thelightningpress.com

SMFLS

SMARTBOOK

Fifth Edition

Sustainment & multifunctional LOGISTICS

Guide to Operational & Tactical Level Sustainment

The Lightning Press
Norman M Wade

he Lightning Press

2227 Arrowhead Blvd.
Lakeland, FL 33813
24-hour Voicemail/Fax/Order: 1-800-997-8827
E-mail: SMARTbooks@TheLightningPress.com

www.TheLightningPress.com

Fifth Edition

(SMFLS5) The Sustainment & Multifunctional Logistics SMARTbook

Guide to Operational & Tactical Level Sustainment

SMFLS5:The Sustainment & Multifunctional Logistics SMARTbook, 5th Ed. (Guide to Operational & Tactical Level Sustainment) has been completely updated for 2021. At 368 pages, SMFLS5 topics and references include the sustainment warfighting function (ADP 4-0); sustainment operations (FM 4-0), sustainment execution (logistics, financial management, personnel services, & health services support); sustainment planning; brigade support (ATP 4-90, Brigade Support Bn); division, corps & field army sustainment (ATP 4-93, Sustainment Brigade); theater support (ATP 4-94, Theater Support Command); joint logistics (JP 4-0); deployment & redeployment (ATP/JP 3-35); and more than a dozen additional new/updated Army sustainment references.

ISBN: 978-1-935886-86-0

Notice of Liability

The information in this SMARTbook and quick reference guide is distributed on an "As Is" basis, without warranty. While every precaution has been taken to ensure the reliability and accuracy of all data and contents, neither the author nor The Lightning Press shall have any liability to any person or entity with respect to liability, loss, or damage caused directly or indirectly by the contents of this book. If there is a discrepancy, refer to the source document. This SMARTbook does not contain classified or sensitive information restricted from public release. "The views presented in this publication are those of the author and do not necessarily represent the views of the Department of Defense or its components."

Photo Credits. Cover photos courtesy of the Dept. of Defense and Dept of the Army.

Printed and bound in the United States of America.

(SMFLS5) Notes to Read

Guide to Operational & Tactical Level Susta

The **sustainment warfighting function** is the related tasks and sy support and services to ensure freedom of action, extended operati prolong endurance. Sustainment determines the depth and duration ovide tions. Successful sustainment enables freedom of action by increasin of options available to the commander. Sustainment is essential for ret exploiting the initiative. The sustainment warfighting function consists o **logistics, financial management, personnel services and health ser**

Sustainment is essential for conducting operations and generating com as the Army performs its **strategic roles**. Sustainment provides the opera commander **freedom of action, operational reach, and prolonged endu** necessary to shape operational environments, prevent conflict, prevail in la ground combat operations, and consolidate gains.

The effectiveness of the sustainment warfighting function is dependent upon ac of units and staffs at the **operational and tactical levels**. Execution is putting a plan into action by applying combat power to accomplish the mission. It focuses on actions to seize, retain, and exploit the initiative. **Sustainment** determines the depth and duration of Army operations. It is essential to retaining and exploiting the initiative and it provides the support necessary to maintain operations until mission accomplishment. Failure to provide sustainment could cause a pause or culmination of an operation resulting in the loss of the initiative.

SMFLS5:The Sustainment & Multifunctional Logistics SMARTbook, 5th Ed. (Guide to Operational & Tactical Level Sustainment) has been completely updated for 2021. At 368 pages, SMFLS5 topics and references include the sustainment warfighting unction (ADP 4-0); sustainment operations (FM 4-0), sustainment execution (logistics, nancial management, personnel services, & health services support); sustainment lanning; brigade support (ATP 4-90, Brigade Support Bn); division, corps & field army stainment (ATP 4-93, Sustainment Brigade); theater support (ATP 4-94, Theater pport Command); joint logistics (JP 4-0); deployment & redeployment (ATP/JP 3-35); d more than a dozen additional new/updated Army sustainment references.

ARTbooks - DIME is our DOMAIN!

ARTbooks: Reference Essentials for the Instruments of National Power (D-I-M-E: omatic, Informational, Military, Economic)! Recognized as a "whole of government" rinal reference standard by military, national security and government professionals nd the world, SMARTbooks comprise a comprehensive professional library ned with all levels of Soldiers, Sailors, Airmen, Marines and Civilians in mind.

books can be used as quick reference guides during actual operations, as ides at education and professional development courses, and as lesson d checklists in support of training. Visit **www.TheLightningPress.com**!

SMFLS5: ...e Sustainment & Multifunctional Logi... SMARTbook, 5th Ed. (G... Operational & Tactical Level Sustainment)

...Sustainment & Multifunctional Logistics SMARTbook (Guide to Operational ...evel Sustainment) has been completely updated for 2021. At 368 pages, ...opics and references include the sustainment warfighting function (ADP ...tainment operations (FM 4-0), sustainment execution (logistics, financial ...ement, personnel services, & health services support); sustainment planning; ...e support (ATP 4-90, Brigade Support Bn); division, corps & field army sustainment ...4-93, Sustainment Brigade); theater support (ATP 4-94, Theater Support ...mmand); joint logistics (JP 4-0); deployment & redeployment (ATP/JP 3-35); and more ...an a dozen additional new/updated Army sustainment references.

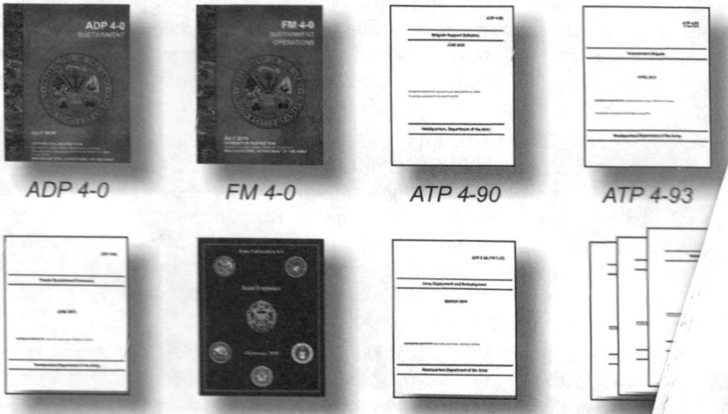

ADP 4-0 | FM 4-0 | ATP 4-90 | ATP 4-93

ATP 4-94 | JP 4-0 | ATP/JP 3-35 | plus 1...

Chap 1: Sustainment Warfighting Function (ADF...

The sustainment warfighting function is the related tasks and system... support and services to ensure freedom of action, extended operati... long endurance. Sustainment determines the depth and duration o... Successful sustainment enables freedom of action by increasing t... available to the commander. Sustainment is essential for retaining... initiative. The sustainment warfighting function consists of four el... financial management, personnel services and health service su...

Chap 2: Sustainment Operations (FM 4-0)

FM 4-0, Sustainment Operations, provides a doctrinal approac... divisions, and brigades to address the challenges of sustainin... four Army strategic roles – Shape Operational Environments, ... Large-Scale Ground Combat and Consolidate Gains. It is the... ment doctrine, detailing how the Army will provide logistics, f... sonnel services and health service support to the force duri...

Chap 3: Sustainment Execution (Logisti... agement, Personnel Services, & Health...

Execution is putting a plan into action by applying comb... mission. It focuses on actions to seize, retain, and explo... determines the depth and duration of Army operations.

exploiting the initiative and it provides the support necessary to maintain operations until mission accomplishment. Failure to provide sustainment could cause a pause or culmination of an operation resulting in the loss of the initiative. It is essential that sustainment planners and operation planners work closely to synchronize all of the warfighting functions, in particular sustainment, to allow commanders the maximum freedom of action.

Chap 4: Sustainment Planning

Planning sustainment support of an operation is vital to mission success. Sustainment commanders and their planning staffs must coordinate and synchronize every stage of the planning process with the operational staff. They must also coordinate, synchronize and integrate the sustainment plan with joint and multinational partners to ensure a continuous linkage with strategic level providers.

Chap 5: Brigade Support (ATP 4-90)

The brigade support battalion (BSB) is the most important sustainment organization in the Army. It supports the brigade combat team and the other brigade formations that constitute the majority of close combat capability in the Army. The BCT area of operations is expansive and its missions diverse. The BSB and its subordinate companies normally operate within the close area of the operational construct closer to the forward line of troops than any other battalion-sized sustainment organization. This places the BSB in operational environments that are highly lethal, rapidly changing, and extremely demanding.

Chap 6: Division, Corps & Field Army Sustainment (ATP 4-93)

The effectiveness of the sustainment warfighting function is dependent upon actions of units and staffs at the **operational and tactical levels**. The **sustainment brigade** is a multifunctional headquarters integrating and employing all assigned and attached units while planning and synchronizing sustainment operations. It is the Army's primary brigade level sustainment headquarters. Sustainment brigades are usually assigned or attached to a sustainment command.

Chap 7: Theater Support (ATP 4-94)

The theater support command (TSC) is responsible for providing sustainment support for an area of responsibility. The TSC executes its mission through human resource sustainment centers, financial management centers, and the use of modular forces, to include expeditionary sustainment commands (ESC), sustainment brigades, combat sustainment support battalions (CSSB), and other modular sustainment formations. The combination of these capabilities gives the sustainment commander the ability to organize and provide tailored support such as theater opening, theater distribution and sustainment support to forces, and the theater closing within an area of responsibility.

Chap 8: Joint Logistics (JP 4-0)

The relative combat power that military forces can generate against a threat is constrained by their capability to plan for, gain access to, and deliver forces and materiel to points of application. Joint logistics is the coordinated use, synchronization, and often sharing of two or more combatant commands (CCMDs) or Military Departments' logistics resources to support the joint force. To meet the wide variety of global challenges, combatant commanders (CCDRs), subordinate commanders, and their staffs must develop a clear understanding of joint logistics, to include the relationship between logistic organizations, personnel, core functions, principles, imperatives, and the operational environment.

Chap 9: Deployment & Redeployment (ATP/JP 3-35)

Force projection is the ability to project the military instrument of national power from the United States or another theater, in response to requirements for military operations. It is a demonstrated ability to alert, mobilize, rapidly deploy, and operate effectively anywhere in the world. The Army, as a key member of the joint team, must be ready for global force projection with an appropriate mix of combat forces together with support and sustainment units.

(SMFLS5) References

The following references were used to compile The Sustainment & Multifunctional Logistics SMARTbook (SMFLS5). All references are considered public domain, available to the general public, and designated as "approved for public release; distribution is unlimited." SMARTbook do not contain classified or sensitive material restricted from public release.

Army Doctrinal Publications (ADPs)

ADP 3-0*	Jul 2019	Operations
ADP 4-0*	Jul 2019	Sustainment
ADP 5-0*	Jul 2019	The Operations Process
ADP 6-0*	Jul 2019	Mission Command

Army Techniques Publications (ATPs)

ATP 1-19	Feb 2015	Army Music
ATP 3-34.40	Feb 2015	General Engineering
ATP 3-35*	Mar 2015	Army Deployment and Redeployment (w/Chg 2)
ATP 4-0.1	Oct 2014	Army Theater Distribution
ATP 4-02.55*	Sept 2015	Army Health System Support Planning
ATP 4-11*	Aug 2020	Army Motor Transport Operations
ATP 4-16	Apr 2013	Movement Control
ATP 4-31*	Nov 2020	Recovery & Battle Damage Assessment and Repair
ATP 4-33*	Jul 2019	Maintenance Operations (w/change 1)
ATP 4-42*	Nov 2020	Materiel Management, Supply, and Field Services Operations
ATP 4-90*	Jun 2020	Brigade Support Battalion
ATP 4-92 (FM 4-92)	Oct 2014	Contracting Support to Unified Land Operations
ATP 4-93	Apr 2016	Sustainment Brigade
ATP 4-94 (FM 4-94)	Jun 2013	Theater Sustainment Command
ATP 5-19	Apr 2014	Risk Management (w/change 1)

Field Manuals (FMs)

FM 1-0	Apr 2014	Human Resources Support
FM 1-04*	Jun 2020	Legal Support to Operations
FM 1-05*	Jan 2019	Religious Support
FM 1-06	Apr 2014	Financial Management Operations
FM 4-01	Apr 2014	Army Transportation Operations
FM 4-02*	Nov 2020	Army Health System
FM 4-30	Apr 2014	Ordnance Operations
FM 6-0	Apr 2016	Commander and Staff Organization and Operations (w/change 2)

Joint Publications (JPs)

JP 3-35*	Jan 2018	Deployment and Redeployment Operations
JP 4-0*	Feb 2019	Joint Logistics

** New/updated references in this edition.*

(SMFLS5) Table of Contents

Chap 1 Sustainment Warfighting Function

Chap 2

Sustainment Operations

Chap 3

Sustainment Execution

Chap 4

Sustainment Planning

Chap 5 Brigade Support

Chap 6

Division, Corps, and Field Army Support

Chap 7

Theater Support

Chap 8

Joint Logistics

Chap 9

Deployment & Redeployment

Chap 1 I. The Sustainment Warfighting Function

Ref: ADP 4-0, Sustainment (Jul '19), chap. 1 and ADP 3-0, Operations (Jul '19), pp. 5-5 to 5-6.

The sustainment warfighting function is the related tasks and systems that provide support and services to ensure freedom of action, extended operational reach, and prolong endurance (ADP 3-0). Sustainment determines the depth and duration of Army operations (ADP 3-0). Successful sustainment enables freedom of action by increasing the number of options available to the commander. Sustainment is essential for retaining and exploiting the initiative. The sustainment warfighting function consists of four elements: logistics, financial management, personnel services and health service support as shown in the sustainment warfighting function logic chart.

See p. 1-3 for discussion of the sustainment warfighting function underlying logic.

Elements of Sustainment

- I **Logistics** *(See pp. 3-5 to 3-42.)*
- II **Financial Management** *(See pp. 3-43 to 3-44.)*
- III **Personnel Services** *(See pp. 3-45 to 3-46.)*
- IV **Health Service Support** *(See pp. 3-47 to 3-50.)*

I. Logistics

Logistics is planning and executing the movement and support of forces. It includes those aspects of military operations that deal with: design and development; acquisition, storage, movement, distribution, maintenance, and disposition of materiel; acquisition or construction, maintenance, operation, and disposition of facilities; and acquisition or furnishing of services. The explosive ordnance disposal tasks are discussed under the protection warfighting function. Army logistics elements are:

- Maintenance
- Transportation
- Supply
- Field Services
- Distribution
- Operational contract support
- General engineering

See pp. 3-5 to 3-42 for further discussion.

Sustainment Overview

Ref: ADP 4-0, Sustainment (Jul '19).

Introduction

ADP 4-0, Sustainment, is the Army's doctrine for sustainment in support of operations. The doctrine discussed in this manual is nested with ADP 3-0, Operations, and describes the sustainment warfighting function. The endurance of Army forces is primarily a function of their sustainment and is essential to retaining and exploiting the initiative. Sustainment provides the support necessary to maintain operations until mission accomplishment. The relationship between sustainment and operation is depicted in introductory figure-1 *(facing page).*

Sustainment must be integrated and synchronized with operations at every level to include those of our joint and multinational partners. Sustainment depends on joint and strategic links for strategic airlift, sealift, intra¬theater airlift, and strategic and theater-level supply support. Sustainment depends on our host nation (HN) partners to provide infrastructure and logistics support necessary to ensure both maneuver forces and follow-on sustainment are delivered to right place, at the right time, and in an operable condition. This dependence on our joint and strategic links and our multinational partners means sustainment is inherently joint. Through joint interdependence, the combatant commander (CCDR) is able to maximize the effect the Army's capabilities in an operational area. The Army's robust sustainment capability assists in providing theater and port opening functions enabling joint forces to conduct strategic and operational reach. Army sustainment capabilities provide the bulk of Army support to other services in the forms of executive agent, common-user logistics, lead service, and other common sustainment resources. Army support to other services enables joint forces with freedom of action and endurance.

Fundamentals of Sustainment

For the Army, sustainment is the provision of logistics, financial management, personnel services, and health service support necessary to maintain operations until successful mission completion. Sustainment is accomplished through the coordination, integration, and synchronization of resources from the strategic level through the tactical level in conjunction with our joint and multinational partners.

Sustainment operations enable force readiness. Sustainment operations maintain Army forces by equipping it with materiel, funding it with required resources, staffing it with trained Soldiers and leaders, and by providing it with the force health protection needed.

Army sustainment is based on an integrated process (people, systems, materiel, health service support, and other support) inextricably linking sustainment to operations. The concept focuses on building an operational ready Army, delivering it to the CCDR as part of the joint force, and sustaining its combat power across the depth of the operational area and with unrelenting endurance.

Principles of Sustainment

The principles of sustainment are essential to maintaining combat power, enabling strategic and operational reach, and providing Army forces with endurance. While these principles are independent, they are also interrelated and must be synchronized in time, space, and purpose. The principles of sustainment and the principles of logistics are the same.

See following pages (pp. 1-4 to 1-5) for further discussion.

Sustainment Underlying Logic

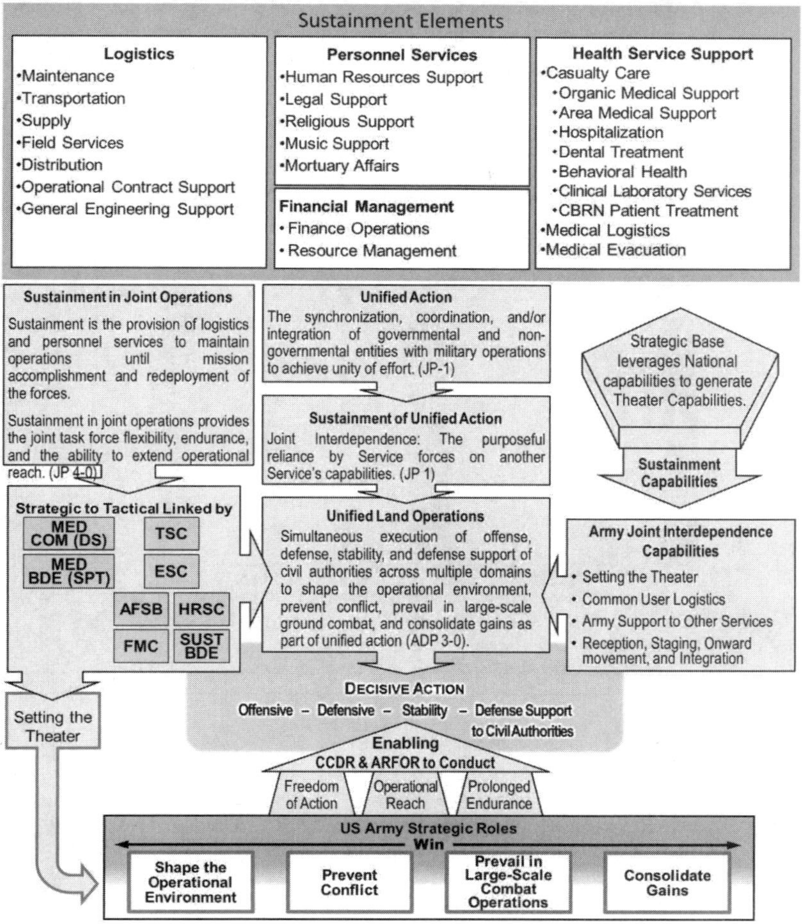

LEGEND:

ADP =	Army doctrinal publication	FMC =	financial management company
AFSB =	Army field service brigade	FMSC =	financial management support center
ARFOR =	Army forces	HRSC =	human resource support center
BDE =	brigade	JP =	joint publication
CCDR =	component commander	MED =	medical
COM =	command	SPT =	support
(DS) =	deployment support	SUST =	sustainment
ESC =	expeditionary sustainment command	TSC =	theater sustainment command

Defense Climate Risk assessment

Principles of Sustainment (& Logistics)

Ref: ADP 4-0, Sustainment (Jul '19), pp. 1-2 to 1-4.

The principles of sustainment shown below are essential to maintaining combat power, enabling strategic and operational reach, and providing Army forces with endurance. While these principles are independent, they are also interrelated and must be synchronized in time, space, and purpose. The principles of sustainment and the principles of logistics are the same.

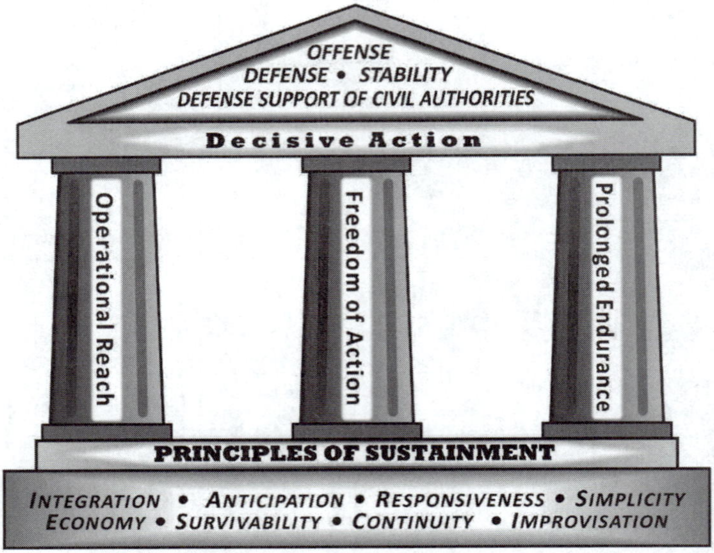

Ref: ADP 4-0 (Jul '19), fig. 1-1. Principles of sustainment.

Integration. Integration is combining all of the sustainment elements within operations assuring unity of command and effort. It requires deliberate coordination and synchronization of sustainment with operations across all levels of war. Army forces integrate sustainment with joint and multinational operations to maximize the complementary and reinforcing effects of each Service component's and other UAPs resources. One of the primary functions of the sustainment staff is to ensure the integration of sustainment with operations plans.

Anticipation. Anticipation is the ability to foresee operational requirements and initiate necessary actions that most appropriately satisfy a response without waiting for operations orders or fragmentary orders. It is shaped by professional judgment resulting from experience, knowledge, education, intelligence, and intuition. Commanders and staffs must understand and visualize future operations and identify appropriate or required support. They must then start the process of acquiring the resources and capabilities that best support the operation. Anticipation is facilitated by automation systems that provide the common operational picture upon which judgments and decisions are based. Commanders integrate risk management into the operations process to identify threats, assess those threats, and emplace controls to mitigate the risk of gaps in support. Anticipation is also a principle of personnel services.

Responsiveness. Responsiveness is the ability to react to changing requirements and respond to meet the needs to maintain support. It is providing the right support in the right place at the right time. It includes the ability to anticipate operational requirements. Responsiveness is facilitated by a common operational picture facilitated by the Army Readiness-Common Operating Picture, and follow-on business intelligence tools associated with Army enterprise resource planning systems. That common operational picture enables commanders to see all supported forces, anticipate requirements based on situational understanding, and provide support when and where needed. Responsiveness involves identifying, accumulating, and maintaining sufficient resources, capabilities, and information necessary to meet rapidly changing requirements. Through responsive sustainment, commanders maintain operational focus and pressure, set the tempo of friendly operations to prevent exhaustion, replace ineffective units, and extend operational reach.

Simplicity. Simplicity relates to processes and procedures to minimize the complexity of sustainment. Unnecessary complexity of processes and procedures leads to confusion. Clarity of tasks, standardized and interoperable procedures, and clearly defined command relationships contribute to simplicity. Simplicity enables economy and efficiency in the use of resources, while ensuring effective support of forces. Simplicity is also a principle of financial management.

Economy. Economy is providing sustainment resources in an efficient manner that enables the commander to employ all assets to the greatest effect possible. Economy is achieved through efficient management, discipline, prioritization, and allocation of resources. Economy is further achieved by eliminating redundancies and capitalizing on joint interdependencies. Disciplined sustainment assures greatest possible tactical endurance and constitutes an advantage to commanders. Economy may be achieved by contracting for support or using HN resources that reduce or eliminate the use of limited military resources. By efficiently and ethically managing Army resources, Army professionals are stewards who act in the best interest of the American people.

Survivability. Survivability is all aspects of protecting personnel, weapons, and supplies while simultaneously deceiving the enemy (JP 3-34). Survivability consists of a quality or capability of military forces to avoid or withstand hostile actions or environmental conditions while retaining the ability to fulfill their primary mission. This quality or capability of military forces is closely related to protection (the preservation of a military force's effectiveness) and to the protection/force protection warfighting function (the tasks or systems that preserve the force). Hostile actions and environmental conditions can disrupt the flow of sustainment and significantly degrade forces' ability to conduct and sustain operations. In mitigating risks to sustainment, commanders often must rely on the use of redundant sustainment capabilities and alternative support plans.

Continuity. Continuity is the uninterrupted provision of sustainment across all levels of war. Continuity is achieved through a system of integrated and focused networks linking sustainment to operations. Continuity is achieved through joint interdependence; linked sustainment organizations; a strategic to tactical level distribution system, and integrated information systems. Continuity assures confidence in sustainment allowing commanders freedom of action, operational reach, and endurance.

Improvisation. Improvisation is the ability to adapt sustainment operations to unexpected situations or circumstances affecting a mission. It includes creating, arranging, or fabricating resources to meet requirements. It may also involve changing or creating methods that adapt to a changing operational environment. Sustainment leaders must apply operational art to visualize complex operations and understand additional possibilities. These skills enable commanders to improvise operational and tactical actions when enemy actions or unexpected events disrupt sustainment operations. While deception is related to survivability in that deception contributes to survivability, improvisation is where logisticians can actively achieve deception of enemy forces. Improvisation is also a principle of financial management.

II. Financial Management

Financial management leverages fiscal policy and economic power across the range of military operations. Financial management encompasses finance operations and resource management.

See pp. 3-43 to 3-44 for further discussion.

III. Personnel Services

Personnel services are those sustainment functions related to Soldiers' welfare, readiness, and quality of life. Personnel services complement logistics by planning for and coordinating efforts that provide and sustain personnel. Personnel services include—

- Human resources support
- Legal support
- Religious support
- Band support

See pp. 3-45 to 3-46 for further discussion.

IV. Health Service Support

Army Health System (AHS) support includes both health service support and force health protection that are critical capabilities embedded within formations across all warfighting functions. The force health protection mission falls under the protection warfighting function and will not be covered in this publication (see ADP 3-37).

Health service support encompasses all support and services performed, provided, and arranged by the Army Medical Department to promote, improve, conserve, or restore the behavioral and physical well-being of Army personnel and as directed, unified action partners (UAPs). Health service support includes the following—

- Casualty care, which encompasses a number of medical functions, including:
  - Medical treatment (organic and area medical support).
  - Hospitalization.
  - Dental care (treatment aspects).
  - Behavioral health/neuropsychiatric treatment.
  - Clinical laboratory services.
  - Treatment of chemical, biological, radiological, and nuclear patients).
- Medical evacuation (including medical regulating).
- Medical logistics (including blood management).

See pp. 3-47 to 3-50 for further discussion.

Refer to AODS6-1 (w/SMARTupdate 1): The Army Operations & Doctrine SMARTbook (Guide to FM/ADP 3-0 Operations & the Elements of Combat Power) for full discussion of each of the six warfighting functions (and elements of combat power): command & control (ADP 6-0), movement and maneuver (ADPs 3-90, 3-07, 3-28, 3-05), intelligence (ADP 2-0), fires (ADP 3-19), sustainment (ADP 4-0), & protection (ADP 3-37).

Chap 1

II. Sustainment of Unified Action

Ref: ADP 4-0, Sustainment (Jul '19), chap. 2.

Unified action is the synchronization, coordination, and/or integration of the activities of governmental and nongovernmental entities with military operations to achieve a unity of effort (JP 1). Unified action requires fully integrating the U.S. military with the efforts of interorganizational and multinational partners to achieve strategic, operational, and tactical success. The sustainment of unified action requires a continuous link, close coordination, and collaboration with other Services, allies, HN forces, and other governmental organizations.

I. Strategic Support Area

In the U.S., sustainment originates in the strategic support area. The strategic support area consists of the defense industrial base that includes DOD, government, and private sector worldwide industrial complexes with capabilities to perform research and development, design, produce, and maintain military weapon systems, subsystems, components, or parts to meet military requirement. The strategic support area generates Army sustainment capabilities, which enable sustained operations through end-to-end processes that drive efficiencies across military Services, agencies, and industry. The continuous effects created in the strategic support area may occur anywhere long the competition continuum, which ranges from peace to war.

The strategic support area describes the area extending from the joint security area within a theater to the continental United States (CONUS) or another combatant's area of responsibility (AOR). It includes a vast array of DOD, government, and private sector agencies that participate in the sustainment enterprise. The support provided includes materiel integration and sealift support conducted by USAMC and United States Transportation Command (USTRANSCOM). Class I, III (B), IV, and VIII support is provided by the Defense Logistics Agency (DLA). United States Army Financial Management Command coordinates, synchronizes, and integrates financial management between strategic partners and the operational and tactical Army levels. The U.S. Army Human Resources Command is the U.S. Army's manning enterprise that distributes personnel in accordance with Army priorities and readiness objectives. The strategic support area also includes the air and seaports that support the force projection, flow, and sustainment of forces into the joint security area within a theater. The joint security area is a specific surface area designated by a joint force commander to facilitate protection of joint bases and their connecting lines of communications (JP 3-10). The theater Army and its assigned TSC operate in and provide support to Army forces operating in the joint security area. The TSC integrates and synchronizes sustainment in the joint security area and across the theater Army's area of responsibility.

Army forces through joint interdependence rely upon joint capabilities, air and maritime, to deliver sustainment to a theater of operations. Through coordination and collaboration between strategic and operational partners, a continuous and uninterrupted flow of sustainment is provided to achieve national military objectives. Also through coordination, collaboration, and agreements with HN, allies and intergovernmental organizations certain sustainment efficiencies are achieved to facilitate a unity of effort.

II. Army Sustainment Responsibilities

Each Service retains responsibility for the sustainment of forces it allocates to a joint force. The Secretary of the Army exercises this responsibility through the Chief of Staff, United States Army (CSA) and the Theater Army assigned to each combatant command. The Theater Army is responsible for the preparation and administrative support of Army forces assigned or attached to the combatant command.

A. Army Title 10 Sustainment Requirements

Title 10, USC and DOD Directive 5100.1, Functions of the DOD and Its Major Components, describe the organization, roles, and responsibilities for the elements of the DOD to include the statutory requirements for each Military Department to provide support to assigned forces.

See facing page for further discussion.

B. Executive Agent (EA)

Executive Agent (EA) is a term used to indicate a delegation of authority by the Secretary of Defense or Deputy Secretary of Defense to a subordinate to act on behalf of the Secretary of Defense (JP 1). An EA may be limited to providing only administration and support or coordinating common functions; or it may be delegated authority, direction, and control over specified resources for specified purposes.

When designated as an EA, the Army is specifically tasked by the Secretary of Defense for certain responsibilities, sometimes limited by geography, sometimes for a particular operation, and sometimes for the entire DOD on a continuing basis. The list below (not all inclusive) is an example of some of the Army's sustainment EA responsibilities:

- DOD Combat Feeding Research and Engineering Program
- Management of Land-based Water Research in Support of Contingency Operations
- Law of War Program
- Defense Mortuary Affairs Program
- Military Postal Service
- Explosive Safety Management

C. Lead Service

The CCDR may choose to assign specific common-user logistics support functions, to include both planning and execution to a lead Service. A lead Service or agency for common-user logistics is a Service component or DOD agency that is responsible for execution of common-user item or service support in a specific combatant command or multinational operation as defined in the combatant or subordinate joint force commander's operation plan, operation order, and/or directives (JP 4-0). These assignments can be for single or multiple common-user logistics and may be based on phases and/or locations within the AOR. The CCDR may augment the lead Service logistics organization with capabilities from another component's logistics organizations as appropriate. The lead Service must plan, issue procedures, and administer sustainment funding for all items issued to other Services as well as a method for collecting items from other Services.

D. Directive Authority for Logistics (DAFL)

The Directive Authority for Logistics is the CCDR's authority to issue directives to subordinate commanders to ensure the effective execution of approved operation plans, optimize the use or reallocation of available resources, and prevent or eliminate redundant facilities and/or overlapping functions among the Service component

Army Title 10 Sustainment Requirements

Ref: ADP 4-0, Sustainment (Jul '19), pp. 2-2 to 2-3.

There are 12 Army Title 10 responsibilities; of the 12, ten are sustainment related responsibilities:

- Recruiting
- Organizing
- Supplying
- Equipping (including research and development)
- Training
- Servicing
- Mobilizing
- Demobilizing
- Administering (including the morale and welfare of personnel)
- Maintaining
- Construction, outfitting, and repair of military equipment
- Construction, maintenance, repairs of building and structures, utilities, acquisition of real property and interests in real property necessary to carry out the responsibilities

The Secretary of the Army exercises this responsibility through the Chief of Staff the Army and the Theater Army assigned to each combatant command. The Theater Army is responsible for the preparation and administrative support of Army forces assigned or attached to the combatant command. However, the purposeful combination of complementary Service capabilities to create joint interdependent forces is often the most effective and efficient means by which to sustain a joint force. Therefore additional authorities to Title 10 have been developed to provide for inter-service and interagency mutual support. The options for executing sustainment of a joint force may include any combination of following:

- Executive Agent
- Lead Service
- Directive Authority for Logistics

All of these authorities may have the same possible impact on Army sustainment headquarters, which is that the Army may be required to provide support to other Services and agencies involved in an operation or that other Services or agencies may provide support to Army units which would normally receive such support from the Army. Army sustainment leaders must be prepared to plan and execute such operations as tasked.

commands (JP1). The CCDRs may delegate directive authority for as many common user commodities and services as required to accomplish the assigned mission. It includes peacetime measures to ensure the effective execution of approved operation plans, effectiveness and economy of operation, prevention or elimination of unnecessary duplication of facilities, and overlapping of functions among the Service component commands.

When the CCDR gives a Service component lead agent responsibility, the responsibility must be specifically defined. When two or more Services have common commodities or support services, one Service may be given responsibility for management based on DOD designations or inter service support agreement. However, the CCDR must formally delineate this directive authority by function and scope. The Army, when directed to provide management of common sustainment functions that include other services, most often establishes and leads joint boards.

E. Army Medical Support to Other Services

The Army Surgeon General is responsible to the Secretary of the Army (under the supervision of the Chief of Staff of the U.S. Army) for execution of assigned responsibilities as outlined in Title 10 U. S. Code Section 3013(b). In this role, The Surgeon General serves as the Army's strategic enabler for globally integrated health services and provides AHS support to other Services through oversight of assigned EA or lead Service responsibilities for a number of medical functions. This requires the synchronization and integration of national medical resources on a global scale to ensure that Soldiers and unified action partners receive the best possible operational medical support regardless of their geographic location. The Surgeon General's EA or lead Service responsibilities include—

- Serving as the DOD sole provider for veterinary public and animal health services in support of all of the military Services (except for the food inspection mission on U.S. Air Force installations).
- Serving as the EA for the Armed Services Blood Program for blood and blood product distribution.
- Serving as the DOD EA for medical research for prevention, mitigation, and treatment of blast injuries.
- Coordination for joint patient movement or strategic evacuation.
- Providing intra-theater aeromedical evacuation (which includes medical evacuation from ship-to¬shore and shore-to-ship for U.S. Navy and Marine personnel).
- Coordinating for medical support to joint special operations forces. *Refer to FM 4-02 for additional information.*

The medical command (deployment support) (MEDCOM [DS]), medical brigade (support) (MEDBDE [SPT]), and subordinate units are responsible to The Surgeon General and the Army Service Component Commander (ASCC) (through coordination with the ASCC surgeon) for regionally focused execution and oversight of AHS support to other Services within the AO. When designated, the MEDCOM (DS) is also responsible for providing operational medical logistics support to other Services. In cases where the ASCC is designated by the combatant commander as the single integrated medical logistics manager, functions are most commonly carried out through a coordinated effort. These functions are carried out between the ASCC surgeon, the Medical Logistics Management Center, the medical logistics company, and the MEDBDE (SPT), supported by the MEDCOM (DS) to provide common user medical logistics support for joint operations in theater.

III. Joint Interdependence

Joint interdependence is the purposeful reliance by one Service's forces on another Service's capabilities to maximize the complementary and reinforcing effects of both. Army forces operate as part of an interdependent joint force. Joint forces support and provide each other with many services. The U.S. Air Force provides airlift capabilities to quickly move Army forces across strategic lines of communication to theater operations. In emergencies, the U.S. Air Force may support aerial delivery of sustainment to forward areas or areas where terrain may be too restrictive for ground operations. The U.S. Air Force through the Air Mobility Command provides worldwide cargo and passenger airlift, air refueling, and aeromedical evacuation. Air Mobility Command also provides Contingency Response Elements that provide en route ground support for airlift operations.

The Naval forces provide critical capabilities to support Army logistics over-the-shore operations. Naval forces provide essential causeway, sealift and logistics over-the-shore support ensuring sustainment is provided to land forces when ports may be austere, damaged, or non-existent. Naval forces may be responsible for removing sustainment from vessels and delivering them to port operations for release to Army forces. The Naval construction capabilities provide port construction such as warehouses, storage facilities. The Navy also provides explosive ordnance disposal support to locate and dispose of mines along ports and channels.

As part of joint interdependence, the Army plays a crucial role in opening and setting the theater. Setting the theater is a continuous shaping activity and is conducted as part of steady-state posture and for contingency or crisis response operations. Set the theater describes the broad range of actions conducted to establish the conditions in an operational area for the execution of strategic plans. It begins with the identification and analysis of host and partner nation resources, and capabilities required to support campaign and contingency plans. Planners leverage whole-of-government initiatives such as bilateral or multilateral diplomatic agreements to allow U.S. forces to have access to ports, terminals, airfields, and bases within the AOR to support future military contingency operations. Setting the joint operations area includes activities such as theater opening, establishing port and terminal operations, conducting reception, staging, onward movement, and integration, and providing Army support to other Services and common-user logistics to Army, joint, and multinational forces operating in the joint operations area.

See pp. 2-9 to 2-10 for discussion setting the theater from FM 4-0.

IV. Role of the Institutional Army

The institutional Army consists of those Army organizations within the strategic support area whose primary mission is to generate and sustain the operational Army's capabilities for employment. The strategic area is the space where friendly strategic forces gain their combat power, project combat power, and sustain operations. Institutional Army activities in the strategic area, include materiel integration, support of readiness, Army force generation, strategic lift, and the routine performance of functions specified and implied in Title 10, USC. Because of its performance of functions specified and implied by law; the institutional Army also possesses operationally useful capabilities for employment by or in direct support of joint force commanders. These organizations enable strategic reach by helping to project Army capabilities into the support, close, and deep areas of combat operations. Institutional Army capabilities include analyzing, understanding and adapting, and generating operational forces tailored to the specific context in which they will be employed.

The institutional Army is responsible for moving Army forces to and from ports of embarkation. They also provide capabilities to assist in the management and operation of ports of embarkation and debarkation and provide capabilities to GCC to conduct reception, staging, onward movement, and integration.

V. Role of Operating Forces

Ref: ADP 4-0, Sustainment (Jul '19), pp. 2-10 to 2-16.

The operating forces are those forces whose primary missions are to participate in combat and the integral supporting elements thereof. Operational Army units are typically assigned to CCDRs. The Army normally executes its responsibilities to organize, train, and equip operational Army units through ASCCs.

Army Service Component Command /Theater Army

When an ASCC is in support of a GCC, it is designated as a theater Army. The theater Army is the primary vehicle for Army support to joint, interagency, intergovernmental, and multinational forces (MNFs). The theater Army headquarters (HQ) performs functions that include reception, staging, onward movement, and integration; logistics over-the-shore operations; and security coordination.

The theater Army is responsible for providing support to Army forces and common sustainment/support to other Services as directed by the CCDR and other authoritative instructions. The TSC is assigned to the theater Army and provides lead Service and executive agency support for designated logistics and services to other government agencies, MNFs, and nongovernmental organizations (NGO). The TSC is the Army's command for the integration and synchronization of sustainment in the AOR. MEDCOM (DS) is also assigned to the ASCC. It is the theater medical command responsible for command and control, integration, synchronization, and execution of AHS support within the AOR. The MEDCOM (DS) commander coordinates with the ASCC surgeon (as the staff proponent with execution through assistant chief of staff, operations channels under the authority of the ASCC commander) to provide AHS support within the AOR.

The theater Army exercises administrative control over all Army forces in the area of responsibility unless modified by DA. This includes Army forces assigned, attached, or OPCON to the combatant command. It coordinates with the TSC for operational sustainment planning and management. The theater Army defines theater policies and coordinates with the TSC for technical guidance and execution of force projection and sustainment.

Theater Army Surgeon's Section

The ASCC surgeon is the theater Army staff proponent responsible for (in coordination with the MEDCOM [DS] commander) the provision of AHS support within the AOR. The ASCC surgeon has staff responsibility for medical planning, coordination, and policy development for AHS support to deployed forces. This officer advises the ASCC commander concerning the health of the command, recommends changes to the theater evacuation policy, and provides input to and personnel in support of the theater patient movement requirements center, as required. Organizations from battalion through Army Service component command level are authorized a surgeon. The Army Medical Department leverages the surgeon's cells (staff channels) at each echelon and medical command channels (through the MEDCOM [DS], medical brigade [support], and medical battalion [multifunctional]) to provide AHS support to the deployed force. Integration of these two chains and other elements of sustainment occur at command headquarters at each echelon and not just between sustainment organizations. The ASCC surgeon and the surgeon cells at each echelon (including the TSC, ESC, and sustainment brigade surgeon cells) work with their staff to conduct planning, coordination, synchronization, and integration of AHS support to plans to ensure that all ten medical functions are considered and included in running estimates, operations plans, and operation orders in coordination with the MEDCOM (DS). The surgeon also prepares a portion of annex E (Protection) and annex F (Sustainment) to the operation order or operation plan. *For additional information, refer to FM 4-02.*

Theater Sustainment Command (TSC)

The TSC is the Army's command for the integration and synchronization of sustainment in the AOR. The MEDCOM (DS) is the theater medical command responsible for command and control, integration, synchronization, and execution of AHS support within the AOR. *For additional information on the TSC, refer to ATP 4-94.*

The TSC is capable of planning, preparing, executing, and assessing logistics and human resource support for Army forces in theater. It provides support to unified land operations. As the distribution coordinator in theater, the TSC leverages strategic partnerships and joint capabilities to establish an integrated theater-level distribution system that is responsive to theater Army requirements. It employs sustainment brigades to execute theater opening, theater sustainment, and theater distribution operations.

The TSC includes units capable of providing multifunctional logistics: supply, maintenance, transportation, petroleum, port, and terminal operations. Other specialized capabilities, such as mortuary affairs, aerial delivery, human resources, sustainment to internment/resettlement operations, and financial management, are available from the force pool. *See p. 7-4.*

Expeditionary Sustainment Command (ESC)

The ESC provides mission command of sustainment units in designated areas of a theater. At the theater echelon, one or more ESC's may be attached to a TSC. An ESC may also be attached to the field Army and one assigned to the corps to act as the headquarters for the integration and synchronization of sustainment at those echelons. The ESC plans, prepares, executes, and assesses sustainment, distribution, theater opening, and reception, staging, and onward movement operations for Army forces in theater. It normally deploys when the TSC determines that a forward command presence is required. This capability provides the TSC commander with the regional focus necessary to provide effective operational-level support to Army or joint task force missions. *See p. 7-6.*

Theater Engineer Command

The theater engineer command is designed to mission command engineer capabilities for all assigned or attached engineer brigades and other engineer units and missions for the joint force land component or theater Army commander. It is the only organization designed to do so without augmentation and can provide the joint force commander with an operational engineer headquarters or augment an engineer staff for a joint task force. The theater engineer command is focused on operational-level engineer support across all three of the engineer disciplines and typically serves as the senior engineer headquarters for a theater Army, land component headquarters, or potentially a joint task force. *Refer to FM 3-34.*

Human Resource Sustainment Center (HRSC)

The HRSC is a multifunctional, modular organization (staff element), and theater-level center assigned to a TSC that integrates and ensures execution of personnel accountability, casualty, and postal functions throughout the theater as defined by the policies and priorities established by the ASCC G-1/AG. The HRSC, in coordination with the TSC, has a defined role to ensure that the theater HR support plan is developed and supported with available resources within the TSC. This includes collaborating with the ASCC G-1/AG and TSC to ensure appropriate HR support relationships are established and properly executed through the operation order process.

Financial Management Center

The Financial Management Center is a modular and tailorable operational financial management unit whose mission is inextricably linked to the theater Army assistant chief of staff, finance. In order to provide adequate theater and national-provider responsiveness and support, the Financial Management Center maintains visibility of all financial

Continued on next page

Operating Forces (Continued)

Ref: ADP 4-0, Sustainment (Jul '19), pp. 2-8 to 2-12.

Continued from previous page

management operations and placement of all operational and tactical financial management units in theater. The primary mission of the Financial Management Center is to provide technical coordination of all theater finance operations and serve as the principal advisor to the theater Army assistant chief of staff, finance and the TSC commander on all aspects of theater finance operations. Technical coordination of theater financial management units (financial management companies and their subordinate detachments) encompasses the provision of recommendations and advice to theater commanders and staff regarding the employment, integration, direction, and control of their financial management forces for the accomplishment of assigned missions. (FM 1-06).

Sustainment Brigade

When deployed, the sustainment brigade is a subordinate command of the TSC, or the ESC. The sustainment brigade is a flexible, multifunctional sustainment organization, tailored and task organized according to mission, enemy, terrain and weather, troops and support available, time available and civil considerations. It plans, prepares, executes, and assesses sustainment operations within an area of operations. It provides mission command of sustainment operations and distribution management. *See p. 6-4.*

Division Sustainment Brigade

The division sustainment brigade is assigned to a division. The division sustainment brigade is a renamed sustainment brigade. The division sustainment brigade's commander is the primary senior advisor to the division commander and the deputy commanding general (support) for the sustainment warfighting function. The commander is responsible for the integration, synchronization, and execution of sustainment operations at echelon. The division sustainment brigade employs sustainment capabilities to create desired effects in support of the division commander's objectives. Its role as the sustainment coordinator does not replace the role of the division assistant chief of staff, logistics as the division sustainment planner responsible for developing the sustainment support concept based on the division assistant chief of staff, operations concept. *See p. 6-6.*

Division Sustainment Support Battalion

The division sustainment support battalion is employed using various task-organizations. The division sustainment support battalion is a renamed combat sustainment support battalion. It is organic to division sustainment support brigades assigned to divisions. The division sustainment support battalion and its subordinate units must be able to move and displace at the pace of large-scale combat operations. The division sustainment support battalion commands and controls all organic, assigned, and attached units. As directed by the division support brigade commander, the division sustainment support battalion conducts maintenance, transportation, supply, field services, and distribution. Division sustainment support battalions organic to division sustainment brigades supporting divisions have an organic composite supply company, composite truck company, and support maintenance company. Other capabilities are task organized by the division commander in accordance with requirements. The division sustainment support battalion synchronizes and executes logistics support to BCTs and multifunctional support brigades attached to the division. *See p. 6-10.*

Continued from previous page

Combat Sustainment Support Battalion (CSSB)

The combat sustainment support battalion (CSSB) is a flexible and responsive unit that executes logistics throughout the depth of an area of operations including transportation, maintenance, ammunition, supply, mortuary affairs, aerial delivery, field services, water, and petroleum. The CSSB is attached to a sustainment brigade and is the building block upon which the sustainment brigade capabilities are developed. The CSSB is tailored to meet specific mission requirements. Employed on an area basis, the CSSB plans, prepares, executes, and assesses logistics operations within an area of operations. The CSSB also supports units in or passing through its designated area.

The CSSB may operate remotely from the sustainment brigade and therefore must maintain communications with the sustainment brigade. The CSSB establishes voice communications to support mission command and convoy operations as well as to monitor, update, and evaluate the logistics posture. *See pp. 6-16 to 6-17.*

Medical Command (Deployment Support)

The MEDCOM (DS) is the senior medical command in support of the ASCC based upon operational and mission variables. The MEDCOM (DS) commander is responsible for maintaining a regional focus in support of the GCC and ASCC theater engagement plan, while providing effective and timely direct health service support and force health protection to tactical commanders and general support (on an area basis) to theater forces at echelons above brigade. The enduring regional focus of the ASCC drives organizational specialization in the supporting MEDCOM (DS) to address unique health threats, specific needs of the local populace, availability of other Service medical capabilities, and geographic factors that are distinctly related to a particular region.

Medical Brigade (Support)

The medical brigade (support) (MEDBDE [SPT]) is a subordinate mission command organization of the MEDCOM (DS). It provides the mission command and planning capabilities necessary to deliver responsive and effective AHS support across the range of military operations. The MEDBDE (SPT) is mission variables driven to ensure the right mix of medical forces and expertise (operational, technical, and clinical) to synchronize AHS support to unified land operations.

Medical Battalion (Multifunctional)

The medical battalion (multifunctional) is an echelon above brigade headquarters. The unit provides mission command, administrative assistance, logistical support, and technical supervision for assigned and attached medical functional organizations (companies, detachments, and teams) task-organized for support to deployed forces operating within the area of responsibility. The medical battalion (multifunctional) can be deployed to provide mission command of medical forces during early entry operations and facilitate the reception, staging, onward movement, and integration of theater medical forces. All echelon above brigade medical companies, detachments, and teams in theater may be assigned, attached, or placed under the OPCON of a medical battalion (multifunctional). It is under the mission command of the MEDBDE (SPT) and/or MEDCOM (DS). *Refer to FM 4-02 for additional information.*

Brigade Support Battalion (BSB)

The BSB is an organic component of BCT, field artillery, and aviation brigades. The BSB is tailored to support the particular brigade to which it is organic. For example, the BSB of an armor BCT has more fuel distribution capabilities and maintenance than does an infantry BCT BSB. The BSB provides supply, maintenance, motor transport, and medical support to the supported brigade. The BSB plans, prepares, and executes, logistics operations in support of brigade operations. Additionally, the BSB plays a significant role in the conduct of post-isolation screening following the recovery of previously isolated personnel. *See p. 5-1.*

Aviation Support Battalion (ASB)

The aviation support battalion is the support battalion for a combat aviation brigade, an expeditionary combat aviation brigade, and a theater aviation brigade. *See p. 5-12.*

VI. Joint Logistics

Ref: JP 4-0, Joint Logistics (Feb '19), chap. II and pp. I-10 to I-12.

Core logistics functions provide a framework to facilitate integrated decision making, enable effective synchronization and allocation of resources, and optimize joint logistics processes. The challenges associated with support cut across all core logistics functions, especially when multiple joint task forces (JTFs) or multinational partners are involved.

Core Logistic Functions

Core Functions	Functional Capabilities
Deployment and Distribution	• Move the force • Sustain the force • Operate the joint deployment and distribution enterprise
Supply	• Manage supplies and equipment • Inventory management • Manage global supplier networks
Maintenance	• Depot maintenance operations • Field maintenance operations • Equipment reset
Logistics Services	• Food service • Water and ice service • Contingency base services • Hygiene services • Mortuary affairs
Operational Contract Support	• Contract support integration • Contracting support • Contractor management
Engineering	• General engineering • Combat engineering • Geospatial engineering
Joint Health Services	• Force health protection • Health service support

The core logistics functions are: deployment and distribution, supply, maintenance, logistics services, operational contract support (OCS), engineering, and joint health services.

See chapter 8, Joint Logistics, for further discussion.

Chap 1 III. Sustainment of Unified Land Operations

Ref: ADP 4-0, Sustainment (Jul '19), chap. 3.

Operational Context

Any operational environment consists of many interrelated variables and sub variables, as well as the relationships among those variables and sub variables. How the many entities and conditions behave and interact with each other within an operational environment is difficult to discern and always results in differing circumstances. Different actor or audience types do not interpret a single message in the same way. Therefore, no two operational environments are the same (ADP 3-0).

Unified Land Operations

Unified land operations require the integration of U.S. military operations with that of multinational partners and our UAPs. The goal of unified land operations is to establish conditions that achieve the joint force commander's end state by applying landpower as part of unified action (ADP 3-0).

The sustainment warfighting function is nested within all four of the Army' strategic roles (shape operational environments, prevent conflict, prevail in large-scale ground combat operations, and consolidate gains). FM 4-0 describes in detail how the sustainment warfighting function supports the strategic roles during operations described in FM 3-0. The sustainment warfighting function is also essential for conducting unified land operations and providing resources for generating and maintaining combat power. Sustainment provides maneuver commanders with operational reach, freedom of action, and operational endurance needed to maintain the initiative in conducting unified land operations.

Sustainment Support of the Army Strategic Roles

Sustainment supports the Army strategic role of shaping operational environments by setting the theater and supporting military engagements. Sustainment activities during shaping operational environments include establishing logistics partnerships, enhancing interoperability, establishing or refining HNS agreements, and gaining access to potential critical infrastructure nodes. Sustainment supports the Army strategic role of preventing conflict by tailoring forces to the type of operation, geographic location, permissiveness of the environment, threat, and a host of other planning considerations determined during continued analysis of the operational environment and mission variables. In addition, refinement of plans and logistics estimates to support expected operations occur during prevent conflict. Sustainment supports the Army strategic role of prevailing in large-scale ground combat operations by providing the freedom of action, prolonged endurance, and extended operational reach required to conduct sustained defensive and offensive operations. During the Army strategic role of consolidate gains, sustainment provides support to combat operations while establishing security, restoring combat power, and preparing for continued operations to destroy remaining enemy forces.

See pp. 2-2 to 2-3 for a detailed discussion of how sustainment supports the Army strategic roles from FM 4-0.

Sustainment Operations (Overview)

Ref: ADP 4-0, Sustainment (Jul '19), chap. 3.

Sustainment is inherently joint and requires a coordinated and collaborated effort between the Services, UAPs, and multinational partners. Sustainment determines the depth and duration of Army operations. It is essential to retaining and exploiting the initiative and it provides the support necessary to maintain operations until mission accomplishment. Failure to provide sustainment could cause a pause or culmination of an operation resulting in the loss of the initiative and failure to exploit tactical successes. It is essential that sustainment planners and operation planners work closely to synchronize all of the warfighting functions, in particular sustainment, to allow commanders the maximum flexibility to conduct decisive action.

When executing sustainment operations, commanders, staffs, and subordinates ensure their decisions and actions comply with U.S., international, and in some cases, HN laws and regulations. Commanders at all levels ensure their Soldiers operate in accordance with the law of war (FM 27-10) and the rules of engagement.

See chap. 2, Sustainment Operations, and chap. 3, Sustainment Execution, for further discussion.

Command and Control of Sustainment Operations

Sustainment staffs provide sustainment commanders with input and feedback that enable sustainment commanders to use the art of command and science of control to array sustainment support within the operational environment. Sustainment commanders use the philosophy of mission command to enable subordinate sustainment commanders to make decisions and adjustments within the supported commander's intent. This ensures supported commanders receive continued sustainment support even as the tactical and operational situations change.

Sustainment staffs monitor and coordinate sustainment functions between the sustainment and supported staffs. The sustainment staffs also coordinate with specialized functional centers (HRSC and Financial Management Support Center) for oversight of those operations. Theater Army, corps, and division headquarters have staff elements aligned with the movement and maneuver, field artillery, intelligence, mission command, and protection warfighting functions. The sustainment cell within these headquarters consists of the assistant chief of staff, personnel, assistant chief of staff, logistics, assistant chief of staff, finance, surgeon staff, and engineer officer. These staff elements assist the sustainment commander in providing command and control.

Mission Command of Sustainment Operations

Sustainment commanders must understand the application of mission command in executing command and control. Sustainment commanders use command in exercising authority and responsibility for the using available assets and resources for organizing, directing, and coordinating forces in executing missions. Resident in command is control. Sustainment staffs support their commanders by receiving communications and providing feedback that allows sustainment commanders to provide direction while coordinating and synchronizing support actions. Mission command is the Army's approach to command and control and allows subordinates to make operational decisions needed to achieve the commander's intent as conditions change. It provides sustainment leaders with the ability to anticipate and improvise methods for providing support based on the tactical situation.

Mission command is the Army's approach to command and control that empowers subordinate decision making and decentralized execution appropriate to the situation (ADP 6-0). It allows sustainment leaders to anticipate and improvise methods for providing support as conditions change and based on the tactical situation. Mission command also helps sustainment commanders exercise better control of sustainment forces as they receive feedback. Successful mission command is enabled by: competence, trust, shared understanding, commander's intent, mission orders, initiative, and risk acceptance.

I. Operational Reach

Operational reach is the distance and duration across which a unit can successfully employ military capabilities (JP 3-0). The limit of a unit's operational reach is its culminating point. Operational reach enables commanders to determine where to engage the enemy by giving them the ability to strike enemy decisive points to achieve decisive force at the appropriate time and place. Extending operational reach is critical to achieving the necessary freedom of action enabling the commander to seize and retain the initiative. Extended operational reach allows the commander to present the adversary with multiple dilemmas, confounding his decision making, and challenging their ability to act effectively.

Sustainment enables operational reach. Extending operational reach is a paramount concern for commanders. Commanders and staff increase operational reach through deliberate, focused operational design, and the allocating appropriate sustainment resources. This requires strategic sustainment capabilities such as materiel, supplies, health services, and other support and global distribution systems to deploy, maintain, and conduct operations over great distances for extended periods of time. Army forces can increase the joint force's ability to extend operational reach by securing and operating bases in the AOR, the use of contracted and local procurements, and the use of aerial delivery.

Topics related to operational reach include:

- Army Pre-positioned Stocks
- Force Projection
- Theater Opening

See p. 1-21 for further discussion of operational reach.

II. Freedom of Action

Freedom of action enables commanders with the will to act, to achieve operational initiative and control and maintain operational tempo. Enabling freedom of action requires that the sustainment commanders synchronize the sustainment plan with the operation plan to ensure supported commanders can operate freely and unencumbered due to limited resources. Sustainment commanders can enable freedom of action through preparing and putting in place sustainment activities.

Topics related to freedom of action include:

- Sustainment Preparation
- Sustainment Execution

See p. 1-28 for further discussion of freedom of action.

III. Endurance

Endurance refers to the ability to employ combat power anywhere for protracted periods. Endurance stems from the ability to create, protect, and sustain a force, regardless of the distance from its base and the austerity of the environment. Endurance involves anticipating requirements and continuity of integrated networks of interdependent sustainment organizations. Prolonged endurance is enabled by an effective distribution system and the ability to track sustainment from strategic to tactical level.

The sustainment principle continuity is paramount for ensuring endurance. Sustainment commanders must ensure the continuous link between strategic to tactical levels are maintained and free flowing. Commanders must be able to track sustainment in near real time and quickly make decisions resulting from changes to missions or operations.

Topics related to endurance include:

- Reconstitution
- Distribution

See p. 1-28 for further discussion of endurance.

Sustainment Planning

Sustainment planning begins with the operational commander's intent and concept of operations. This single, unifying idea provides direction for the entire operation. Based on a specific idea of how to accomplish the mission, commanders refine the concept of operations during planning. They adjust it throughout the operation as subordinates develop the situation or conditions change.

Sustainment planning continues with an analysis of the conditions in the operational environment with emphasis on the enemy and operational variables mission, enemy, terrain and weather, troops and support available, time available, and civil considerations. It involves understanding and framing the problem and envisioning the set of conditions that represent the desired end state.

Planning sustainment support of an operation is vital to mission success. Sustainment commanders and their planning staffs must coordinate and synchronize every stage of the planning process with the operational staff. They must also coordinate, synchronize and integrate the sustainment plan with joint and multinational partners to ensure a continuous linkage with strategic level providers.

The sustainment staff's role in synchronizing sustainment planning with operations is necessary to assist operational commanders and their staffs establish the conditions for what is possible. To ensure maximum freedom of action sustainment planners must understand the commander's intent, be able to visualize the operation and articulate the operational risks. Limitations like, insufficient infrastructure or the availability of a key class of supply or replacement weapon systems has bearing on the commander's ability to execute the mission. Sustainment commanders and staffs must present credible courses of action commensurate with sustainment capabilities to allow as much freedom of action as possible to accomplish the operational end state. While sustainment should not be an impediment to an operation, poor planning, lack of coordination, and understanding could severely influence the success of the operation.

See chap. 4, Sustainment Planning, for further discussion.

Doctrinal Linkage

Sustainment commanders must ensure their staffs are conversant in operational and sustainment doctrine as well as joint and multinational doctrine. Doctrine is a guide and establishes a basic framework for the conduct of operations that facilitates the planning process. Sustainment is a complex operation with many branches and sequels. Doctrine, based on past experiences and knowledge, helps leaders understand the broad context of what is possible, and then allow commanders and staffs to use judgment in the application of doctrine to adjust to the unique circumstances facing them.

To conduct sustainment in unified land operations, sustainment planners must fully understand the operations process—planning, preparing, executing, and continuously assessing the operation; operational art and mission command. Additionally, sustainment planners must understand the mechanics of joint operations and the necessary links that ensure strategic level support.

See pp. 4-2 to 4-3 for further discussion of the operations process.

A firm doctrinal grasp enables sustainment staffs to use and apply the planning tools of the operations process. In addition to a firm foundation in planning, operational, and sustainment doctrine, planners must understand maneuver doctrine in order to arrange sustainment actions in a manner to effectively support the operation.

Planning Considerations

Sustainment staffs create viable plans that are well coordinated and synchronized, facilitate operational tempo and support the commander's priorities before, during, and after operations. Sustainment planners in an operational headquarters generally do not drive the planning process, but must be fully integrated throughout planning to help understand, visualize, and describe solutions. Sustainment planners use the commander's intent, planning guidance, and the military decision making process to develop the sustainment concept of support.

The concept of support is derived from running estimates developed using a variety of planning tools. These running estimates project consumption rates for key classes of supply, casualty figures, maintenance requirements, and other sustainment requirements. (See ADP 5-0 for additional information.) Sustainment planners participate in all aspects of the military decision making process to ensure synchronization and unity of effort.

Planning in a sustainment headquarters requires planners to take an active role in the maneuver planning process. They assist in the development of the commander's understanding of the operational environment, identify shortfalls, articulate risks, and articulate the sustainment commander's vision. Sustainment planners must have the most current products from the organizations they support as well as planning products from their higher headquarters to ensure proper nesting and synchronization. Developing effective plans facilitates well-synchronized transitions between operational phases.

A comprehensive analysis of HN capabilities and plans incorporating these resources provides sustainment commanders with an array of options. For example, the availability of reliable contracting resources could reduce the burden on military resources and an already strained distribution system. Contracted resources could enable military resources to be focused on high priority operations that are unsuitable for civilian personnel. The use of contractors and HNS are often directly tied to the level of violence and threat in the operational environment.

I. Operational Reach

Operational reach is the distance and duration across which a unit can successfully employ military capabilities (JP 3-0). The limit of a unit's operational reach is its culminating point. Operational reach enables commanders to determine where to engage the enemy by giving them the ability to strike enemy decisive points to achieve decisive force at the appropriate time and place. Extending operational reach is critical to achieving the necessary freedom of action enabling the commander to seize and retain the initiative. Extended operational reach allows the commander to present the adversary with multiple dilemmas, confounding his decision making, and challenging their ability to act effectively.

See p. 2-5 for related discussion of operational reach from FM 4-0.

Sustainment enables operational reach. Extending operational reach is a paramount concern for commanders. Commanders and staff increase operational reach through deliberate, focused operational design, and the allocating appropriate sustainment resources. This requires strategic sustainment capabilities such as materiel, supplies, health services, and other support and global distribution systems to deploy, maintain, and conduct operations over great distances for extended periods of time. Army forces can increase the joint force's ability to extend operational reach by securing and operating bases in the AOR, the use of contracted and local procurements, and the use of aerial delivery.

A. Army Pre-Positioned Stocks (APS)

Pre-positioning of stocks in potential theaters provides the capability to rapidly supply and resupply forces until air and sea lines of communication are established. Army pre-positioned stocks are located at or near the point of planned use or at other designated locations. This reduces the initial amount of strategic lift required for power projection, to sustain the war fight until the line of communication with CONUS is established, and industrial base surge capacity is achieved (ATP 3-35.1).

Force Projection

Force projection is the ability to project the military instrument of national power from the United States or another theater, in response to requirements for military operations (JP 3-0). Force projection includes the processes of mobilization, deployment, employment, sustainment, and redeployment of forces. These processes are a continuous, overlapping, and repeating sequence of events throughout an operation. Force projection operations are inherently joint and require detailed planning and synchronization.

Sustainment of force projection operations is a complex process involving the GCC, strategic and joint partners such as USTRANSCOM, and transportation component commands like Air Mobility Command, military sealift command, SDDC, USAMC, DLA, Service Component Commands, and Army generating forces. ATP 3-35 provides greater detail on the force projection processes, but a general summarization is provided—

- **Mobilization** is the process of assembling and organizing national resources to support national objectives in time of war or other emergencies. The process by which the Armed Forces of the United States, or part of them, are brought to a state of readiness for war or other national emergency. (JP 4-05).
- **Deployment** is the movement of forces into and out of an operational area (JP 3-35). Sustainment is crucial to the deployment of forces. Joint transportation assets including air and sealift provide the movement capabilities for the Army.
- **Employment** is the strategic, operational, and tactical use of forces (JP 5-0).
- **Sustainment** provides logistics, financial management, personnel services, and health service support to maintain forces until mission completion. It gives Army forces its operational reach, freedom of action and endurance.
- **Redeployment** is the transfer of forces and materiel to the home and/or mobilization station for reintegration and/or out-processing (ATP 3-35). It requires retrograde of logistics, personnel services, and health service support and reuniting unit personnel and equipment at their home station.

See pp. 2-47 and 4-10 for related discussion of force projection, deployment operations, and RSOI.

B. Theater Opening

Theater opening is the ability to establish and operate ports of debarkation (air, sea, and rail), to establish a distribution system and sustainment bases, and to facilitate port throughput for the reception, staging, onward movement and integration of forces within a theater of operations. Preparing for theater opening operations requires unity of effort among the various commands and a seamless strategic¬to-tactical interface. It is a complex joint process involving the GCC and strategic and joint partners such as USTRANSCOM and DLA. Theater opening functions set the conditions for effective support and lay the groundwork for subsequent expansion of the theater distribution system.

See following pages (pp. 1-24 to 1-25) for further discussion.

C. Theater Closing

Theater closing is the process of redeploying Army forces and equipment from a theater, the drawdown and removal or disposition of Army non-unit equipment and materiel, and the transition of materiel and facilities back to host nation or civil authorities. Theater closing begins with the termination of joint operations.

See pp. 1-26 to 1-27 for further discussion.

Basing

Ref: ADP 4-0, Sustainment (Jul '19), pp. 3-8 to 3-9. See also AODS6 p. 1-32.

A base camp is an evolving military facility that supports military operations of a deployed unit and provides the necessary support and services for sustained operations (ATP 3-37.10). Basing directly enables and extends operational reach, and involves the provision of sustainable facilities and protected locations from which units can conduct operations. Army forces typically rely on a mix of bases and/or base camps to deploy and employ combat power to operational depth. Options for basing range from permanent basing in CONUS to permanent or contingency (non-permanent) basing OCONUS. A base camp is an evolving military facility that supports military operations of a deployed unit and provides the necessary support and services for sustained operations.

Bases or base camps may have a specific purpose (such as serving as an intermediate staging base, logistics base, or forward operating base) or they may be multifunctional. A base or base camp has a defined perimeter and established access controls, and should take advantage of natural and man-made features.

Bases or base camps may be joint or single service and will routinely support both U.S. and multinational forces, as well as interagency partners, operating anywhere along the range of military operations. Commanders often designate a single commander as the base or base camp commander that is responsible for protection, terrain management, and day-to-day operations of the base or base camp. This allows other units to focus on their primary function. Units located within the base or base camp are under the tactical control of the base or base camp commander for base security and defense.

Within large echelon support areas, controlling commanders may designate base clusters for mutual protection and mission command. Within a support area, a designated unit such as a BCT or maneuver enhancement brigade provides area security, terrain management, movement control, mobility support, clearance of fires, and required tactical combat forces. Operational area security operations focus on the protected force, base, base camp, route, or area. This allows sustainment units to focus on their primary function. Sustainment commanders and planners must constantly coordinate with supported operational staffs to synchronize sustainment operations to include all activities of the base camp life cycle and the basing strategy.

Refer to ATP 3-37.10 for more information on base camps.

Intermediate Staging Bases (ISB)

An intermediate staging base is a tailorable, temporary location used for staging forces, sustainment and/or extraction into and out of an operational area (JP 3-35). While not a requirement in all situations, the intermediate staging base may provide a secure, high-throughput facility when circumstances warrant. The commander may use an intermediate staging base as a temporary staging area en route to a joint operation, as a long-term secure forward support base, and/or secure staging areas for redeploying units, and noncombatant evacuation operations. An intermediate staging base is task organized to perform staging, support, and distribution functions as specified or implied by the CCDR and the theater Army operations order.

Forward Operating Bases (FOB)

Forward operating bases extend and maintain the operational reach by providing secure locations from which to conduct and sustain operations. They not only enable extending operations in time and space; they also contribute to the overall endurance of the force. Forward operating bases allow forward deployed forces to reduce operational risk, maintain momentum, and avoid culmination. Forward operating bases are generally located adjacent to a distribution hub. This facilitates movement into and out of the operational area while providing a secure location through which to distribute personnel, equipment, and supplies.

B. Theater Opening

Ref: ADP 4-0, Sustainment (Jul '19), pp. 3-6 to 3-9. See also p. 3-26.

Theater opening is the ability to establish and operate ports of debarkation (air, sea, and rail), to establish a distribution system and sustainment bases, and to facilitate port throughput for the reception, staging, onward movement and integration of forces within a theater of operations. Preparing for theater opening operations requires unity of effort among the various commands and a seamless strategic¬to-tactical interface. It is a complex joint process involving the GCC and strategic and joint partners such as USTRANSCOM and DLA. Theater opening functions set the conditions for effective support and lay the groundwork for subsequent expansion of the theater distribution system.

When given the mission to conduct theater opening, a sustainment brigade, and a mix of functional battalions and multi-functional CSSBs are assigned based on mission requirements. The sustainment brigade will participate in assessing and acquiring available HN infrastructure capabilities and contracted support and coordinating with military engineers for general engineering support.

See pp. 6-24 to 6-25 for additional discussion of theater opening from ATP 4-93.

Port Opening

Port opening is a subordinate function of theater opening. Port opening is the ability to establish, initially operate and facilitate throughput for ports of debarkation to support unified land operations.

The port opening process is complete when the port of debarkation and supporting infrastructure is established to meet the desired operating capacity for that node. Supporting infrastructure can include the transportation needed to support port clearance of cargo and personnel, holding areas for all classes of supply, and the proper in-transit visibility systems established to facilitate force tracking and end-to-end distribution.

Port opening and port operations are critical components for preparing and conducting theater opening. Commanders and staffs coordinate with the HN to ensure seaports and aerial ports possess sufficient capabilities to support arriving vessels and aircraft. USTRANSCOM is the port manager for deploying U.S. forces (ATP 4-0.1).

Joint Task Force Port Opening (JTF-PO)

The joint task force-port opening is a joint capability designed to rapidly deploy and initially operate aerial and seaports of debarkation, establish a distribution node, and facilitate port throughput within a theater of operations (JP 4-0). It is a standing task force that is a jointly trained, ready set of forces constituted as a joint task force at the time of need. The Army contribution to the Joint Task Force-Port Opening is the rapid port opening element which deploys within hours to establish air and seaports of debarkation in contingency response operations. The rapid port opening element also provides in-transit visibility and cargo clearance.

The joint task force-port opening facilitates joint reception, staging, onward movement, and integration and theater distribution by providing an effective interface with the theater joint deployment and distribution operations center and the sustainment brigade for initial aerial port of debarkation (APOD) operations. It is designed to deploy and operate for up to 60 days. As follow-on theater logistics capabilities arrive, the joint task force-port opening will begin the process of transferring mission responsibilities to arriving sustainment brigade forces or contracted capabilities to ensure the seamless continuation of airfield and distribution operations.

Seaports

USTRANSCOM's SDDC is the single port manager for all common user seaports of debarkation (SPODs) and as the single port manager it develops policy and advises the GCC on port management, recommends ports to meet operational demands, and is primarily responsible for the planning, organizing, and directing the operations at the seaport. The TSC and its subordinate sustainment brigades, terminal battalions and seaport operating companies perform the port operator functions at SPODs. These functions can include port preparations and improvement, cargo discharge and upload operations, harbor craft services, port clearance and cargo documentation activities. If the operational environment allows, SDDC may have the ability to contract locally for port operator support eliminating or decreasing the requirement for the TSC and its subordinate units.

The single port manager may have OPCON of a port support activity that is an ad hoc organization consisting of military and/or contracted personnel with specific skills to add in port operations. The TSC and SDDC will coordinate the port support activity requirements. It assists in moving unit equipment from the piers to the staging/marshaling/loading areas, assisting the aviation support element with movement of helicopters in preparation for flight from the port, providing limited maintenance support for equipment being offloaded from vessels, limited medical support, logistics support, and security for port operations.

Ideally, the SPOD will include berths capable of discharging large medium speed roll-on/roll-off ships. The SPOD can be a fixed facility capable of discharging a variety of vessels, an austere port requiring ships to be equipped with the capability to conduct their own offloading, or beaches requiring the conducting of joint logistics over-the-shore operations. Whatever the type of SPOD, it should be capable of accommodating an armored BCT.

When vessels arrive at the SPOD, the TSC and/or SDDC is responsible for discharging the unit equipment, staging the equipment, maintaining control and in-transit visibility, and releasing it to the unit. This includes minimum standards that are critical for the physical security/processing of DOD sensitive conventional arms, ammunition, and explosives, including non-nuclear missiles and rockets.

The theater gateway personnel accounting team and supporting HR company and platoons will normally operate at the SPOD as well as movement control teams to facilitate port clearance of personnel and equipment. The movement control team that has responsibility for the SPOD, coordinates personnel accounting with the supporting CSSB or sustainment brigade for executing life support functions for personnel who are transiting into or out of the theater.

Aerial Ports

Airfields supporting strategic air movements for deployment, redeployment, and sustainment are designated aerial ports. Aerial ports are further designated as either an aerial port of embarkation for departing forces and sustainment, or as an APOD for arriving forces and sustainment. Reception at the APOD is coordinated by the senior logistics commander and executed by an Air Force contingency response group/element and an arrival/departure airfield control group. It is an ad hoc organization established to control and support the arrival and departure of personnel, equipment, and sustainment cargo at airfields and must be a lead element when opening an APOD. Elements of a movement control team and an inland cargo transfer company typically operate the arrival/departure airfield control group however; any unit can perform the mission with properly trained personnel and the appropriate equipment. USTRANSCOM's Air Mobility Command is the single port manager for all common user APODs

Basing

A base camp is an evolving military facility that supports military operations of a deployed unit and provides the necessary support and services for sustained operations (ATP 3-37.10). Basing directly enables and extends operational reach, and involves the provision of sustainable facilities and protected locations from which units can conduct operations. Army forces typically rely on a mix of bases and/or base camps to deploy and employ combat power to operational depth. Options for basing range from permanent basing in CONUS to permanent or contingency (non-permanent) basing OCONUS. A base camp is an evolving military facility that supports military operations of a deployed unit and provides the necessary support and services for sustained operations.

See previous page (p. 1-23) for further discussion of basing.

C. Theater Closing

Ref: ADP 4-0, Sustainment (Jul '19), pp. 3-15 to 3-17. See also 3-32.

Theater closing is the process of redeploying Army forces and equipment from a theater, the drawdown and removal or disposition of Army non-unit equipment and materiel, and the transition of materiel and facilities back to host nation or civil authorities. Theater closing begins with the termination of joint operations.

Terminating Joint Operations

Terminating joint operations is an aspect of the CCDR's functional or theater strategy that links to achievement of national strategic objectives (JP 5-0). Based on the President's strategic objectives that compose a desired national strategic end state, the supported CCDR can develop and propose termination criteria. The termination criteria describe the standards to be met before conclusion of a joint operation. These criteria help define the desired military end state, which normally represents a period in time or set of conditions beyond which the President does not require the military instrument of national power as the primary means to achieve remaining national objectives. Termination criteria should account for a wide variety of operational tasks that the joint force may need to accomplish, to include disengagement, force protection (including force health protection support to conduct retrograde cargo inspections and pest management operations), transition to post-conflict operations, reconstitution, and redeployment. While there may be numerous terminating tasks the Army must achieve, the discussion below is deliberately broad and not all-inclusive. The discussion focuses on redeployment, drawdown of non-unit materiel, and transitioning of materiel, facilities and capabilities to HN or civil authorities.

Redeployment

Redeployment involves the return of personnel, equipment, and materiel to home and/or demobilization stations and is considered as an operational movement critical in reestablishing force readiness (ATP 3-35). Under the sustainable readiness model deployment and redeployment of forces in support of extended operations is a cyclic process. However, for terminating joint operations, Army forces may be completely redeployed from the joint operational area. Many of the procedures used in the initial deployment of forces to theater apply during redeployment. Unlike cyclic deployment where units fall in on positioned unit equipment and sets, termination redeployment efforts require movement of unit sets to aerial or seaports of embarkation for shipment to home station or other designated locations. After completion of military operations, redeploying forces move to designated assembly areas or directly to redeployment assembly areas. The same elements that operate and manage the theater distribution system during deployment and sustainment will usually perform support roles during redeployment.

Two critical aspects of equipment and materiel redeployment are property book accountability and asset visibility. Furthermore, the identification of how much equipment is on the ground, location of the equipment, type of equipment, condition of the equipment, and reporting procedures will allow for timely planning, as it will impact mode of transportation, resources, timeline, personnel, storage capabilities, and the like. Moreover, the accurate reporting of equipment on property books, locations, and condition will influence strategic-level decision making in terms of funding, field or sustainment reset, and disposal of equipment.

Drawdown

Planning for drawdown of non-unit equipment and materiel should occur early in the operational and strategic planning process. Drawdown planning entails more than returning equipment to CONUS. At the strategic level, the requirement for specific types of equipment may necessitate the redistribution of equipment to another AOR.

Even though equipment drawdown is an important mission in the redeployment operation, it may not be the Army's or the GCC main priority; thus, prioritization of equipment redistribution/disposition must be established early on to maximize distribution capacity and velocity. A challenge is visibility of strategic-level materiel requirements synthesized into the already established priority timeline. Overcoming this challenge requires strategic-level collaboration between partners including Service Headquarters, GCCs, USAMC, DLA, and USTRANSCOM to effectively, efficiently and strategically reset both Joint and Army forces. Through this partnership, the nation's resources are preserved for other security needs.

As planners begin the process of reducing forces in a theater of operations, they must develop a balance between operational capability and sustainment capability. There is a natural tendency to eliminate the sustainment and enabler forces first because they do not provide an inherent capability to engage with the population or enemy. However, as the sustainment and enabling force are withdrawn, there is a direct impact on the operational forces in the form of reduced operational reach and requirements for assumption of additional missions.

To provide unity of effort and ensure operational freedom of action through rapid return, repair, redistribution, and combat power regeneration for the Army, a USAMC Responsible Reset Task Force provides a comprehensive solution for drawdown. Reset is a coordinated effort to methodically plan and execute the timely, repair, redistribution, and/or disposal of non-unit equipment, non-consumable and materiel identified as excess to theater requirements, to home station, sources of repair, or storage or disposal facilities. Through the phased redeployment of forces, the Responsible Reset Task Force mission will reset the Army in the shortest time possible.

The TSC and ESC work closely with DLA in the close out of materiel in the theater. The DLA support team serves as the single point of contact to the TSC and ESC. The DLA support teams are tasked to provide support to the theater closure plans, and are focused on providing support to all echelons based on the priorities of effort. During theater closure, DLA provides support in the form of adjusting the flow of supply classes I, II, III (B) (P), IV, VIII and IX to ensure support to the warfighter. DLA could, if requested, provide a theater consolidation and shipping point for departing forces. Additionally, DLA assists Army forward operating base closure by providing expeditionary disposal remediation teams to provide expert advice and oversight to U.S. Forces on the preparation for and the closure of Army units.

Closing Operational Contracts

The supporting contracting organization will be required to terminate and close out existing contracts and orders and includes the redeployment of contractor personnel. Ratifications and claims must be processed to completion. Contracting for life support services and retrograde support may continue until the last Soldiers and U.S. personnel depart, but standards of support should be reduced as much as possible prior to final contract closeout. In some operations, the supporting contracting organization may be required to assist in the transition of contracted support (the contracts themselves are not transferable) to the Department of State, a multi-national partner, or to the HN. This transition of contract support may include limited continuation of existing contracts in support of high priority Department of State operations. Because of the nature of contract support transition and/or closeout during termination operations, contingency contracting officers will often be some of the last Soldiers to leave the area of operations.

Port Closing

USTRANSCOM, through SDDC is responsible for providing and managing strategic common-user sealift and terminal services in support GCC's drawdown or termination requirements. As the single port manager for USTRANSCOM, it is SDDC's responsibility to integrate and synchronize strategic and theater re-deployment execution and distribution operations within each CCDR's area of responsibility. Ensures drawdown and/or termination requirements are met through the use of both military and commercial transportation assets based on the supported commander business rules and Joint Deployment and Distribution Enterprise best business practices.

II. Freedom of Action

Preparation for the sustainment of operations consists of activities performed by units to improve their ability to execute an operation. Preparation includes but is not limited to plan refinement, rehearsals, information collection, coordination, inspections, and movements. For sustainment to be effective, several actions and activities are performed across the levels of war to properly prepare forces for operations.

A. Sustainment Preparation

Sustainment preparation of the operational environment is the analysis to determine infrastructure, physical environment, and resources in the operational environment that will optimize or adversely impact friendly forces means for supporting and sustaining the commander's operations plan. The sustainment preparation of the operational environment assists planning staffs to refine the sustainment estimate and concept of support. It identifies friendly resources (HNS, contractible, or accessible assets) or environmental factors (endemic diseases, climate) that impact sustainment.

See facing page for further discussion.

B. Sustainment Execution

Execution is putting a plan into action by applying combat power to accomplish the mission (ADP 5-0). It focuses on actions to seize, retain, and exploit the initiative.

Sustainment determines the depth and duration of Army operations. It is essential to retaining and exploiting the initiative and it provides the support necessary to maintain operations until mission accomplishment. Failure to provide sustainment could cause a pause or culmination of an operation resulting in the loss of the initiative. It is essential that sustainment planners and operation planners work closely to synchronize all of the warfighting functions, in particular sustainment, to allow commanders the maximum freedom of action.

Sustainment plays a key role in enabling the simultaneous offensive, defensive, and stability or defense support of civil authorities tasks that occur as part of unified land operations. For example, general engineering support provides construction support to protect key assets such as personnel, infrastructure, and bases. Horizontal and vertical construction enables assured mobility of transportation networks and survivability operations to alter or improve cover and concealment to ensure freedom of action, extend operational reach, and endurance of the force. Legal personnel supporting rule of law activities may find themselves working closely with HN judicial, law enforcement, and corrections systems personnel.

See chap. 2, Sustainment Operations, and chap. 3, Sustainment Execution.

III. Endurance

Endurance refers to the ability to employ combat power anywhere for protracted periods. Endurance stems from the ability to create, protect, and sustain a force, regardless of the distance from its base and the austerity of the environment. Endurance involves anticipating requirements and continuity of integrated networks of interdependent sustainment organizations. Prolonged endurance is enabled by an effective distribution system and the ability to track sustainment from strategic to tactical level.

Reconstitution operations are extraordinary actions that commanders plan and implement to restore attritted units' combat effectiveness commensurate with the mission requirements and available resources. Reconstitution restores combat power to the levels necessary to maintain endurance and continue operations.

Distribution is the primary means to prolong endurance. Distribution is the operational process of synchronizing all elements of the logistic system to deliver the "right things" to the "right place" at the "right time" to support the geographic combatant commander (JP 4-0).

See following page (p. 1-30) for an overview of distribution.

Sustainment Preparation

Ref: ADP 4-0, Sustainment (Jul '19), pp. 3-12 to 3-13.

Preparation for the sustainment of operations consists of activities performed by units to improve their ability to execute an operation. Preparation includes but is not limited to plan refinement, rehearsals, information collection, coordination, inspections, and movements. For sustainment to be effective, several actions and activities are performed across the levels of war to properly prepare forces for operations.

Negotiations and Agreements

Negotiating and establishing agreements with HN resources is an important task of mission command for sustainment operations. Through negotiation and agreements, Army forces are able to reduce the military sustainment footprint and/or all military sustainment resources to focus on higher priority operations that may not be conducive to civilian support functions. Negotiation of agreements enables access to HNS resources identified in the requirements determination phase of planning. This negotiation process may facilitate force tailoring by identifying available resources (such as infrastructure, transportation, warehousing, and other requirements) which if not available would require deploying additional sustainment assets to support.

Host nation support agreements may include pre-positioning of supplies and equipment, OCONUS training programs, and humanitarian and civil assistance programs. These agreements are designed to enhance the development and cooperative solidarity of the HN and provide infrastructure compensation should deployment of forces to the target country be required. The pre-arrangement of these agreements reduces planning times in relation to contingency plans and operations.

Sustainment Preparation of the Operational Environment

Sustainment preparation of the operational environment is the analysis to determine infrastructure, physical environment, and resources in the operational environment that will optimize or adversely impact friendly forces means for supporting and sustaining the commander's operations plan. The sustainment preparation of the operational environment assists planning staffs to refine the sustainment estimate and concept of support. It identifies friendly resources (HNS, contractible, or accessible assets) or environmental factors (endemic diseases, climate) that impact sustainment. Some of the factors considered (not all-inclusive) are as follows:

- **Geography.** Information on climate, terrain, and endemic diseases in the AO to determine when and what types of equipment are needed. For example, water information determines the need for such things as early deployment of well-drilling assets and water production and distribution units.
- **Supplies and Services.** Information on the availability of supplies and services readily available in the AO. Supplies (such as subsistence items, bulk petroleum, and barrier materials) are the most common. Common services consist of bath and laundry, sanitation services, and water purification.
- **Facilities.** Information on the availability of warehousing, cold-storage facilities, production and manufacturing plants, reservoirs, administrative facilities, hospitals, sanitation capabilities, and hotels.
- **Transportation**. Information on road and rail networks, inland waterways, airfields, truck availability, bridges, ports, cargo handlers, petroleum pipelines, materials handling equipment, traffic flow, choke points, and control problems.
- **Maintenance**. Availability of host nation maintenance capabilities.
- **General Skills**. Information on the general skills such as translators and skilled and unskilled laborers.

See pp. 2-12 and 4-1 for additional discussion.

Distribution

Ref: ADP 4-0, Sustainment (Jul '19), pp. 3-13 to 3-14.

Distribution is the primary means to prolong endurance. Distribution is the operational process of synchronizing all elements of the logistic system to deliver the "right things" to the "right place" at the "right time" to support the geographic combatant commander (JP 4-0).

The distribution system consists of a complex of facilities, installations, methods, and procedures designed to receive, store, maintain, distribute, manage, and control the flow of military materiel between point of receipt into the military system and point of issue to using activities and units.

The joint segment of the distribution system is referred to as global distribution. Global distribution is defined as the process that coordinates and synchronizes fulfillment of joint force requirements from point of origin to point of employment (JP 4-09). It provides national resources (personnel and materiel) to support the execution of joint operations.

The Army segment of the distribution system is referred to as theater distribution. Theater distribution is the flow of personnel, equipment, and materiel within theater to meet the GCC's missions (JP 4-09). The theater segment extends from the ports of debarkation or source of supply (in theater) to the points of need.

Distribution management synchronizes and optimizes transportation, its networks, and materiel management with the warfighting functions to move personnel and materiel from origins to the point of need in accordance with the supported commander's priorities. Distribution management includes the management of transportation and movement control, warehousing, inventory control, order administration, site and location analysis, packaging, data processing, accountability for equipment (materiel management), people, and communications.

See pp. 3-38 to 3-39 for further discussion. *Refer to ATP 4-0.1, Army Theater Distribution, for additional information on distribution and distribution management.*

A support team from the Medical Logistics Management Center accomplishes the distribution management of medical materiel. The Medical Logistics Management Center support team collocates with the DMC of the TSC and ESC to provide the MEDCOM (DS) with visibility and control of all class VIII.

In-Transit Visibility

In-transit visibility is the ability to track the identity, status, and location of DOD units, and non-unit cargo (excluding bulk petroleum, oils, and lubricants) and passengers; patients and personal property from origin to consignee, or destination across the range of military operations (JP 4-01.2). This includes force tracking and visibility of convoys, containers/pallets, transportation assets, other cargo, and distribution resources within the activities of a distribution node.

Retrograde of Materiel

Another aspect of distribution is retrograde of materiel. Retrograde of materiel is an Army logistics function of returning materiel from the owning or using unit back through the distribution system to the source of supply, directed ship-to location, and/or point of disposal (ATP 4-0.1). Retrograde includes turn-in/classification, preparation, packing, transporting, and shipping. Contractor equipment is considered when planning retrograde. To ensure these functions are properly executed, commanders must enforce supply accountability and discipline and utilize the proper packing materials. Retrograde of materiel can take place as part of theater distribution operations and as part of redeployment operations. Retrograde of materiel must be continuous and not be allowed to build up at supply points/nodes.

Early retrograde planning is essential and necessary to preclude the loss of materiel assets, minimize environmental impact, and maximize use of transportation capabilities. Planners must consider environmental issues when retrograding hazardous material.

Sustainment Operations

Ref: FM 4-0, Sustainment Operations (Jul '19), forward and preface.

> *"Logistics sets the campaign's operational limits."*
> *– Joint Pub 1, Joint Warfare of the U.S. Armed Forces, November 1991*

Today's operational environment presents threats to the Army and joint force that are more dangerous in terms of capability and magnitude than those we faced in Iraq and Afghanistan. Major regional powers — Russia, China, Iran, and North Korea seek to gain strategic positional advantage. These nations, and other competitors, are fielding capabilities to deny U.S. freedom of action in the air, land, maritime, space, and cyberspace domains and reduce U.S. influence in critical areas of the world. In some domains, they already have parity or overmatch, a challenge the joint force has not faced in twenty-five years.

The proliferation of advanced technologies; competitor emphasis on training, modernization, and professionalization; combined with extremist ideologies; and the increasing speed of human interaction makes large-scale combat more lethal. As the Army and the joint force focused on counter-insurgency and counter-terrorism at the expense of other capabilities, our competitors watched, learned, adapted, modernized, and devised strategies that placed us at a position of relative disadvantage.

Combat power may win battles, but sustainment wins wars. In light of current threats, simply acquiring more resources is not enough to succeed. Sustainment doctrine is critical to more than just the sustainment community; it lays the foundation for all Army leaders to plan and execute unified land operations. The Army and joint force must adapt and prepare for large-scale combat operations in highly contested, lethal environments where the enemy employs long-range fires and other capabilities that rival or surpass our own. The key to all successful military campaigns throughout history is the ability to sustain itself. Sustainment units must be able to operate effectively and survive in expeditionary environments across all contested domains. The ability to sustain ourselves ensures that the Army, with its joint and multi-national partners, has the operational reach, freedom of action, and the endurance to execute campaigns.

FM 4-0, Sustainment Operations, provides a doctrinal approach for our armies, corps, divisions, and brigades to address the challenges of sustaining operations across all four Army strategic roles – Shape Operational Environments, Prevent Conflict, Prevail in Large-Scale Ground Combat and Consolidate Gains. It is the cornerstone of all sustainment doctrine, detailing how the Army will provide logistics, financial management, personnel services and health service support to the force during unified land operations. FM 4-0 is applicable across the range of military operations during competition and conflict. Its logic and format aligns with FM 3-0 to illustrate how sustainment is central to all military operations, tasks, and activities. FM 4-0 also includes critical planning considerations and examples that illustrate the volume and scope of materiel required to maintain an expeditionary army.

> *FM 4-0, Sustainment Operations*, augments the Army's principal doctrine on providing sustainment support as found in *ADP 4-0*. It describes how Army sustainment forces, as part of a joint team, provide support to Army and other forces with particular emphasis on support to large-scale combat operations. It is a companion manual to *FM 3-0* on Army operations and, together with *ADP 4-0*, provides the foundation for how Army sustainment forces support prompt and sustained large-scale combat operations.

I. The Army in Joint Operations

Ref: FM 3-0 (w/Chg 1), Operations (Dec '17), pp. 1-14 to 1-16.

Joint operations are military actions conducted by joint forces and those Service forces employed in specific command relationships with each other, which of themselves, do not establish joint forces. Traditionally, campaigns are the most extensive joint operations, in terms of the amount of forces and other capabilities committed and the duration of operations. In the context of large-scale combat operations, a campaign is a series of related major operations achieving strategic and operational objectives within a given time and space. A major operation is a series of tactical actions, such as battles, engagements, and strikes, and it is the primary building block of a campaign. Army forces conduct supporting operations as part of a joint campaign.

The Army's primary mission is to organize, train, and equip its forces to conduct prompt and sustained land combat to defeat enemy ground forces and seize, occupy, and defend land areas. The Army accomplishes its mission by supporting the joint force in four strategic roles: shape operational environments, prevent conflict, conduct large-scale ground combat, and consolidate gains. The strategic roles clarify the enduring reasons for which the U.S. Army is organized, trained, and equipped.

U.S. Army Strategic Roles

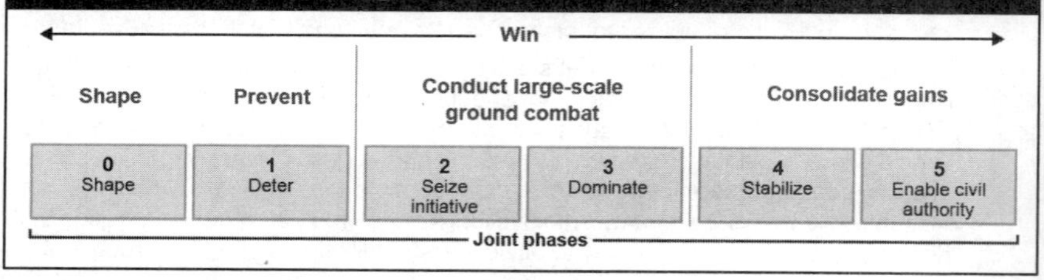

Ref: FM 3-0 (Oct '17), fig. 1-4. Army strategic roles and their relationships to joint phases.

I. Shape Operational Environments

Army operations to shape bring together all the activities intended to promote regional stability and to set conditions for a favorable outcome in the event of a military confrontation. Army operations to shape help dissuade adversary activities designed to achieve regional goals short of military conflict. As part of operations to shape, the Army provides trained and ready forces to geographic combatant commanders (GCCs) in support of their theater campaign plan. The theater army and subordinate Army forces assist the GCC in building partner capacity and capability and promoting stability across the AOR. Army operations to shape are continuous throughout a GCC's AOR and occur before, during, and after a joint operation within a specific operational area.

Shaping activities include security cooperation and forward presence to promote U.S. interests, developing allied and friendly military capabilities for self-defense and multinational operations, and providing U.S. forces with peacetime and contingency access to a host nation. Regionally aligned and engaged Army forces are essential to achieving objectives to strengthen the global network of multinational partners and preventing conflict.

See pp. 2-9 to 2-18 for discussion of sustaining operations to shape.

II. Prevent Conflict

Army operations to prevent include all activities to deter an adversary's undesirable actions. These operations are an extension of operations to shape designed to prevent adversary opportunities to further exploit positions of relative advantage by raising the potential costs to adversaries of continuing activities that threaten U.S. interests. Prevent activities are generally weighted toward actions to protect friendly forces, assets, and partners, and to indicate U.S. intent to execute subsequent phases of a planned operation. As part of a joint force, Army forces may have a significant role in the execution of directed FDOs. Additionally, Army prevent activities may include mobilization, force tailoring, and other predeployment activities; initial deployment into a theater to include echeloning command posts; employment of intelligence collection assets; and development of intelligence, communications, sustainment, and protection infrastructure to support the JFC's concept of operations.

See pp. 2-19 to 2-26 for discussion of sustaining operations to prevent.

III. Conduct Large-Scale Ground Combat

During large-scale combat operations, Army forces focus on the defeat and destruction of enemy ground forces as part of the joint team. Army forces close with and destroy enemy forces in any terrain, exploit success, and break their opponent's will to resist. Army forces attack, defend, conduct stability tasks, and consolidate gains to attain national objectives. Divisions and corps are the formations central to the conduct of large-scale combat operations, organized, trained and equipped to enable subordinate organizations. The ability to prevail in ground combat is a decisive factor in breaking an enemy's will to continue a conflict.

See pp. 2-27 to 2-40 for discussion of large-scale combat operations sustainment.

See chap. 3 for related discussion of sustainment execution (logistics, personnel services, and health service support).

IV. Consolidate Gains

Army operations to consolidate gains include activities to make enduring any temporary operational success and to set the conditions for a sustainable environment, allowing for a transition of control to legitimate civil authorities. Consolidation of gains is an integral and continuous part of armed conflict, and it is necessary for achieving success across the range of military operations. Army forces deliberately plan to consolidate gains during all phases of an operation. Early and effective consolidation activities are a form of exploitation conducted while other operations are ongoing, and they enable the achievement of lasting favorable outcomes in the shortest time span. Army forces conduct these activities with unified action partners. In some instances, Army forces will be in charge of integrating forces and synchronizing activities to consolidate gains. In other situations, Army forces will be in support. Army forces may conduct stability tasks for a sustained period of time over large land areas. While Army forces consolidate gains throughout an operation, consolidating gains becomes the focus of Army forces after large-scale combat operations have concluded.

See 2-41 to 2-44 for discussion of sustaining operations to consolidate gains.

II. Sustainment of Unified Land Operations

Ref: FM 4-0, Sustainment (Jul '19), pp. 1-5 to 1-6.

Multi-Domain Extended Battlefield

The interrelationship of air, land, maritime, space, and cyberspace requires a cross-domain understanding of an OE. Sustainment commanders and staffs must understand friendly and enemy capabilities that reside in each domain and the potential impacts to sustainment operations. Space and cyberspace operations that degrade sustainment information system capabilities as well as the peer threat employment of conventional, special operations, guerilla, and insurgent forces to interdict friendly air, land, and maritime operations, can all degrade sustainment operations. Since many friendly capabilities are not organic to sustainment formations, commanders and staffs plan, coordinate for, and integrate joint and other unified action partner capabilities in a multi-domain approach to operations.

During large-scale combat against peer threats, all friendly forces, including those conducting sustainment tasks, are in contact and under observation in the space and cyberspace domains, as well as the information environment. In light of potential adversaries' capabilities, Soldiers must be prepared to operate in denied, degraded, and disrupted communication environments. FM 6-99 includes standardized report and message formats. The formats in the field manual are for manual and voice use. The report and message formats help users prepare and manually transmit written and voice reports and messages. Each format provides an organized template to record, pass, and store information. Operations on the multi-domain battlefield will require Soldiers maintain the ability to operate in a contested cyber or space environment, without the use of enterprise systems. Sustainment organizations should maintain manual reporting skills and should also be prepared to use alternate methods of reporting such as, telephone, radio transmission, messenger, or hard copy.

Sustainment is inherently joint. Joint interdependence is the purposeful reliance of all the services upon each other's capabilities to maximize the complementary and reinforcing effects of both. The United States Air Force provides airlift capabilities to move Army forces quickly across strategic LOCs. The United States Navy provides strategic sealift into deep draft ports and land capabilities to supplement Army theater opening forces entering areas where ports are austere, damaged, or non-existent. The joint force enables Army sustainment and provides the services and capabilities needed to sustain unified land operations.

See chap. 8, Joint Logistics, for further discussion.

Army sustainment enables unified land operations by providing the support required to keep the Army and its unified action partners engaged in operations across the OE. Lack of sufficient sustainment support slows operational tempo, contributes to early culmination, and could lead to the defeat of friendly forces on the ground. Since most Army sustainment capabilities reside within the Reserve Component, contracted support, high readiness reserve units and Army prepositioned stocks play critical roles during initial stages of a crisis response. It is essential that sustainment and operational planners be inseparable in the planning, preparation, execution, and assessment of operations.

Freedom of Action, Operational Reach, and Prolonged Endurance

The sustainment warfighting function is essential for conducting operations and generating combat power as the Army performs its strategic roles. Sustainment provides the operational commander freedom of action, operational reach, and prolonged endurance necessary to shape operational environments, prevent conflict, prevail in large-scale ground combat operations, and consolidate gains.

Freedom of Action

Freedom of action enables commanders to achieve operational initiative and maintain operational tempo. Successful sustainment commanders facilitate freedom of action by planning and executing sustainment activities able to support the greatest possible number of courses of action for the supported commander in a particular situation. Effective sustainment execution provides continuous and responsive sustainment that allows supported commanders to operate freely.

Sustainers enable freedom of action by conducting sustainment preparation of the OE, synchronizing the sustainment plan with the scheme of maneuver, and maintaining flexibility and disciplined initiative while supporting maneuver commanders. The proper application of the mission command philosophy to sustainment operations encourages the greatest possible freedom of action from subordinates. Appropriate sustainment unit command and support relationships equips subordinate commanders with the authority and flexibility to act boldly and achieve desired support effects in the absence of continuous direction from higher headquarters (HQ).

See p. 1-28 for discussion of freedom of action from ADP 4-0.

Operational Reach

Operational reach is the distance and duration across which a force can successfully employ military capabilities (JP 3-0). Depth is the extension of operations in time, space, or purpose to achieve definitive results (ADP 3-0). Army forces require significant sustainment to strike enemy forces and disrupt or destroy throughout their depth, preventing the effective employment of reserves, command and control nodes, sustainment, and other capabilities not in direct contact with friendly forces. Operations in depth disrupt the enemy's decision cycle and contribute to protecting the force by destroying enemy capabilities before the enemy can use them. Commanders balance their forces' tempo and momentum to produce simultaneous results throughout their operational areas. To achieve simultaneity, commanders establish a higher tempo of operations to target enemy capabilities located at the limit of a force's operational reach. Generally, the higher the tempo of operations, particularly during large-scale combat, the greater the demand for sustainment.

Sustainment operations are critical to enabling operational reach. The Army's sustainment capability assists in providing crucial theater and port-opening functions that further enable the strategic and operational reach of the joint force. Effective operational reach requires gaining and maintaining operational access in the face of enemy anti-access and area denial capabilities and actions. (JP 3-0). Army sustainment capabilities continue to provide the bulk of Army support to other Services, CUL, and other common sustainment resources to prevent friendly forces from culminating.

See p. 1-21 for discussion of operational reach from ADP 4-0.

Prolonged Endurance

Endurance refers to the ability to employ combat power anywhere for protracted periods. It stems from the ability to organize, protect, and sustain a force, regardless of the distance from its base and the austerity of the environment. Endurance involves anticipating requirements and making the most effective, efficient use of available resources. Endurance gives Army forces their campaign quality. It makes permanent the transitory effects of other capabilities.

Sustainment provides the support necessary to ensure operations continue until mission accomplishment. Sustainment also provides the capabilities necessary for maintaining the personnel and materiel replacements essential to maintaining combat power. Sustainment support that enables prolonged endurance requires the synchronization of the elements of sustainment.

See p. 1-28 for discussion of endurance from ADP 4-0.

III. Command and Support Relationships by Echelon

Ref: FM 4-0, Sustainment (Jul '19), pp. 2-63 to 2-64.

The command and support relationships most commonly used vary by echelon. The relationships described below are typical examples. However, relationships may vary based on mission and operational variables.

See pp. 2-36 to 2-40 for related discussion of support areas.

Joint Security Area

Discussion of each type of Army headquarters assumes employment in the joint security area. All units may or may not be employed. Within the joint security area command relationships will normally be the following:

- The TSC is assigned to the theater Army.
- The MEDCOM (DS) is assigned or attached to the theater Army.
- Task-organized medical formations in the joint security area from the MEDBDE (SPT) are attached to the MEDCOM (DS). The MLMC forward team is assigned to the MEDCOM (DS) and coordinates class VIII distribution with the DMC of the TSC and ESC.
- The ESC is attached to the TSC.
- Sustainment brigades are attached to the ESC. The sustainment brigades will have the same relationship with the TSC if an ESC is not employed in the joint security area.
- CSSBs are attached to the sustainment brigade.
- Functional logistics and HR companies are attached to the CSSB or special troops battalion.

Corps Area

The task organization of sustainment units in the corps area may be complex. Multiple sustainment headquarters may be operating in the area. Some of the sustainment headquarters are assigned or attached to the corps headquarters while others are not.

An ESC is assigned to the corps. The MEDCOM (DS) provides direct or GS to the corps through the MEDBDE (SPT). The theater Army will also attach a sustainment brigade, or multiple sustainment brigades, to the ESC. Furthermore, CSSBs, functional logistics battalions, and functional companies will be attached to the sustainment brigades as required to support the corps mission. All command relationships will be designated in an OPORD issued by the theater Army. The units attached to the corps receive mission orders from the corps headquarters, normally in support of specific corps operations.

Other ESCs operating in the corps area will remain attached to the TSC and will also have sustainment brigades, CSSBs, functional logistics battalions, and functional logistics companies attached. MEDBDEs (SPT) will have hospital centers and medical battalions (multifunctional) attached. Sustainment units not attached to the corps receive mission orders from the TSC to support broader theater operations.

The normal support relationship for all sustainment units in the corps area is GS. However, other support relationships may be designated based on mission requirements.

See pp. 6-2 to 6-5 for related discussion.

Division Area

The task organization of sustainment units in the division area is subject to the same complexities as the corps area. Multiple echelons above division sustainment units and elements of the MEDBDE (SPT) may be operating in the area alongside the DSB with its organic DSSB. Medical elements of the MEDBDE (SPT) are normally OPCON to the division commander and their parent medical organization retains ADCON.

Other sustainment brigades operating in the division area will remain attached to an ESC in the corps area. Sustainment brigades attached to the ESC will have CSSBs, functional logistics battalions, and functional logistics companies attached. The CSSBs attached to the sustainment brigade will have functional companies attached. Sustainment units not assigned to the division receive mission orders from the ESC to support broader theater operations.

The DSB and its organic DSSB are assigned to the division. The division's task-organized DSB provides direct support to the division and GS for all units in or passing through their geographic area. The DSB commander remains responsible for integration and synchronization of sustainment in the division area.

See pp. 6-5 to 6-10 for related discussion.

Brigade Combat Team Area

The BSB is organic to the BCT headquarters; it is an integral part of the BCT table of organization and equipment. The distribution, field maintenance, brigade medical support, and forward support companies are organic to the BSB. The distribution, field maintenance, and brigade medical support companies have a GS relationship with the BCT units to include the forward support companies. The forward support companies have a direct support relationship with the supported maneuver battalion. The forward support companies may be attached to the supported maneuver battalion for limited periods of time based on mission requirements. This attachment is normally limited to the duration of a specific mission or phase of an operation.

See chap. 5, Brigade Support, for related discussion.

Refer to AODS6-1 (w/SMARTupdate 1): The Army Operations & Doctrine SMARTbook, pp. 1-50 to 1-53, for related discussion of the Army's operational framework. An operational framework organizes an area of geographic and operational responsibility for the commander and provides a way to describe the employment of forces. The framework illustrates the relationship between close operations, operations in depth, and other operations in time and space across domains.

IV. Training for Large-Scale Combat Operations

Ref: FM 4-0, Sustainment (Jul '19), pp. 2-69 to 2-73.

Sustainment units train to support large-scale combat operations and a wide range of operations that include regular and irregular warfare, humanitarian assistance operations, security force assistance, and support to civil authorities. In today's complex and uncertain strategic environment, it is imperative that sustainment support remains globally responsive and supports not only the Army but also its joint and multinational partners. Meeting the challenges of an uncertain, complex, and interconnected strategic environment requires sustainment formations that are adaptive, innovative, flexible, and agile in training and operations.

Support of large-scale combat operations requires deployment of sustainment forces from all components on short notice. Therefore, the increased training readiness of sustainment units is key to the success of our Army; the majority of which are in the Reserve Component. Since the Army is largely a CONUS-based force, leaders and units must practice the ability to deploy into an austere theater through contested ports, conduct RSOI, and immediately support large-scale combat operations.

Training Sustainment Formations

Unit commanders at all levels are the critical link for implementing unit training management to ensure progressive readiness and unit proficiency within their formations. At all echelons, sustainment formations play a critical role in enabling military decision making within mission command, and sustainment organizations train collectively with our war-fighting partners to sustain this proficiency.

With the majority of EAB sustainment force structure being in the Reserve Component, commanders ensure these formations are well trained and integrated into the multi-component sustainment force structure prior to deployment. Commanders also leverage opportunities to strengthen sustainment training relationships with other Services, interagency organizations, private industry, and multinational partners whenever feasible.

Sustainment units at all levels integrate training with the units that the sustainment units support. Commanders do not train sustainment units in isolation. Rather, commanders develop organizational proficiency as part of a combined arms or joint team, supporting other war-fighting functions to achieve specified outcomes. Maneuver units and their supporting sustainment elements routinely train on resupply (including delivery of logistics packages), vehicle recovery, convoy operations, unit maintenance, casualty collection, HR support, and financial management team missions. Establishing the forward arming and refueling points (FARP) and conducting refuel on the move operations to support extended moves for operations like attacks, mobile defenses, and defensive retrograde are other examples of key sustainment activities integrated into maneuver unit collective training events.

Commanders maximize institutional and operational training opportunities at home station and ensure use of current tools, such as the Individual Critical Task List. The technical nature of sustainment core functions requires constant engagement and maintenance to prevent skill atrophy. Sustainment units strive for mastery-level proficiency of these essential tasks through multiple iterations of training events. That concept applies to not only the actual technical functions, but also staff activities and basic field craft.

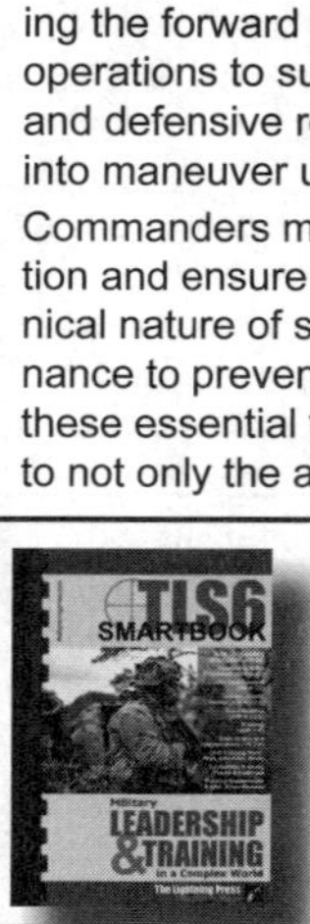

Refer to TLS6: The Leader's SMARTbook, 6th Ed. (Military Leadership & Training in a Complex World) for complete discussion of Military Leadership (ADP 6-22); Leader Development (FM 6-22); Counsel, Coach, Mentor (ATP 6-22.1); Army Team Building (ATP 6-22.6); Military Training (ADP 7-0); Train to Win in a Complex World (FM 7-0); Unit Training Plans, Meetings, Schedules, and Briefs; Conducting Training Events and Exercises; Training Assessments, After Action Reviews (AARs); and more!

(Sustainment Operations)

I. Operations to Shape

Ref: FM 4-0, Sustainment Operations (Jul '19), chap. 3.

Sustaining Operations to Shape

Examples of shaping activities are set the theater, military engagements, security cooperation, combined training and exercises, and sustainment preparation of the OE. The following paragraphs describe these activities and provide examples of sustainment support. There are a number of sustainment considerations that typically apply to specific types of shaping activities. Shaping activities incorporate a large portion of conventional force and SOF core activities, requiring planning for both.

I. Set the Theater

Setting the theater is a continuous shaping activity and is conducted as part of steady-state posture and for contingency or crisis response operations. Set the theater describes the broad range of activities conducted to establish the conditions in an operational area for the execution of strategic plans. The GCC has overall responsibility for this activity, but executes many of its responsibilities through the TSC of the ASCC. The purpose of setting a theater is to establish favorable conditions for the rapid execution of military operations and the support requirements for a specific OPLAN during crisis or conflict.

Setting the theater involves all of the warfighting functions.

- The focus of the command and control warfighting function is the organization and the command and control of forces to accomplish missions. The ASCC tailors, and controls Army forces in the AOR. The TSC provides command and control of assigned and attached sustainment forces in the AOR.
- The movement and maneuver warfighting function focuses on mobilization, deployment, employment, and redeployment of forces. The ASCC conducts theater opening and is responsible for RSOI of Army forces. The TSC provides TACON for movement of Army of forces into theater, conducts RSOI, and provides command and control for theater distribution.
- The intelligence warfighting function focuses on planning, collecting, producing, and disseminating intelligence. The ASCC provides Army intelligence capabilities to support CCMD operations. Part of that intelligence support is setting the theater from an intelligence perspective that is bigger than supporting sustainment operation (see FM 2-0). Supporting theater sustainment intelligence requirements is a significant and complex mission. To meet theater sustainment intelligence requirements the TSC G-2 directs, plans, collects, and disseminates intelligence to subordinate units. However, the TSC G-2 also depends on national to tactical intelligence and the theater intelligence architecture for access to critical intelligence.
- The fires warfighting function plans and directs Army fires in coordination with joint fires. Sustainment supports joint and Army fires.
- Protection focuses on establishing force protection measures for Army forces. The ASCC is responsible for Army forces protection in the AOR. Sustainment plans, coordinates, and executes protection plans for sustainment forces.

Sustainment Operations in Support of Military Engagement & Security Cooperation

Ref: FM 4-0, Sustainment (Jul '19), pp. 3-3 to 3-6.

II. Military Engagement

Military engagement is the routine contact and interaction between individuals or elements of the Armed Forces of the U.S. and those of another nation's armed forces, or foreign and domestic civilian authorities or agencies to build trust and confidence, share information, coordinate mutual activities, and maintain influence (JP 3-0). Military engagements are primarily State Department-led engagements. These occur as part of security cooperation, but also extend to interaction with domestic civilian authorities. Military engagements can reduce tensions and may preclude conflict; if conflict is unavoidable, these engagements may allow the U.S. to enter into it with stronger alliances or coalitions.

A military-to-military engagement requires less sustainment support than large-scale combat operations. Host-nation support is civil and/or military assistance rendered by a nation to foreign forces within its territory during peacetime, crises or emergencies, or war based on agreements mutually concluded between nations (JP 4-0). Sustainers leverage contacts during this shaping activity to facilitate logistics partnerships, enhance interoperability, establish or refine HNS agreements, and gain access to potential critical infrastructure nodes. Routine interactions during operations to shape will help establish agreements and partnerships that may be mutually beneficial during follow-on operations. Sustainers are involved in all these engagements primarily to facilitate sustainment agreements and coordinate planning for future operations.

III. Security Cooperation

Security cooperation is all DOD interactions with foreign security establishments to build security relationships that promote specific U.S. security interests, develop allied and partner nation military and security capabilities for self-defense and multinational operations, and provide U.S. forces with peacetime and contingency access to allied and partner nations (JP 3-20). The Department of State leads and provides oversight for security cooperation efforts.

Sustainment commands simultaneously plan and synchronize sustainment operations for theater security cooperation activities. The Army approach to supporting the larger DOD security cooperation effort is either indirect or direct.

Indirect approach activities involve the U.S. supporting a nation with security cooperation programs, given legitimate authorities, designed to enhance its capability and capacity. The sustainment command supports the following programs and activities typical of the indirect approach:

- International military education and training.
- Multinational and joint exercises and exchange programs.

Direct approach activities involve U.S. forces assisting the host nation by conducting operations for the mutual benefit of the host nation and U.S. interests. These operations either provide a capability the host nation does not possess or increase the capacity of the host nation to conduct the operation.

Establishing a host nation coordination center is a means to enhance stability and interaction between nations. The lead element for coordination center is security cooperation (G-9), with representation from the logistics and medical (G-4, Surgeon), financial management (G-8, FMSC), engineers (G-7), G-3, host nation representatives, Department of State and any other governmental agencies or non-governmental organizations as required. A multinational commander, especially one that operates under a parallel command structure, establishes a coordination center during the shape role of an operation. It organizes and controls functional areas including logistics, and civil-military operations. A coordination center is the initial focal point for support issues such as force sustainment, medical support, infrastructure engineering, HNS, and movement control. As a multinational force matures, the center's role includes activities such as force provision or force deployment. Member nations provide action officers who are familiar with its activities when a coordination center is activated. Multinational forces are encouraged to maintain contact with parent headquarters. For more information, see FM 3-16.

Security assistance is an element of security cooperation that is authorized by the Department of State and administered by the DOD (ATP 3-93). It is a group of programs the U.S. Government uses to provide defense articles, military training, and other defense-related services by grant, loan, credit, or cash sales. Security assistance programs are typically focused on the transfer of defense articles and services to eligible foreign governments, the provision of training and education to foreign military personnel, and the sale of construction services in support of partner nations' military establishments. Sustainers are frequently required to provide support and logistics training to support these activities.

Security force assistance are the DOD activities that contribute to unified action by the U.S. Government to support the development of the capacity and capability of foreign security forces and their supporting institutions (JP 3-22). Army sustainers interact with sustainers of partner countries in these operations. These partnerships assist in future interoperability and enhance partner nation militaries to be more self-sufficient in logistics capabilities.

Foreign internal defense is participation by civilian and military agencies of a government in any of the action programs taken by another government or other designated organization to free and protect its society from subversion, lawlessness, insurgency, terrorism, and other threats to its security. Sustainment support operations are limited by applicable U.S. law without an acquisition and cross servicing agreement (ACSA). Such support usually consists of transportation or limited maintenance support, although an ACSA can allow additional support.

In security cooperation, support considerations include support to U.S. forces and support to host nation forces based on a variety of authorizations. Support to host or partner nations is primarily driven by already established ACSAs. If ACSAs are not in place, local State Department officials act as lead for determining what support can be granted to host nations.

Combined Training and Exercises

Army forces build partner combat readiness and set conditions for future contingencies through training and exercises. Combined exercises familiarize both forces with the capabilities and shortfalls of the other force and develop procedures to leverage capabilities and mitigate shortfalls. These serve to sustain and/or develop interoperability between nations as well as build partnership capacity.

These exercises are extremely diverse in size, participation, duration, and sustainment requirements. The requirements may be a few aircraft being sustained by HNS or OCS with minimal DOD sustainment to large-scale training operations requiring a combination of HNS, OCS, and home station sustainment activities. These exercises may be with one U.S. military agency and another country or with joint agencies and multiple nations.

Refer to TAA2: Military Engagement, Security Cooperation & Stability SMARTbook (Foreign Train, Advise, & Assist) for further discussion. Topics include the Range of Military Operations (JP 3-0), Security Cooperation & Security Assistance (Train, Advise, & Assist), Stability Operations (ADRP 3-07), Peace Operations (JP 3-07.3), Counterinsurgency Operations (JP & FM 3-24), Civil-Military Operations (JP 3-57), Multinational Operations (JP 3-16), Interorganizational Cooperation (JP 3-08), and more.

Setting the theater during operations to shape (from a sustainment perspective) involves actions to provide sustainment (logistics, financial management, personnel services, and health service support) to SOF, SFAB, military engagements, and security cooperation as well as conduct sustainment preparation of the OE, and other activities in support of daily Title 10 requirements, contingency operations, and crisis response operations.

Sustainment support to setting the theater in operations to shape involves theater opening, receiving initial forces, equipment, and supplies and assembling them into mission-tailored units; and transporting them to their final destination. LOGCAP ensures a rapid response to emerging crises. The Army should also utilize HNS to the maximum extent possible to give sustainment units time to close in theater. It places set the theater as a priority and provides for holistic operational planning across the Army's strategic roles. LOGCAP planners are placed at the ASCC, TSC and corps with LOGCAP decision authority being at the ASCC.

Set the theater activities during operations to shape may also include the establishment of a distribution network, identify local procurement requirements, employing the theater gateway PAT, and setting the conditions for medical operations. The broad range of setting the theater activities also includes synchronization and integration of sustainment through the establishment of boards, bureaus, centers, cells, and working groups.

IV. Sustainment Preparation of the Operational Environment (OE)

Sustainment preparation of the OE is a continuous shaping activity involving analysis to determine infrastructure, environmental, or resource factors in the OE that impact the Army's ability to sustain a commander's operations plan. Sustainment planners in the TSC and/or ESC use the sustainment preparation of the OE analysis to update and refine sustainment estimates and the concept of support. Analysis products cover such topics as selection of LOC, determination of operational stock assets, and design of a distribution network and information technology infrastructure for the theater.

The sustainment preparation of the OE identifies resources available to friendly forces and factors that impact sustainment for example, endemic diseases, and climate. Representative questions planners try to answer include—

- Are any agreements with host nation or multinational partners in place?
- Does the host nation have supply resources available to support U.S. forces?
- Are health threats present in the OE and have mitigation strategies been identified?
- What host nation facilities are available (airports, sea ports, military facilities, hospitals, barracks, water systems, electrical networks, warehouses, and other storage facilities are available?
- Has geospatial information gathering, site reconnaissance and surveys of ports been conducted?
- What are the contracting options?
- What resources have other U.S. Government departments and agencies or unified action partners identified?
- What skills does the local labor force have?
- What host nation communications capabilities are available?
- What is the entry environment—permissive or non-permissive?
- Are host nation banking infrastructure and e-commerce capabilities available?
- Are considerations in place on host nation postal limitations and restrictions?

See following pages (pp. 2-14 to 2-15) for an overview and further discussion. See also p. 1-29 and 4-1.

Army Health System Support to Set the Theater

Ref: FM 4-0, Sustainment (Jul '19), pp. 3-2 to 3-3.

The MEDCOM (DS) maintains a regional focus that encompasses the GCC's entire AOR and is critical for the successful provision of AHS support to set the theater. The medical commander's ability to assess host nation medical capability/capacity and the presence of health threats prevalent in the AOR, facilitates the planning and execution of regional strategies for establishment of the theater joint trauma system and mitigation of identified threats. The MEDCOM (DS) provides the GCC an effective tool to assist in maximizing the use of scarce medical resources, shaping the security environment by building partner medical capacity, and alleviating health conditions that not only impact U. S. military forces, but multinational partners, and particular challenges faced by the host nation. Efforts must also be made to understand the roles and responsibilities of all agencies involved (to include the Department of State, World Health Organization, partner nations, and others) for integration and synchronization of all medical capabilities in the region. The MEDCOM (DS) also provides AHS support to set the theater through coordination and integration of strategic medical capabilities from the U.S. sustaining base. It also provides support through global health engagements, establishment and maintenance of medical support agreements, deploying medical technical expertise for consultation services and other support, and military medical training exercises, as well as the following:

- Executing AHS support to other Services.
- Ensuring adherence to eligibility criteria for treatment in U.S. military MTFs.
- Recommending theater evacuation policy adjustments.
- Providing theater food protection support.
- Coordinating with USTRANSCOM for patient movement plans.
- Ensuring integration and interoperability of theater medical capabilities.
- Providing AHS support to foreign humanitarian assistance and disaster relief.
- Conducting medical preparation of the OE.
- Maximizing use of host nation medical capabilities.
- Establishing and executing occupational and environmental health surveillance programs and countermeasures.
- Coordinating with the National Center for Medical Intelligence, Centers for Disease Control and Prevention, and other strategic partners for identification and mitigation of regional health threats.
- Planning and coordination for AHS support to—
  - Noncombatant evacuation operations.
  - Detainee operations.
  - RSOI and theater opening.
  - Large-scale casualty events and prolonged care.
  - Other Services.

The MEDCOM (DS) is assigned to the ASCC and serves as the theater medical command responsible for synchronization and execution of AHS support operations within the AOR. The MEDCOM (DS) commander coordinates with the ASCC surgeon to provide AHS support within the AOR.

Sustainment Preparation of the OE (Overview)

Ref: FM 4-0, Sustainment (Jul '19), pp. 3-8 to 3-13.

Geography and Environmental Factors. Information on climate, terrain, flooding, precipitation, and endemic diseases in the AO is used to determine when and what types of equipment are needed. For example, water information determines the need for such things as early deployment of well-digging assets and water purification and distribution units. Endemic disease conditions, vector-borne illnesses, and local medical resources greatly influence requirements for Army medical assets. Terrain, weather, lengths of LOCs, and other factors influence maintenance and fuel requirements. Sustainment planners must consider assistance through veterinary services to reduce risk and prevent disease during sustainment preparation of the operational environment. Methods to prevent disease include-improving sanitation practices, waste management controls, and pest and vector control. These are crucial to disease prevention. Regional spraying and insect repellent application to guard against hazardous flora and fauna are examples of prevention methods.

Supply and Services. Planners require information on the availability and characteristics of supplies and services readily available in the operational area. Supplies such as subsistence items, water and ice, bulk petroleum, and barrier materials are the most common. Compatibility of commodities and services for example, electrical and fuel connectors and classes of supply, should be considered. Services that may be available consist of shower and laundry, sanitation services, and water purification.

See p. 3-18. For additional information, refer to ATP 4-42.

Facilities. Identifying infrastructure and capacity to receive personnel, commodities, equipment and other resources is an imperative. Assessing availability of warehouses, cold-storage facilities, production and manufacturing plants, reservoirs, administrative facilities, hospitals, sanitation capabilities, hotels, barracks, military bases, airfields, and rail systems can greatly reduce the requirement for the deployment of U.S. assets.

Transportation. Development of any distribution plan depends on information regarding road and rail networks, inland waterways, airfields, truck availability, bridges, ports, cargo handlers, petroleum pipelines, materials-handling equipment, traffic flow, choke points, and potential movement control complications. Movement of cargo in theater must comply with laws, regulations, and rules for transporting materiel as established by the international agreements, the U.S. and the host nation. Cargo restrictions vary from theater to theater. The CCDR has the authority to set cargo restrictions. Non-materiel cargo restrictions include weight limits at APODs and SPODs, vehicle weight and dimension limits on routes, and certain airspace controls. Materiel cargo restrictions may include explosives, pyrotechnics, POLs, compressed gases, corrosives, and batteries. Planners also need to plan for any host nation administrative requirements for activities such as border crossings, customs, and diplomatic clearances.

See p. 3-11. For additional information, refer to ATP 4-11.

Maintenance. Key planning considerations for maintenance support include the availability of host nation maintenance capabilities, information on contract maintenance assets, the commonality or standardization of major end items and repair parts across the force, and the host nation's internal capacity for additive manufacturing and fabricating repair parts.

See p. 3-5. For additional information, refer to ATP 4-33.

General Skills. Sustainment planners at all echelons gather information on the general population with a focus on local personnel who can function as translators and skilled and unskilled laborers. Sustainment plans are influenced by availability of drivers, administrative clerks, dockworkers, materials-handling equipment operators, food service personnel, security guards, and mechanics.

Army Pre-Positioned Stocks. APS is equipment and supplies configured in unit sets forward-positioned afloat and ashore located at or near the point of planned use or at other designated locations. This allows units identified for early entry to deploy with limited to-accompany-troops equipment and draw unit equipment at an RSOI location. This reduces the initial amount of strategic lift required for power projection to sustain the warfighter until the LOC from the strategic base is established.

Refer to ATP 3-35.1 for additional information on APS.

Banking and Economy. Financial management leaders/planners analyze the economic impact of the use of currency (U.S. and foreign) on the local economy. This economic analysis is a detailed report provided to commanders to make decisions on applying the economic instrument of power, identifying all financial aspects of a specific geographical area, and the effect of a U.S. force presence will have on the specific AOR. Banking support includes negotiating with host nation banking facilities, advising unit commanders on the use of local currency, and coordinating with strategic providers. Some examples are Office of the Under Secretary of Defense (Comptroller), the U.S. Treasury, Defense Finance and Accounting Service, Federal Reserve Bank, and the United States Army Financial Management Command.

Non-Organic Support. Approximately eighty percent of Army sustainment capabilities reside in the Reserve Component. The Soldiers and units that provide those capabilities may require several weeks to months to mobilize, train, and deploy before becoming available to support military operations in an AOR. The delayed arrival of this organic capability requires sustainment planners to coordinate and synchronize sustainment from non-organic sources, including HNS, ACSAs, and OCS.

Operational Contract Support. The OCS process enables commanders to acquire services, commodities, and construction support from commercial sources, thereby reducing sealift requirements, accelerating the deployment of combat power, and extending the commander's operational reach. Because of the long lead-time required to coordinate commercial support, planners need to clearly identify capabilities and limitations as part of the deliberate planning process. OCS augments sustainment capabilities through the integration of commercial sector support activities into military operations. OCS plays an important role in shaping operations by aiding in the establishment of favorable conditions for rapid execution of military operations. During operations to shape, sustainment planners at the theater level align allocated military resources against forecasted requirements and employ OCS to mitigate risk. OCS consists of three complementary functions: contract support integration, contracting support, and contractor management.

See p. 3-40. Refer to ATP 4-10 for information regarding OCS roles and responsibilities.

Agreements with other Nations. Sustainment preparation of the OE considers whole-of-government initiatives, including bilateral or multilateral diplomatic agreements. These agreements allow U.S. forces to access ports, terminals, airfields, and bases within the AOR to support future military contingency operations. The Department of State and the appropriate U.S. diplomatic mission negotiate bilateral or multilateral agreements. Positive U.S. relations and successful bilateral engagement in one nation can impact U.S. interests in other regional locations. HNS is civil and military assistance rendered by a nation to foreign forces within its territory during peacetime, crises or emergencies, or war based on agreements mutually concluded between nations. Negotiating HNS and theater support contracting agreements may include pre-positioning of supplies and equipment, civilian support contracts, OCONUS training programs, and humanitarian and civil assistance programs.

V. Analysis, Planning and Coordination

Ref: FM 4-0, Sustainment (Jul '19), pp. 3-13 to 3-15.

Two key concepts in sustaining shaping operations are the requirement for a comprehensive sustainment preparation of the operational environment and the supporting role other agencies perform in sustaining these operations. Analysis, planning, and coordination are critical to successful sustainment operations.

Theater Analysis and Planning

DLA planners conduct mission analysis and develop the DLA concept of logistics support for the theater. As part of the logistics supportability analysis process DLA works with the CCMD and Services to evaluate time-phased requirements against DLA capabilities (industrial base, stock status, contracts, etc.) to identify potential risks to mission execution. Requirements are translated into a sustainment wedge and provided to USTRANSCOM during time-phased force deployment conferences. Mitigation strategies for known shortfalls are developed for potential sourcing as part of setting the theater. Additionally, contracts are reviewed to ensure appropriate surge clauses are in place; market research is conducted to identify local procurement opportunities, readiness rates of deployable DLA capabilities are assessed, and DLA works closely with the Services to support readiness rates of critical weapon systems.

The ASCC conducts theater sustainment analysis as part of setting the theater. This analysis begins as part of shaping activities and continues as part of operations to prevent. This analysis identifies risk in terms of access, capabilities and capacities across the AOR. It consists of the actions taken by sustainment planners to optimize means (force structure, resources, and strategic lift) for supporting the commander's plan. These actions include identifying and preparing intermediate staging bases and forward operating bases; selecting and improving LOC; and forecasting and building operational stock assets forward and afloat. Sustainment preparation of the OE— part of setting the theater—identifies potential risks in terms of access, infrastructure capabilities, and capacities in theater so planners can develop alternatives and mitigating measures. Logistics planners use the sustainment preparation of the operational environment analysis to optimize the distribution system. The analysis products may include identifying forward operating bases, selecting LOC, determining operational stock assets, and designing a distribution and automatic information technology infrastructure for the theater.

See chap. 7, Theater Support. Chapter 4 of ATP 4-94, and FM 4-95 provide detailed information for sustainment preparation of the OE.

The theater campaign plan is the GCC's vehicle for operationalizing the theater strategy. The theater campaign plan provides a framework within which the GCC conducts security cooperation activities and military engagement with regional partners through cooperative security and development. A theater campaign plan's main function is to provide guidance to coordinate steady-state components of contingency planning by conducting security cooperation activities across the AOR.

The theater Army develops the force structure required to support the theater campaign plan. The theater Army requests Army forces and the resources required to support them. These resources include sustainment capabilities required. The theater Army in conjunction with the sustainment command (TSC or ESC) provides support to forces participating in exercises to support the theater security cooperation plan. The theater Army in conjunction with the sustainment command (TSC or ESC) designs effective and efficient movement plans for land forces into and out of the theater of operations. The theater Army in conjunction with the sustainment command (TSC or ESC) also requests forces to support ongoing Army responsibilities for sustainment.

Interagency Cooperation

During operations to shape, it is critical that sustainers work with other agencies to achieve integrated whole-of-government operations and synchronization of interagency activities, such as information sharing, security cooperation, and foreign assistance. This requires Army sustainers to develop and share detailed time-phased logistics requirements/estimates with interagency partners. During competition, the Department of State will play a large role in maintaining or establishing a path toward greater stability, with elements from the intelligence community and DOD in support. Other agencies that may provide crucial logistics support during shaping operations are DLA and USAMC. Liaison officers are often the first sustainment planners on ground. Liaison officers begin sustainment preparations and discussions with the ASCC, TSC, State Department officials, and country teams. The liaison officers may assist with expediting movement of equipment and logistics planning. USTRANSCOM may also send liaison officers to facilitate transportation operations.

Department of State

The Department of State leads and provides oversight for security cooperation efforts through its bureaus, offices, and overseas missions. Security cooperation activities are conducted and coordinated throughout the AOR by, with, or through the ASCC to—

- Build defense relationships that promote specific U.S. security interests.
- Develop multinational and friendly military capabilities for self-defense and multinational operations.
- Provide U.S. forces with peacetime and contingency access to a host nation.

The TSC is heavily involved in the operations to shape and provide the sustainment support that makes shaping operations possible. Since the boundaries for Department of State are different from the boundaries of the GCCs the TSC through the ASCC may have to coordinate operations through multiple Department of State representatives.

Senior Defense Officials

The senior defense official or defense attaché is the principal military advisor on defense and national security issues. The senior defense official is also the senior diplomatically accredited DOD military officer assigned to a U.S. diplomatic mission, and the single point of contact for all DOD matters involving the embassy or DOD elements assigned to or working from the embassy. This individual acts as the in-country focal point for planning, coordinating, supporting, and/or executing U.S. defense issues and activities in the host nation, including theater security cooperation programs under the oversight of the GCC. Army sustainers may be tasked to assist in establishing and maintaining sustainment foundations to enhance nation partnerships. For additional information regarding DOD operations at U.S. embassies see DODD 5105.75(D).

Defense Security Cooperation Agency

The Defense Security Cooperation Agency plays a critical role in shaping activities. Defense Security Cooperation Agency's mission is to advance U.S. national security and foreign policy interests by building the capacity of foreign security forces to respond to shared challenges. Defense Security Cooperation Agency leads the broader U.S. security cooperation enterprise in its efforts to train, educate, advise, and equip foreign partners. Defense Security Cooperation Agency administers security cooperation programs that support U.S. policy interests and objectives identified by the White House, DOD, and Department of State. These objectives include developing specific partner capabilities, building alliances and partnerships, and facilitating U.S. access. Defense Security Cooperation Agency integrates security cooperation activities in support of a whole-of-government approach, provides execution guidance to DOD entities that implement security cooperation programs, exercises financial and program management for the foreign military sales system and many other security cooperation programs, and assists in the long-term development of the security cooperation workforces.

VI. Army Health System Support during Shape

AHS support activities identified for planning and coordination to set the theater are implemented and executed during operations to shape. The MEDCOM (DS), MEDBDE (SPT), and their subordinate units provide AHS support to theater opening, RSOI of early entry forces, integration of joint and multinational medical capabilities for establishment and execution of the theater trauma system, and regional health engagements in support of security cooperation and deterrence missions to build partner medical capacity, which promotes regional stability. Other key AHS activities during shape include the provision of AHS support for maintenance and execution of medical support agreements, home station medical readiness and training activities, force tailoring for generation of medical combat power, providing Army medical support to other Services and unified action partners, as well as assessment and release of theater APS and other medical logistics support. Efforts are also made to identify capability gaps and determine fills.

The role of the AMEDD in support of the GCC is to preserve the fighting strength by providing medically ready forces and trained, ready, and rapidly deployable medical forces. Preserving the fighting strength also focuses on maximizing the number of troops available for employment by preventing or mitigating health threats, maximizing return to duty rates, minimizing morbidity and mortality, and clearing the battlefield. To accomplish these tasks, the AMEDD leverages the surgeon cells (staff channel) at each echelon and medical command channels to provide AHS support of U.S. national objectives, the GCC's theater campaign plan, and the Unified Command Plan across the full range of military operations.

Medical staff channels (surgeon cells) conduct planning, coordination, synchronization, and integration of AHS support to plans. The chain of medical commanders execute AHS support to established plans. Starting with the Surgeon General, the surgeon cells at echelon identify, assess, counter and/or mitigate health threats throughout the range of military operations. The surgeon cells advise commanders on the optimal placement and coordinate use of medical assets to support operations (i.e. forward positioned forces, APS, adjacent/ supporting region assets).

During operations to shape, institutional medical organizations within the AMEDD (in coordination with the Defense Health Agency) conduct home station medical activities to maintain health readiness, support contingencies, and project medical forces in support of the GCC. At EAB, sustainment organizations without organic medical capabilities receive AHS support on an area basis. For METT-TC dependent medical augmentation support requirements, medical assets may be OPCON to echelon commanders. Organic medical elements at BDE and below provide DS to parent units.

Refer to AODS6-1 (w/SMARTupdate 1): The Army Operations & Doctrine SMARTbook, pp. 2-37 to 2-42, for discussion of operations to shape from FM 3-0. While the U.S. Army must be manned, equipped, and trained to operate across the range of military operations, large-scale ground combat against a peer threat represents the most significant readiness requirement. FM 3-0 provides doctrine for how Army forces, as part of a joint team, and in conjunction with unified action partners, do this.

Chap 2

(Sustainment Operations) II. Operations to Prevent

Ref: FM 4-0, Sustainment Operations (Jul '19), chap. 4.

Sustaining Operations to Prevent

The purpose of operations to prevent is to deter adversary actions contrary to U.S. interests (FM 3-0). It is typically conducted in response to activities that threaten unified action partners and require the deployment or repositioning of forces in a theater to demonstrate the willingness to fight if deterrence fails. Furthermore, prevent activities enable the joint force to gain positions of relative advantage prior to future combat operations. As part of crisis response or limited contingency operations, operations to prevent are tailored in scope and scale to achieve a strategic or operational-level objective. It may be conducted as a stand-alone response to a crisis, as in a noncombatant evacuation operation, or as part of a larger joint operation such as large-scale combat.

Operations to prevent are characterized by actions to protect friendly forces and indicate the intent to execute subsequent phases of a planned operation. With the transition from shaping to deterrence, the theater Army shifts to refining contingency plans and preparing estimates for land power based on GCC guidance.

Sustainment of operations to prevent require a force array tailored to the type of operation, geographic location, permissiveness of the environment, threat, and a host of other considerations determined during the planning phase of prevent operations. Commanders must also consider possible sequels and branches during planning to ensure optimal command and support relationships are established, and the right mix of forces are identified for potential follow-on operations. The phased arrival of Army sustainment forces will require planners to integrate non-Army solutions, including acquisitions cross-service agreements and OCS, to enable the rapid deployment of combat forces.

The ASCC is responsible for all Army operations, to include reception of forces, sustaining forces, and preparing to redeploy forces. The ASCC also interacts with strategic and operational commands and organizations essential to the theater distribution network.

Assessment of the sustainment preparation of the OE and analysis conducted during operations to shape is a key activity for sustainers during operations to prevent. The outcomes of sustainment preparation of the OE informed OPLANs and time-phased force deployment data (TPFDD). It is during operations to prevent that plans and estimates are refined, the theater distribution network is expanded, and actions are taken to deploy forces as required.

Refer to AODS6-1 (w/SMARTupdate 1): The Army Operations & Doctrine SMARTbook, pp. 2-43 to 2-60, for full discussion of operations to prevent from FM 3-0. While the U.S. Army must be manned, equipped, and trained to operate across the range of military operations, large-scale ground combat against a peer threat represents the most significant readiness requirement. FM 3-0 provides doctrine for how Army forces, as part of a joint team, and in conjunction with unified action partners, do this.

I. Refining Plans and Developing Estimates

Ref: FM 4-0, Sustainment (Jul '19), pp. 4-1 to 4-4.

DLA, in coordination with the CCMD and Services, continues to refine its mission analysis and logistics estimates to identify joint logistics requirements and continue to set the theater. Actions may include:

- Changes in stock positioning (for example-locations, quantities).
- Contract reviews for appropriate surge clauses and operational contingency zones.
- Market research to identify local procurement opportunities.
- Discussions with industrial base partners regarding production capacities.
- Identification of potential acquisition policy waivers that may be required.
- Close coordination with the Services on readiness rates and critical weapon systems.
- A readiness review of DLA deployable capabilities.

DLA may provide additional liaison officers or deploy a DLA Assessment Team with forward deployed units to coordinate support.

The TSC, ESC, DSB, and sustainment brigade refinement of sustainment plans and estimates is a continuous process. Sustainment planners will continue to conduct detailed analysis and assessment to update support requirements and availability of resources based on the ever changing OE. These estimates are provided to the CCMD and interagency partners for planning purposes. The availability of commercial support will play a particularly important role during early stages of operations to prevent and may change rapidly due to enemy activity or host nation policy.

See pp. 4-15 to 4-22 for related discussion of running estimates and mission analysis.

Logistics Estimation

Sustainers employ tools such as the Operational Logistics Planner that produces class of supply consumption estimates for units at all levels from team, to platoon, to company, to Army headquarters. Operational Logistics Planner enables staffs at all levels to estimate mission requirements for all class I, II, III bulk, III packaged, IV, V, VI, VII, VIII, and IX as well as water, ice, and mail for their units. It also provides an analysis of the estimated transportation assets needed to get the supplies to the units. Operational Logistics Planner uses the latest Army approved planning rates and force structures. Units can be customized to add or remove equipment and personnel.

See pp. 4-18 to 4-19. For additional information, refer to FM 4-0 appendix C.

Casualty Estimation and Replacement Requirements

The Army G-1 is the functional proponent for overall casualty estimation (killed in action, captured, missing in action, and wounded in action, disease and non-battle injury) and must coordinate closely with the Army Surgeon General for Army evacuated rates, in support of projected manning requirements. Replacement requirements consist of the killed in action and evacuated Soldiers not return-to-duty, prisoner of war and missing in action. Casualty estimation is conducted at ASCC level and above as part of the planning process for contingency operations and approved by the CCDR.

The Medical Planners' Toolkit (discussed below) integrates several tools including the Casualty Rate Estimation Tool, which provides the capability for planners to emulate the plan to calculate the combat and noncombat injuries and illnesses that would be expected during military operations. Casualty estimates can be generated for ground combat, ship attacks, fixed facilities, and natural disasters. The Casualty Rate Estimation Tool generated patient streams are based on the casualty estimate and the user selected Patient Condition Occurrence Frequency distribution. It uses stochastic processes to allow users to evaluate medical planning risk.

See pp. 4-20 to 4-22.

Medical Planning Tools

The Joint Medical Planning Tool, the Medical Planners' Toolkit, and the Medical Contingency Requirements Workflow are planning tools approved for calculation of medical requirements. The Joint Medical Planning Tool and Medical Planners' Toolkit are fully integrated for versatility and enhanced medical planning efficiency.

The Joint Medical Planning Tool is a computer-based simulation tool developed by the Naval Health Research Center that models patient flow from the point of injury through more definitive care. It supports research, medical systems analysis, operational risk assessment, and field medical services planning. The Medical Planners' Toolkit is a suite of tools developed to support the joint medical planning community. The Medical Planners' Toolkit combines the Patient Conditions Occurrence Frequency tool, the Casualty Rate Estimation Tool, and the Expeditionary Medicine Requirements Estimator into a single desktop application. The Medical Contingency Requirements Workflow is the authoritative source for estimating medical materiel contingency requirements by time and role of care.

See pp. 4-27 to 4-30 for discussion of Army Health Service (AHS) planning. Refer to JP 4-02, ATP 4-02.55, and ATP 4-02.7 for additional information.

Financial Management Planning

During operations to prevent, FM forces are continuing to assist Commanders with setting the theater, tailoring and projecting FM support to the force. Some initial tasks include—

- Identifying funding authority for deployment operations and funding ISO operations, to include special funding authorities, throughout the strategic roles.
- Determining the requirement for multiple disbursing station symbol numbers (DSSNs) and dispersion throughout theater.
- Securing initial funding for central funding and cash management operations (for example establish a limited depositary account).
- Establishing financial management policies and procedures to address finance and resource management operations (for example cash holding authority, fund certification thresholds).
- Analyzing current banking infrastructure throughout the theater and determining the need to expand synchronization.
- Identifying the mobility the financial management distribution network between strategic and operational echelons for replenishment (strategic lift, surface transportation) in the absence of host nation capabilities.
- Identifying existing and potential contracting vehicles for requirements.

Postal Operations Planning

Postal operations rely on a well-coordinated network to process mail and provide postal services within a theater of operations. It is an important combat multiplier and an essential consideration for military operations. It must be included in all planning and OPORDs. Information contained in the OPORD should describe all information to support efficient postal operations that will require planning for significant logistics given limited air and ground transportation in large-scale combat operations. Postal units must have assured real-time communications by voice and data as well as securable transportation. The HRSC must determine postal support requirements well before activities to prevent. Close coordination with the Military Postal Service Agency and corps/division planners is key. It is the theater commander's responsibility to determine at which point, in the Army roles for large-scale combat operations, postal operations should begin. Additionally, planners should anticipate and prepare for the lethal effects of large-scale combat operations battlefields and adverse weather conditions affecting postal operations (for example; delayed transportation, increased backlog, and possible battle/weather damage) in large-scale combat operations environments.

Refer to FM 4-0 appendix A and FM 1-0 for more details.

II. Distribution Network

Sustainment commands establish the initial distribution network in the operational area, synchronize, and integrate intra-theater deployment and distribution operations. These commands monitor all segments of the distribution network that begins at the strategic source of support and extends to the point of employment. The TSC manages the strategic to operational links of the global distribution network. Figure 4-1 depicts key agencies involved in the distribution network. The ESC manages the operational to tactical links of the global distribution network. This includes the segment of the distribution network that begins at the POD or theater source and ends at the point of employment. Both the TSC and ESC have distribution management responsibilities executed through a DMC. The TSC or ESC will establish movement boards to manage transportation policies, priorities, LOC status, convoy protection and synchronization, and transportation asset allocation to support theater distribution operations. The sustainment commands also provide command and control of organizations that execute distribution and distribution management and control capabilities.

Establishment of the network is a complex joint effort. The GCC typically establishes a JDDOC. It is an integral component of the GCC's staff, normally under the staff supervision of the GCC J-4. A JDDOC may be co-located with the DMC. Some initial tasks of the JDDOC include:

- Monitor airlift and sealift flow.
- Provide movement control of arriving supplies, personnel, equipment, and units.
- Establish theater-wide capabilities required to meet anticipated transportation and throughput capacities.

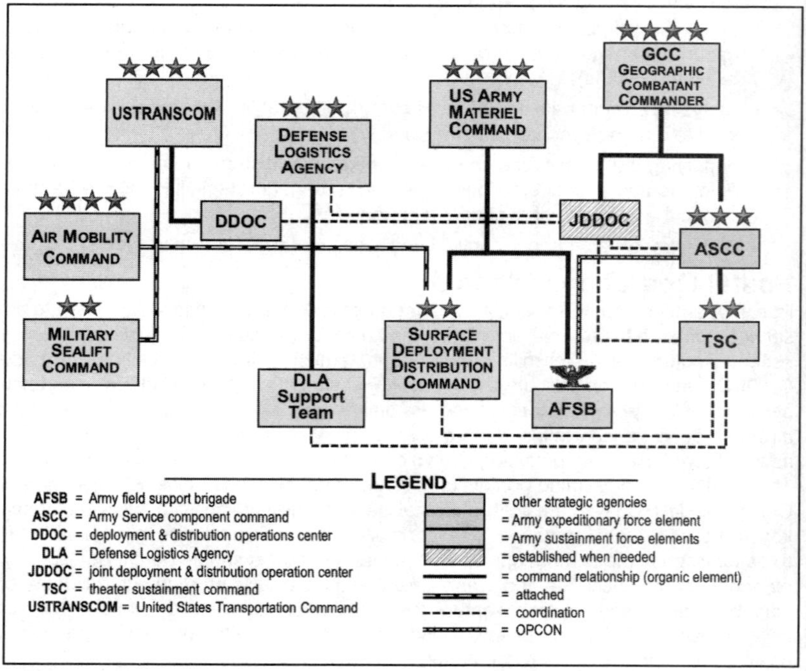

Ref: FM 4-0, fig. 4-1. Strategic providers to Army theater distribution.

III. Sustainment of Prevent Activities

Army special operations conducted during operations to shape and operations to prevent conflict can hasten or delay the employment of a task force and provide more time for the JFC to make a decision or explore alternative options. More importantly, optimal use of special operations could eliminate the threat completely or the need for a large-scale conventional force deployment.

Sustainment of SOF in operations to prevent conflict includes supporting the deployment, RSOI, and sustainment of a special operations JTF. The 528th Sustainment Brigade supports crisis response and initial entry/ RSOI of a special operations JTF during this Army strategic role. The 528th Sustainment Brigade is augmented with Army sustainment enablers to conduct RSOI and establish initial theater distribution for SOF operations for a limited time.

Refer to ATP 3-05.40 for more information on ARSOF sustainment and FM 6-05 for more information on CF and SOF integration, interoperability and interdependence.

Sustainment commanders at all levels have critical roles in each of the major activities that Army forces perform during operations to prevent:

A. Execute Flexible Deterrent Options and Flexible Response Options

A flexible deterrent option is a planning construct intended to facilitate early decision making by developing a wide range of interrelated responses that begin with deterrent-oriented actions carefully tailored to create a desired effect. A flexible response option is the capability of military forces for effective reaction to any enemy threat or attack with actions appropriate and adaptable to the circumstances existing. Flexible deterrent options and flexible response options serve three basic purposes. First, providing a visible and credible message to adversaries about U.S. will and capability to resist aggression. Second, positioning U.S. forces in a manner that facilitates implementation of the operations plan or contingency plan if hostilities become unavoidable. Third, providing options for decision makers during crises. Flexible deterrent options and flexible response options allow for measured increases in pressure to avoid unintentionally provoking full-scale combat and to enable decision makers to develop the situation and gain a better understanding of an adversary's capabilities and intentions.

Flexible deterrent options and flexible response options are only credible as options when coordinated across all warfighting functions, including sustainment. Sustainment planners must work with ARFOR commanders to anticipate potential flexible deterrent options and flexible response options. Sustainment operations can be a flexible deterrent option. Examples include deploying sustainment capabilities and building logistics infrastructure which would present the adversary with multiple dilemmas that may change the intended course of action.

If the force cannot sustain the operation, it will fail in the goals outlined above. Sustainers need to know the size and composition of the forces to be supported, distances involved, threat capabilities, and other factors to estimate support requirements and prepare to support these operations. Sustainers also account for the requirement to rapidly escalate/deescalate from flexible deterrent option to large-scale combat and back based on changing conditions in the OE.

B. Tailor Army Forces

Force tailoring combines two complementary requirements—selecting the right forces and deploying the forces in the optimal sequence. Much of the effort for selection and deployment order occurs prior to deploying forces to a theater.

Throughout the process, sustainment commands work with the theater Army sustainment staff to track force composition and order of deployment to ensure support is available to sustain the force package. The right types and number of sustainment units should be integrated early into the force flow.

Selecting the right force involves identifying, selecting, and sourcing required Army capabilities and establishing their initial task organization to accomplish the mission. The result is an Army force package matched to the needs of the CCDR. The theater Army works with force providers (FORSCOM, U.S. Army Materiel Command, etc.) to match the composition of the force with the forces identified in theater security cooperation plans or contingency plans. In the deliberate sustainment planning process it is critical the Center for Army Analysis analyzes contingency plans and provides battle damage attrition estimates to the theater Army, TSCs, and USAMC LCMCs to ensure that adequate depot maintenance Forward Repair Activities support is identified and integrated into the force flow for planning purposes.

Force tailoring establishes the order of deployment for the force package, given the available lift and the GCC's priorities. The organization established in force tailoring is not necessarily the same as the task organization for combat. It is a macro-level organization established to control the forces through deployment and RSOI.

C. Project the Force

Force projection is the ability to project the military instrument of national power from the United States or another theater in response to requirements for military operations (JP 3-0). Five processes (mobilization, deployment, employment, sustainment, and redeployment) are integral to force projection. In support of operations to prevent, speed becomes a critical planning factor to consider to project combat-ready forces that can enable a position of advantage against an adversary.

See chapter 9, Deployment & Redeployment, for further discussion from ATP 3-35.

Mobilization is the process of bringing the armed forces to a state of readiness in response to a contingency. Sustainment builds and maintains force readiness. Army generating sustainment forces in the execution of its Title 10 mission prepare Army forces for unified land operations. Upon alert for deployment generating force sustainment organizations ensure Army forces are manned, equipped, and meet all Soldier readiness criteria. Army active and Reserve Component units mobilize from Army installations that ensure Soldiers, equipment, materiel, and medical readiness are verified prior to deployment

Deployment as a process that enables force projection involves activities required to prepare and move forces, supplies, and equipment to a theater. The objective of deployment planning is to synchronize deployment activities to facilitate effective execution in the operational area. The steps used in deployment planning are:

- Analyze the mission.
- Structure forces.
- Refine deployment data.
- Prepare the force.
- Schedule movement.

The key output from the deployment planning steps is the TPFDD that incorporates commander's intent, available forces, and designated strategic lift. The TPFDD is both a force and a transportation requirements document. It is the Joint Planning and Execution System data-based portion of the operational plan. It contains time-phased force data, non-unit-related cargo and personnel data, and movement data for the operational plan including:

- In-place units.
- Units to be deployed to the JOA with a priority indicating the desired sequence for their arrival at the POD, including mobilized augmentees or units.
- Routing of forces deployed.

Reception, Staging, Onward Movement, and Integration (RSOI)

Ref: FM 4-0, Sustainment (Jul '19), pp. 4-12 to 4-14.

A critical portion of the deployment process often associated with operations to prevent is RSOI. It is this process that delivers combat power to the JFC in an AOR. RSOI is also a process the Army has not been challenged with to the extent required for large-scale combat operations in recent history. The increased challenge will stem from the initial deployment of large forces, and will continue with a heavy flow of replacements. Accurate casualty estimates and predetermined casualty shelves are a critical part of RSOI planning. RSOI goals are to link units with their equipment while maintaining a balanced flow of supplies, personnel, equipment, and units consistent with strategic lift capabilities and CCDR priorities.

Sustainment commands are assigned responsibilities for RSOI operations by the ASCC. The TSC will normally be assigned the responsibility of RSOI. Subordinate units of the TSC from an ESC, sustainment brigade, CSSB, down to a movement control team can be assigned specific tasks in support of RSOI. Specific tasks include feeding, billeting, limited supply, finance, personnel accountability, health service support, battlefield orientation, and transportation of replacements to their assigned units. Support for execution of RSOI is provided by some combination of theater support contracts, external support contracts (primarily the LOGCAP), regionally available commercial HNS, military assets, and Component 2 or 3 units deployed early. It requires sufficient capabilities and capacity to provide the support required by arriving units at POD. For example heavy vehicle operators and mechanics are required to move and repair vehicles being downloaded from ships for onward movement. The TSC and/or ESC directs the operation of the theater's PODs and the AOR distribution networks to minimize bottlenecks that may impede the flow of cargo and forces into and throughout the theater.

Effective RSOI consists of the processes that transform arriving personnel and equipment into forces capable of meeting operational requirements in accordance with established timelines. Army units are attached from the supporting commander (usually FORSCOM) to the ASCC. The ASCC subsequently passes TACON of the unit to the sustainment command managing RSOI activities. RSOI is an ongoing series of four interrelated and overlapping processes that ensure synchronized clearance of nodes and assembly of combat power that is then rapidly transported to point of need to support operational requirements.

The role of the DSB during RSOI operations is to complete the RSOI process themselves and then provide support to its division at and beyond the tactical assembly area. The sustainment command responsible for RSOI must consider the DSB a tactical unit and not part of the RSOI support structure. The DSB must complete the RSOI process and move forward to a tactical assembly area with its supported division.

Success of RSOI is measured by force closure. Force closure is the point in time when a supported JFC determines that sufficient personnel and equipment resources are in the assigned operational area to carry out assigned tasks. It is essential during RSOI operations that data relating to arriving unit personnel, unit equipment, contractors, containers, and the like be captured at each point of entry into the theater, i.e. APODs, SPODs, railheads, and border crossings to facilitate personnel/asset accountability and force closure reporting.

See chapter 9, Deployment & Redeployment, for further discussion of force projection from JP 3-35 and ATP 3-35.

- Movement data associated with deploying forces.
- Estimates of non-unit-related cargo and personnel movements to be conducted concurrently with the deployment of forces.
- Estimates of transportation requirements, which are fulfilled by common-user lift resources, as well as those requirements that can be fulfilled by assigned or attached transportation resources.

Army Power Projection

The Army Power Projection Program consists of four movement phases, PPPs and installation activities, movement of people and assets to ports of embarkation, port operations, and movement of personnel and equipment from ports of debarkation to the RSOI locations in theater. During all phases of force projection, sustainers play a role in planning and execution. Force projection requires detailed support and sustainment planning.

The Army Power Projection Program begins at a PPP or an installation, to include depots, armories, and other facilities. PPPs are located where one or more brigade size element is deploying. All movements from these activities hinge on out-load support at each site. Each site's infrastructure and capacity to receive strategic assets and perform out-load support in support of personnel and unit equipment is vital to meeting any RSOI timeline associated with a deploying unit. Critical players during this phase of RSOI include the deploying unit, installation senior commander, installation commander, ASC support elements, the LRC including its installation transportation office, SDDC and SDDC support elements including deployment support teams and/or rail operating teams. Force projection operations at an activity may consist of rail, air, line haul, and/or convoy operations that are necessary to begin movement of personnel and equipment to a port of embarkation.

The movement of personnel or equipment from a PPP to a port of embarkation requires close coordination with the installation transportation office and SDDC deployment support teams. Units that are not self-deploying to a port of embarkation require strategic lift. Installation transportation offices are the conduit to receive buses, trains, and line haul assets through USTRANSCOM contracts and are ordered based on the unit's deployment list and timelines in the Joint Operation Planning and Execution System.

Force projection further requires planning for the arrival of personnel and equipment at ports of embarkation to include port support activities, size and make up, staging areas, super cargo, joint inspections, frustrated cargo, and general life support for any supporting activity. Key players at APODs and SPODs include arrival/departure airfield control groups, movement control teams, and PATs.

USAMC provides support to Army power projection through its subordinate LRCs. LRCs are not deployable and provide support to Army installations across the Army's strategic roles.

Editor's note: See chap. 9, Deployment and Redeployment, for more complete discussion of force projection and RSOI from ATP 3-35 (Mar '15) for more complete discussion of additional topics discussed in chap. 4 of FM 4-0 to include: Aerial Port of Embarkation (APOE), Seaport of Embarkation (SPOE), Aerial Port of Debarkation (APOD), Seaport of Debarkation (SPOD), and Intermediate Staging Bases (ISBs).

(Sustainment Operations) III. Large-Scale Combat

Ref: FM 4-0, Sustainment Operations (Jul '19), chap. 5.

Large-Scale Combat Operations Sustainment

As a nation, the U.S. wages war by employing all instruments of national power-diplomatic, informational, military, and economic. The President employs the Armed Forces of the U.S. to achieve national strategic objectives. The nature and scope of some missions may require Army forces to conduct large-scale combat operations to achieve national strategic objectives or protect national interests.

Large-scale combat operations are characterized by simultaneous, geographically dispersed operations that occur in multiple domains. In large-scale combat operations against a peer threat, maneuver commanders conduct decisive action to seize, retain, and exploit the initiative. Maneuver commanders strive to achieve superiority across multiple domains (air, maritime, land, space, and cyberspace) early to allow the Army forces to conduct land operations without prohibitive enemy interference. This involves the orchestration of many simultaneous unit actions in the most demanding of OEs.

Characteristics of sustaining large-scale combat operations include volume, lethality, precision, and tempo. Large-scale combat operations will require a volume of reinforcements, materiel, and equipment significantly greater than other types of operations. It will be more lethal than other types of operations generating mass casualties and replacement of personnel and equipment on a large scale. Large-scale combat operations will require greater precision in our distribution network than other types of operations. It will be executed at a higher operational tempo than other types of operations that require flexible and adaptable sustainment structure to meet mission requirements.

Sustainment Challenges in Large-Scale Combat Operations

Large-scale combat operations are characterized by simultaneous, geographically dispersed operations that occur in various OEs and are challenged across multiple domains. It requires greater sustainment than other types of operations because of the higher operational tempo, greater lethality, and significantly increased consumption of supplies, and equipment. The lethal nature of large-scale combat operations increases the propensity for mass casualties, requirements for mortuary affairs, increased requirements for a robust medical architecture, and large-scale personnel and equipment replacements. Large-scale combat operations will require the distribution system to move a greater volume of personnel and equipment than in other types of operations. Increased velocity and precision will be required to sustain operations.

I. Sustainment Considerations during Large-Scale Combat Operations

Ref: FM 3-0 (w/Chg 1), Operations (Dec '18), pp. 2-48 to 2-49.

Logistics during Large-Scale Combat Operations

Logisticians support operational tempo by delivering supplies and materiel as far forward as possible. They use throughput distribution and preplanned and preconfigured packages of essential items to do this. Logisticians maintain constant contact with operational units to determine requirements for supporting operations. Operational units also provide logisticians with support estimates for contingencies and requirements for cross-loading of supplies to prevent all of one type of supply from being destroyed by the loss of a single system.

Supplies and material should remain close to the maneuver force to ensure short response times for supplies and services. This includes uploading critical materiel—such as water, petroleum, oils, and lubricants and ammunition—in order to anticipate attempted occupation of a piece of terrain by more than one unit. Commanders must make risk decisions regarding logistics preparations and avoidance of enemy detection, since logistic preparations may give indications of friendly tactical plans.

The availability of supplies and materiel to sustain tactical unit operations becomes critical to extend operational reach as large-scale combat operations progress. Operational reach is reduced when supplies fail to keep up with the demand of tactical units. Slow or limited resupply may require commanders to use controlled supply rates for various classes of supply to reduce unit expenditures. When those controlled supply rates are not sufficient to continue operations, the force culminates.

During large-scale combat operations supply lines of communication are strained, and requirements for repair and replacement of weapon systems increase. Requirements for petroleum, oils, and lubricants increase during the offense. Conversely, requirements for munitions tend to be higher in the defense than in the offense. Sustainment units must be as mobile as the forces they support. One way to provide continuous support is to task organize elements of sustainment units or complete sustainment units with their supported maneuver formations as required by the mission.

The variety and complexity of possible situations arising during an attack requires sustainment operators to establish a flexible and tailorable distribution system in support of tactical commanders. There may be a wide dispersion of forces and lengthening of lines of communication. Required capabilities to support longer lines of communications include movement control, in-transit visibility, terminal operations, and mode operations.

Field maintenance assets move as forward as possible to repair inoperable and damaged equipment to return it to service as quickly as possible. Crews perform preventive maintenance checks and services as modified for the climate and terrain in which they find themselves. Battle damage assessment and repair restores the minimum essential combat capabilities necessary to support a specific combat mission or to enable the equipment to self-recover. Crews and maintenance and recovery teams conduct battle damage assessment and repair to rapidly return disabled equipment to battlefield service using field expedient components and means.

Establishing aerial resupply and forward logistics base camps may be necessary to sustain operations. This is especially true in the offense, if an attack transitions to exploitation and pursuit conducted at great distances from unit sustaining bases. Aerial resupply, either by rotary-wing or parachute, delivers critical supplies to the point of need during an entry operation, deep inland operation, or to a rapidly moving unit. The unit or support activity at the airlift's point of origin is responsible for obtaining the required packing, shipping, and sling-load equipment. It prepares the load for aerial transport, prepares the pickup zone, and conducts air-loading operations. The unit located at the airlift destination is responsible for preparing the landing zone to accommodate aerial resupply and for receiving the load.

See pp. 3-5 to 3-42 for discussion of logistics (sustainment execution).

Personnel Services during Large-Scale Combat Operations

During large-scale combat operations, the key subordinate functions of staffing the force that are important are personnel accountability and strength reporting. The subordinate functions of provide human resource services including postal operations, finance services, and casualty operations, continue during large-scale combat operations. Human resource planning and operations are the means by which human resources are addressed in the military decision-making process and in the attack operations plan. This includes casualty forecasts necessary to inform commanders and staffs.

See pp. 3-45 to 3-46 for discussion of personnel services (sustainment execution).

Health Service Support during Large-Scale Combat Operations

Large-scale combat operations place an incredible burden on medical resources due to the magnitude and lethality of the forces involved. Medical units can anticipate large numbers of casualties in a short period of time due to the capabilities of modern conventional weapons and the possible employment of weapons of mass destruction. These mass casualty situations can exceed the capabilities of organic and direct support medical assets without careful planning and coordination. Casualty evacuation must occur concurrently with operations.

Effective management of mass casualty situations depends on established and rehearsed unit-level mass casualty plans and detailed medical planning. There are a number of other variables which can ensure the success of a unit's mass casualty response plan. These include, but are not limited to—

- Coordination and synchronization of additional medical support and augmentation and their dispositions and allocations, such as medical evacuation support, forward resuscitative and surgical teams, combat support and field hospitals, casualty collection points, ambulance exchange points, and established Class VIII resupply.
- Predesignating casualty collection points.
- Quickly locating the injured and clearing them from the battlefield.
- Providing effective emergency medical treatment for the injured.
- Accurate triage and rapid medical evacuation of the injured to medical treatment facilities at the next higher role of care.
- Use of alternative assets when the number of casualties overwhelms the capacity of available medical evacuation systems.

See pp. 3-47 to 3-50 for discussion of health service support (sustainment execution).

II. Sustainment Synchronization

Logistics, financial management, personnel services, and health service support require coordination and synchronization at every stage of the planning process. This synchronization is crucial in large-scale combat operations with its inherent distributed nature. Only by integrating and synchronizing sustainment functions can the sustainment system achieve required effects at the speed, the volume, velocity, and lethality of large-scale combat operations.

Sustainment commanders and staffs present synchronized courses of action commensurate with sustainment capabilities to allow as much freedom of action as possible. Limitations such as insufficient infrastructure or non-availability of key classes of supply have a bearing on the commander's ability to execute the mission and are accounted for in the planning process. Sustainment leaders also coordinate, synchronize, and integrate the sustainment plan with joint and other unified action partners to ensure continuous linkage with strategic-level providers. A successful sustainment plan will extend operational reach, prevent culmination or loss of the initiative, manage transitions, exploit possible opportunities, and mitigate risk.

Identifying and accepting prudent risk is a principle of mission command. Throughout the operations process, commanders and staffs use risk management to identify and mitigate risks associated with all hazards that have the potential to injure or kill friendly and civilian personnel, damage or destroy equipment, or otherwise impact mission effectiveness. For sustainment commanders and staffs, identifying and mitigating risk must always include not only risk to finite and limited sustainment capabilities, but also how those capabilities are employed to enable freedom of action and extended operational reach.

Sustainment synchronization remains the focus as sustainment commanders plan for and coordinate support through such continuing activities as rhythm of military operations events, information collection, liaison, meetings, protection efforts, and reporting. For the purposes of sustaining large-scale combat operations, two of these—liaison and reporting—require special emphasis.

Liaison refers to contact or intercommunication maintained between elements of military forces or other agencies to ensure mutual understanding and unity of purpose and action. Most commonly used for establishing and maintaining close communications, liaison continuously enables direct, physical communications between commands.

Sustainment commanders and staffs have the continuous requirement to coordinate with higher, lower, adjacent, supporting, and supported units and civilian organizations. The sustainment liaisons participate in boards, bureaus, cells, centers, and working groups, especially in the case of the TSC with the ASCC, the ESC with the corps, the DSB with the division, the BSB with the BCT, and ASB with the CAB. While the use of liaisons taxes organic staff manpower in sustainment organizations, their presence and active participation is essential to sustaining large-scale combat operations and mitigating the effects of dispersion, threat disruption of communications, and accelerated tempo.

Both maneuver and sustainment commanders rely on logistics and personnel status reports to identify support requirements and capabilities to enable large-scale combat operations. Sustainment staffs use data from sustainment estimation tools, higher headquarters orders, and documents such as country studies to develop running estimates. A running estimate is the continuous assessment facts, assumptions, constraints and limitations concerning the current situation and operational environment used to determine if the current operation is proceeding according to the commander's intent and if planned future operations are supportable. Using sustainment information systems, commodity managers include information in running estimates such as quantity on-hand, quantity consumed, expected quantity on-hand, expected consumption to anticipate requirements and assist in synchronization. Each staff element and CP functional cell maintains a running estimate focused on how its specific areas of expertise are postured to support future operations.

III. Threats to Sustainment Units

Ref: FM 4-0, Sustainment (Jul '19), pp. 5-5 to 5-6.

A threat is any combination of actors, entities, or forces that have the capability and intent to harm U.S. forces, U.S. national interest, or the homeland (ADP 3-0). Threats may include individuals, groups of individuals, paramilitary or military forces, nation-states, or national alliances. In general, a threat can be categorized as an enemy or an adversary.

FM 4-0 is focused on sustaining large-scale combat operations against peer threats. A peer threat is an adversary or enemy with capabilities and capacity to oppose U.S. forces across multiple domains world-wide or in a specific region where it enjoys a position of relative advantage. Peer threats present credible challenges to sustainment forces through the use of information warfare, isolation, systems warfare, preclusion and sanctuary. Other considerations such as contested LOC and anti-access/area denial techniques coupled with challenges across multiple domains (air, land, maritime, space and cyberspace) challenge sustainment support to operations. Sustainment commanders must consider the OE and all the factors that affect their ability to sustain operations during planning.

Peer threats have the ability to influence and direct irregular forces, criminal elements, and hostile populations. Peer threats have the ability to impose disruptive effects in cyberspace that will challenge Army sustainment during pre-deployment, deployment, employment, and redeployment. These disruptive effects may occur at unit home-stations, ports of embarkation, while in transit to the theater, and upon arrival at ports of debarkation as well as within the theater.

Peer threats employ their capabilities across multiple domains to attack U.S. vulnerabilities, including sustainment facilities, networks and formations. Peer threats use their capabilities to create lethal and nonlethal effects throughout an OE. During combat operations, threats seek to inflict significant damage across multiple domains in a short period of time. Peer threats seek to delay U.S. forces long enough to achieve their goals and end hostilities before U.S. forces reach culmination. One effective way to delay U.S. forces, which generally operate on very long LOCs, is to disrupt sustainment operations and nodes.

Refer to FM 3-0 for additional information on peer threats.

Threat forces may employ tactics that force the U.S. Army into conducting large-scale combat operations in urban areas. Currently more than 50 percent of the world's population lives in urban areas, and this is likely to increase to 70 percent by 2050, making large-scale combat operations in cities likely. Large-scale combat operations in urban terrain is complex and resource intensive. In most urban operations, the terrain, the dense population, military forces, and unified action partners will further complicate sustainment operations.

For additional information on operations in urban areas, refer to ATP 3-06.

Threat use of subterranean spaces and structures (any space or structure located below ground) as a means to covertly maintain the initiative against a more powerful military opponent may occur during large-scale combat operations. Such spaces and structures can be used for command and control, defensive networks, operations, storage, production, or protection. Continued improvements in the construction of subterranean environments have increased their usefulness and their proliferation.

For additional information, refer to ATP 3-21.51.

IV. Large-Scale Defensive Operations

Ref: FM 4-0, Sustainment (Jul '19), chap. 6.

A **defensive task** is a task conducted to defeat an enemy attack, gain time, economize force, and develop conditions favorable for offensive or stability tasks (ADP 3-0). As one of the four decisive action tasks, the defense is a major, complex operation conducted to defeat an enemy attack. Although a defensive task normally does not achieve decisive victory, it sets the conditions necessary to regain the initiative through counterattacks.

There are three defensive tasks executed during large-scale combat operations: mobile defense, area defense, and retrograde. A **mobile defense** defeats the attacking forces by permitting the enemy to advance into a position that exposes them to counterattack. An **area defense** orients on retaining terrain and denying enemy forces access to desired terrain while not destroying the enemy outright. A **retrograde** moves the friendly forces away from the enemy to gain time, preserve forces, place the enemy in unfavorable positions, or avoid combat under undesirable conditions.

As a component of large-scale combat operations, the defense is a combination of highly complex tasks that place tremendous and continuous demands on Army sustainment organizations. Situational awareness, mission analysis, and detailed planning are keys to successful support operations. Commanders take advantage of the time available during a defense to build combat power. However, the time available is likely to be unknown since the enemy typically has the initiative. As a result, sustainment organizations and the functions executed play a critical role in supporting the defense and the success of subsequent future operations.

Defending commanders combine the three types of defensive tasks to fit the situation. All three types of defense use mobile and static elements. In mobile defenses, static positions help control the depth and breadth of the enemy penetration and retain ground from which to launch counterattacks. In area defenses, commanders closely integrate patrols, security forces, and reserve forces to cover gaps among defensive positions. Commanders reinforce positions as necessary and counterattack, as directed. In retrograde operations, some units conduct area or mobile defenses or security operations to protect other units that execute carefully controlled maneuver or movement rearward. These units use static elements to fix, disrupt, turn, or block the attackers. Mobile elements are used to counterattack and destroy the enemy.

Commanders execute defensive tasks for various reasons, such as to retain decisive terrain or deny a vital area to the enemy, weaken or fix the enemy as a prelude to offensive actions, or increase the enemy's vulnerability by forcing the enemy to concentrate subordinate forces. The ultimate purpose of the defense is to create conditions for a counteroffensive whereby it allows Army forces to regain the initiative.

The transition to offensive operations will normally be as a planned operation but may also be a hasty operation conducted to capitalize on tactical opportunities or an identified enemy weakness. Because of this, sustainment commanders and leaders must stay cognizant of the status of the operation. Leaders must also use all available time to execute required sustainment functions knowing the mission may change quickly and frequently.

Sustainment Fundamentals

Enemy commanders look for opportunities to counter corps and division defensive tasks. The enemy will seek to employ special purpose forces, irregular forces, electronic warfare, long-range artillery, rockets, missiles, information capabilities, and cyberspace electromagnetic activities to disrupt sustainment activities. The enemy, to assist in targeting of sustainment units and locations, may exploit use of electronic signals, such as cell phones and geotagged photos. Sustainment commanders must be aware of these unintended threats and focus on those efforts that would help set the conditions necessary to regain the initiative during defensive operations.

All sustainment functions are planned and executed to support defensive operations and build combat power to prepare for future offensive operations. The exact type and extent of support operations and the organizations executing them will vary by echelon based on the support requirement. Even though defensive operations may be the main effort, simultaneous offensive operations with their support requirements are also likely to be ongoing.

Sustainment commanders and staffs plan for increased requirements in class IV, V, and IX items to support the defensive effort and build class III (bulk and package) stocks to prepare to transition to offensive operations. Sustainment planners anticipate where the greatest need might occur during operations and consider pre-positioning sustainment stocks far forward to reduce response times for critical support. Planners also consider alternative methods for delivering sustainment in emergencies. Sustainment of defensive tasks requires a coordinated planning effort designed to maximize synchronization, integration, and continuity of support at all echelons. Commanders and staffs at every echelon must anticipate operational requirements, be responsive in requisitioning and distributing resources, and be prepared to improvise tactics and techniques for execution that ensures responsiveness even in unexpected situations.

Planning Considerations

Sustainment planning is both a continuous and a cyclical activity of the operations process. For sustainment planning, the most important factors are requirements, capabilities, and shortfalls. As outlined in the paragraphs below, planning considerations assist planners in identifying specific support or operational requirements based upon available information. Many planning considerations affect the ability to execute large-scale defensive operations. These considerations must be recognized, analyzed in the time available, and prioritized based on the commander's intent. Planning considerations must encompass all warfighting functions to ensure the plan is integrated across all functions and domains. A planning consideration may have various levels of effects that drive support requirements across all warfighting functions.

Sustaining Defensive Operations

- Plan class IV for transitions from offense to defense
- Expect Increases in class V
- Plan for pre-positioning of supplies
- Plan retrograde support
- Increased demand for class VIII
- Mass casualties
- Large scale personnel replacements in a short period

Refer to AODS6-1 (w/SMARTupdate 1): The Army Operations & Doctrine SMARTbook, pp. 2-69 to 2-82, for full discussion of large-scale defensive operations from FM 3-0. While the U.S. Army must be manned, equipped, and trained to operate across the range of military operations, large-scale ground combat against a peer threat represents the most significant readiness requirement. FM 3-0 provides doctrine for how Army forces, as part of a joint team, and in conjunction with unified action partners, do this.

Sustainment Operations

V. Large-Scale Offensive Operations

Ref: FM 4-0, Sustainment (Jul '19), chap. 7.

An **offensive task** is a task conducted to defeat and destroy enemy forces and seize terrain, resources, and population centers (ADP 3-0). The intent of an offensive task is to impose the commander's will on the enemy. Against a capable, adaptive enemy, the offense is the most direct and sure means of seizing, retaining, and exploiting the initiative to gain physical, temporal, and cognitive advantages and achieve definitive results.

There are four offensive tasks executed during large-scale combat operations: **movement to contact, attack, exploitation, and pursuit**. These tasks enable commanders to impose their will on the enemy and deprive the enemy of resources, seize decisive terrain, deceive or divert the enemy, develop intelligence, or hold an enemy position.

Sustainment commanders and their staffs prepare to support each offensive tasks. Sustainment determines the depth, duration, and endurance of Army operations, and plays ... enabling decisive action. Failure to provide adequate sustainment during of-

Initial Guidance ↗ why do we do MA
See p. 4-15
As you conduct analysis
Spend time on Dev good estimates

* see p. 4-16

– Issue warning order to subordinate

tainment systems, interdicted LOCs, and ... overmatch capabilities. Sustainment commanders and planners prepare to push forward critical supplies in an OE where degraded systems and communications exist.

If the force is to maintain the initiative and combat power necessary for the successful performance of offensive tasks, the continued forward movement of units and sustainment support is critical. Maintaining the initiative in the close area often results in significant numbers of bypassed enemy forces and remnants of defeated units as friendly forces maneuver deep into enemy areas by avoiding enemy units in well prepared positions. The fluidity and rapid tempo of operations pose challenges when planning for the area security of support and consolidation areas.

Enemy commanders look for opportunities to counter or at least hinder the performance of corps and division offensive tasks. Enemy commanders attempt to strike deeply into friendly support and consolidation areas using multiple combinations of lethal and nonlethal effects from multiple domains. The enemy will seek to employ special purpose forces, irregular forces, electronic warfare, long-range artillery, rockets, missiles, information capabilities, and cyberspace electromagnetic activities to disrupt sustainment activities. Sustainment commanders remain aware of conventional enemy units and other elements bypassed during the advance of friendly forces and the threat presented by their presence in support and consolidation areas.

Sustainment units synchronize with maneuver units to ensure security of support and consolidation areas. Corps and division headquarters must plan to keep CPs operating, sustainment capabilities functional, respective LOCs open, and supply stocks at an acceptable level. The conduct of noncontiguous operations increases the difficulty of these tasks, as does the lack of friendly host nation security forces.

Sustainment of offensive tasks is a high-intensity operation. Sustainment commanders and staffs plan for increased requirements in class III (B), IX items, and personnel replacements to sustain the pace and tempo of operations. Plan and rehearse command and control, forward positioning, orders issuance, personnel accounting, logistical support, processing, and transportation of replacements, and most critically maneuver unit rapid integration of replacements. Sustainment planners anticipate where the greatest need might occur during offensive operations. Planners consider positioning sustainment units in close proximity to operations to reduce response times for critical support. Planners also consider alternative methods for delivering sustainment in emergencies. Extended LOCs require analysis of how to best emplace forward sustainment elements to support the commander. It is important to clearly lay out key actions for rehearsing offensive operations for example casualty evacuation routes, logistics release points, support area displacement times and locations, detainee collection points and holding areas, and fuel and ammunition resupply points to foresee potential problems and means to mitigate them.

Sustaining Offensive Operations

- Continually update running estimates
- Support offense and consolidate gains simultaneously
- Understand enemy threat and challenges
- Increased class III (bulk and package) and class IX requirements
- Increased casualties and personnel replacements over extended battlefield

Refer to AODS6-1 (w/SMARTupdate 1): The Army Operations & Doctrine SMARTbook, pp. 2-83 to 2-100, for full discussion of large-scale offensive operations from FM 3-0. While the U.S. Army must be manned, equipped, and trained to operate across the range of military operations, large-scale ground combat against a peer threat represents the most significant readiness requirement. FM 3-0 provides doctrine for how Army forces, as part of a joint team, and in conjunction with unified action partners, do this.

Current sustainment systems possess vulnerabilities and connectivity requirements that may make them susceptible to disruption and deliberate targeting by threat forces, both lethally and non-lethally. To mitigate this vulnerability and maintain an accurate readiness COP, organizations develop the rhythm of military operations, data cut-off times, as-of times, and reporting times. Commanders and staffs also balance the timeliness and potential latency of reporting with the amount of time needed to analyze data when evaluating unit readiness and combat capability.

Sustainment enterprise resource planning systems and associated decision support tools help provide near real-time status with minimal staff effort required to gather and display information from multiple databases. Integrating this information with command and control systems is crucial to give the sustainment leaders and supported commanders and staffs the identical current COP. The value of integrated sustainment information systems and command and control systems is that everyone on the network can see and use the same reported information to plan and control operations.

Sustainment Rehearsals

Sustainment rehearsals are critical to synchronization and the success and accomplishment of the mission. Conducting sustainment rehearsals immediately after the combined arms rehearsal ensures understanding and synchronization of the unit's maneuver and sustainment plan as it traverses the battlefield. It is critical that the combined arms team and all elements of sustainment are represented and participate in sustainment rehearsals to ensure all sustainment commodities understand how these integrate with other elements of sustainment to accomplish the mission. The sustainment rehearsal helps synchronize the sustainment warfighting functions with the other warfighting functions to create a common understanding of the plan.

VI. Support Area

The support area is a smaller, subordinate AO inside the commander's overall AO. The support area is normally, but not always, positioned within and surrounded by the consolidation area. It is where most of an echelon's sustaining operations occur. The geographic size of a support area is based on mission and operational variables and is difficult to quantify. These variables include the number of units assigned to the support area, the existing threat, and the amount of terrain that can be influenced by the unit assigned support area responsibility. As an example for a division support area, if it is assumed to be a brigade-sized area, it will be approximately 10 square kilometers. This number is for general planning consideration and to give readers an idea of the geographic scope of a division support area and the impact it has on command and control and protection. It should be understood that division support area size may vary widely. The corps support area will be significantly larger.

See related discussion on pp. 2-6 to 2-7.

Within the joint security area, strategic enablers such as USTRANSCOM, USAMC, DLA and each of their individual subordinate components link strategic support activities with theater support activities. Examples of these activities include synchronizing strategic and operational distribution of equipment, supplies and personnel; managing materiel and establishing contracts, establishing theater fuel farms and managing excess property turn-in. USASOC coordinates operational support requirements while monitoring SOF activities within the theater. The TSC, ESC and its attached sustainment brigade conduct RSOI for units arriving in theater and support the movement of those units forward to corps and division areas. MEDCOM (DS) provides command and control of all EAB medical units providing direct or GS to the corps and division areas. Other sustainment forces in the joint security area support activities including -classes I and III (Bulk) distribution, APOD and SPOD operations, personnel services, financial management activities, and other support tasks.

Within a division support area, a MEB is normally designated to control the area. If a MEB is not available, a BCT must be designated to control the area. For the corps support area, a MEB is normally designated to control the area. If a MEB is not available, a corps MP BDE with augmentation can control the corps support area. Support area control responsibilities include area security, terrain management, information collection, integration, and synchronization, civil affairs operations, movement control, mobility support, and clearance of fires, personnel recovery, airspace control, and minimum-essential stability tasks. This allows sustainment units to focus on their primary functions.

The corps headquarters is likely to position assets in the division support area to facilitate division operations and enable freedom of action. The division headquarters orchestrates the sustainment and protection tasks essential to ensuring freedom of action in the division close and deep areas. Planning in the support area largely influences current and future operations in the deep, close, and consolidation areas. The support area is not a single large base. It is a base cluster comprised of multiple bases, each established by units assigned to the support area. The MEB is responsible for terrain management to include placement and integrated protection of the bases.

Depending on the situation, including the threat, size of the support area, and number of units within the support and consolidation areas, division and corps commanders may employ a support area command post (SACP) in the support area to assist in controlling operations. The SACP enables division and corps commanders to exercise command and control over disparate functionally focused elements operating within the support and consolidation areas that may exceed the effective span of control of the MEB or division and corps main CPs.

At the corps level, the SACP is not a separate section in the corps' table of organization and equipment. The corps commanders form the SACP in the support area from the equipment and personnel from the main and tactical CPs when required. Normally, the deputy corps commander leads the corps SACP.

For the division, the SACP is organic to the division headquarters. The SACP consists of 14 personnel and has a command group, a movement and maneuver section, and a sustainment section. The DSB has a limited role in establishing the SACP. The DSB must however coordinate with the SACP for support prioritization. The primary role of the SACP is to provide command authority and general officer oversight of division support area operations and sustainment, medical and other division support activities. The SACP performs tasks and functions, as defined by the commander, based on operational & mission variables. The division SACP in the support area normally co-locates with the MEB, which provides the CP with signal connectivity, life support, security and workspace. Functions of the CP include planning and directing sustainment, terrain management, movement control, and area security. When augmented by the MEB staff, the CP may also plan and control combined arms operations with units under division or corps control, manage airspace, and employ fires.

Sustainment forces prepare for the various threats in the support area. Threats in the division support area are categorized by the three levels of defense required to counter them. Any or all threat levels may exist simultaneously in the division support area. All threats pose potential risks to sustainment and other support operations. Emphasis on base defense and security measures may depend on the anticipated threat level. A Level I threat is a small enemy force that can be defeated by those units normally operating in the echelon support area or by the perimeter defenses established by friendly bases and base clusters. A Level I threat for a typical base consists of a squad-sized unit or smaller groups of enemy soldiers, agents, or terrorists. Typical objectives for a Level I threat include supplying themselves from friendly supply stocks, disrupting friendly command and control nodes and logistics facilities, and interdicting friendly LOCs.

VII. Reconstitution Operations

Ref: FM 4-0, Sustainment (Jul '19), pp. 5-35 to 5-37.

The scale, complexity, and increased destructive power of large-scale combat operations assumes an all-encompassing multi-domain fight, potentially resulting in the greater loss of personnel, weapon platforms, supplies, and equipment of our warfighting formations. Restoring combat power to the levels necessary, within a limited window of time to continue the fight, is the objective of normal, day-to-day sustainment actions, specifically personnel replacement operations, rebuild and maintain units at strength. Under exceptional conditions, with severely degraded units, constrained time, and limited or no personnel replacements, commanders may make the decision to execute reconstitution. Reconstitution operations are extraordinary actions that commanders take to restore degraded units to an acceptable level of combat effectiveness as determined by the commander, commensurate with mission requirements and available time and resources. Reconstitution must be planned and resourced during operations to shape and prevent. The commander directing the reconstitution mission uses assets under their control, along with those provided by higher echelons.

Reconstitution is not a sustainment operation, although sustainment plays an integral part. Reconstitution is an operational event enabled by all warfighting functions. Activities such as personnel and equipment restoration, health service support to restore combat power, and collective training, are involved to ensure combat readiness. The reconstitution effort should be thoroughly planned and understood by all involved to ensure success. Any maneuver, maneuver support, or sustainment unit may require reconstitution. Therefore, planners at all levels of command must have a contingency plan prepared to execute reconstitution operations, when necessary.

Two Types of Reconstitution

Reconstitution operations consists of two elements: reorganization and regeneration.

Reorganization involves the cross leveling of forces and resources following combat operations to increase combat effectiveness of an attrited unit.

Regeneration involves the large-scale replacement of personnel, weapon platforms, equipment, and supplies; as well as the reestablishment and/or replacement of the chain of command, and the conduct of mission essential training. Regeneration operations take place at EAB, using corps, theater and strategic capabilities and resources to enable restoration of combat power.

For additional information, refer to FM 4-0, appendix C.

Regeneration, being the more complex of the reconstitution elements, returns degraded Army units into large-scale combat operations. Planning for reconstitution operations is necessary prior to beginning of large-scale combat operations. A critical aspect of regeneration is that of decision authority to direct units to conduct regeneration operations. Regeneration will require the removal of attrited units from the battlefield to an area that allows regeneration activities out of contact. Commanders must balance the resources required to move attrited units out of combat versus the overall mission, time constraints, and resources required.

Operational readiness thresholds that trigger both entry and exit criteria for regeneration are critical in determining the scope of operations. Operational readiness thresholds are measures used to determine a units capability to perform the missions or functions for which it is organized or designed. Thresholds to enter or exit regeneration must be determined by an operational. At the theater level, the operational plans should address combat power degradation thresholds to aid the commander in determining regeneration operations. Commanders must also consider further degradation of a unit during retrograde operations.

Regeneration Task Force (RTF)

During large-scale combat operations, regeneration operations are resource intensive and METT-TC dependent. Designation of a specialized regeneration task force (RTF) allows for the rapid (days to weeks) execution of large-scale replacement operations. The RTF provides the synchronization of command and control, security, and sustainment support activities to a specified location, either within the corps consolidation area, joint security area, or outside of the theater, to rebuild a degraded unit's combat power.

The RTF predetermines maintenance collection points suitable for Heavy Equipment Transporter System on and/or off-loading during hours of darkness with limited time available. Equipment towed or carried from the battle area goes to a central location that facilitates sorting of equipment. The RTF determines special requirements for vehicles that require special road clearances prior to beginning regeneration.

Sustainment commanders must consider time constraints required to support regeneration versus sustainment of large-scale combat operations. Utilization of sustainment assets to support regeneration operations may affect the execution of distribution operations of classes I, III, and V supplies. The RTF may need to request echelon above brigade transportation assets to support regeneration.

Other sustainment planning considerations are as follows—

- Regeneration of units is an exceptionally intensive form of reconstitution requiring the direction of maneuver commanders two levels up, and substantial sustainment support.
- Attrited units are not able to support any sustainment requirements.
- Planners and the RTF should request equipment from theater reserve stocks.
- Movement control support should be requested by the RTF to assist with planning the attrited unit's move.
- Planners should standardize information required by the RTF from an attrited unit.
- Reconstitution plans should reflect procedures for managing replacements of Soldiers in attrited units.
- Reconstitution plans should include combat stress counselors and psychologists in the RTF.
- Planners should plan for the recovery of an attrited unit's disabled vehicles and crews from forward areas to the regeneration site.
- Planners should prepare for attrited units arriving to the regeneration site with a form of chain of command. However, regardless of how the attrited unit is organized, the RTF regenerates all the entities within the attrited unit.
- Planners should ensure the RTF should has a 24-hour operating capability.
- The regeneration site should be as close as possible to the attrited unit location to minimize amount of time and difficulty reaching the site.
- The site should be far enough away to prevent easy enemy interdiction and provide the safest feasible location for personnel executing regeneration operations.
- The site should contain enough space to conduct regeneration operations.
- Planners should be prepared to coordinate with the unit's parent organization. For example, the RTF may be required to coordinate a division property book office to establish correct hand receipts.
- Planners and the RTF should plan for signal support when conducting reconstitution in austere locations.
- Planners and the RTF should request decontamination support and prepare to handle decontaminated personnel and equipment.

Units must also employ cover and concealment to prevent observation and detection of sustainment equipment and bases by reconnaissance elements, reconnaissance aircraft, drones, or attack aircraft. Cover and concealment includes signature management and emissions control. Dispersion aids in concealment and limits destruction and losses in the event of an attack. Units at all echelons must conduct CBRN defense preparation.

At higher echelons military police units from the MEB enhance protection capabilities by performing area security within the support areas. These units perform response-force operations to defeat Level II threats against bases and base clusters located in that support area. A Level II threat is an enemy force or activities that can be defeated by a base or base cluster's defensive capabilities when augmented by a response force (ATP 3-91). These units maintain contact with Level III threats in the division support area until the tactical combat force under the MEB's control can respond. Level III threat is an enemy force or activities beyond the defensive capability of both the base and base cluster and any local reserve or response force.

Corps and division commanders designate close, deep, support, and consolidation areas to describe the physical arrangement of forces in time, space, and focus. Commanders should always designate a close area and a support area, and designate a deep area and consolidation area as required. The consolidation area does not necessarily need to surround, nor contain, the support area base clusters, but typically, it does. The consolidation area requires a purposefully task organized, combined arms unit to conduct area security and stability tasks and employ and clear fires.

The consolidation area is the portion of the commander's area of operations that is designated to facilitate the security and stability tasks necessary for freedom of action in the close area and to support the continuous consolidation of gains (ADP 3-0). Corps and division commanders may establish a consolidation area, particularly in the offense as the friendly force gains territory, to exploit tactical success while enabling freedom of action for forces operating in the other areas. When designated, a consolidation area refers to an AO assigned to an organization that extends from its higher headquarters boundary to the boundary of forces in close operations where forces have established a level of control and large-scale combat operations have ceased.

Army forces consolidate gains as part of a combat operation to enable combat power for continued action against remaining enemy forces in support of a host nation and its civilian population, or as part of the pacification of a hostile state. These gains may include the establishment of population security temporarily by using the military as a transitional force, the relocation of displaced civilians, re-establishment of law and order, and restoration of key infrastructure. Concurrently, corps and divisions must be able to accomplish these activities while sustaining, repositioning, and reorganizing subordinate units to continue operations in the close and deep area.

Consolidation of gains activities may encompass a lengthy period of post-conflict operations prior to redeployment. Consolidation of gains may occur even if large-scale combat operations are occurring in other parts of an AO to exploit tactical success. Anticipation and early planning for activities after large-scale combat operations ease the transition process.

Commanders address the decontamination, disposal, and destruction of war materiel. Commanders must also address the removal and destruction of unexploded ordnance and the responsibility for demining operations. (The consolidation of friendly and available enemy mine field reports is critical to this mission.) In addition, if support areas are surrounded by a consolidation area, the higher echelon headquarters must clearly articulate the roles and responsibilities for controlling supply routes and clearance procedures. Additionally, the theater Army is prepared to provide medical support, emergency restoration of utilities, support to social needs of the indigenous population, and other humanitarian activities.

(Sustainment Operations) IV. Consolidate Gains

Ref: FM 4-0, Sustainment Operations (Jul '19), chap. 8.

Sustaining Operations to Consolidate Gains

Consolidate gains are those activities to make enduring any temporary operational success and set the conditions for a stable environment allowing for a transition of control to legitimate authorities (ADP 3-0). Commanders continuously consider activities necessary to consolidate gains and achieve the end state. Operations to consolidate gains present unique challenges to sustainment planners. While remnants of enemy formations may pose a continuing threat to U.S. forces during these operations, U.S. forces will also face threats from criminal and insurgent forces.

Deployed sustainment headquarters continue to provide sustainment support to maneuver forces while enabling combat power for continued action against remaining enemy forces. As with sustainment for other missions, successful support for operations to consolidate gains requires sustainment commanders to anticipate shifting priorities and integrate simple and responsive solutions to dynamic battlefield conditions.

Activities to consolidate gains may occur over a significant amount of time and involve transitions in both focus and partners. Emphasis will shift from actions to ensure the defeat of remaining threat forces to measures that address the needs of the population and eventually to the transfer of responsibility from Army forces to a host nation government, interagency partner, or other organizations.

Consolidation of gains activities are conducted by a separate maneuver force in the designated corps or division consolidation areas. A corps or division commander typically assigns responsibility for the consolidation area to a task-organized division or BCT. Consolidation activities include security and stability tasks and will likely involve combat operations against bypassed enemy forces.

Successful operations to consolidate gains employ a combination of offensive, defensive, and stability tasks to achieve continued operational success. Operations to consolidate gains assume that enemy forces will employ all means available to protract the conflict, even as organized enemy resistance continues in other sectors. To defeat these efforts, commanders take into account both the purpose of an operation and the capacity for the enemy to resist. Success depends on the commander's ability to exploit tactical success by establishing security and stability in a manner sufficiently decisive to achieve national strategic aims.

Refer to AODS6-1 (w/SMARTupdate 1): The Army Operations & Doctrine SMARTbook, pp. 2-101 to 2-110, for full discussion of operations to consolidate gains from FM 3-0. While the U.S. Army must be manned, equipped, and trained to operate across the range of military operations, large-scale ground combat against a peer threat represents the most significant readiness requirement. FM 3-0 provides doctrine for how Army forces, as part of a joint team, and in conjunction with unified action partners, do this.

II. Planning Considerations

Ref: FM 4-0, Sustainment (Jul '19), pp. 8-5 to 8-6.

In addition to sustaining combat operations, sustainment of operations to consolidate gains may involve the provision of essential government services, emergency infrastructure reconstruction, and humanitarian relief for the local population. Typically, it involves support for both U.S. forces and other unified action partners.

Forces conducting operations to consolidate gains may require the full complement of sustainment capabilities, especially if those operations are prolonged. During the initial stages, however, sustainment planners should prioritize the following functions:

- Ammunition.
- Medical evacuation.
- Explosive ordnance disposal.
- Maintenance.
- Transportation.
- Supply and distribution operations.
- General engineering support.
- Operational contract support.
- Personnel accountability.
- Postal operations.
- Essential personnel services.
- Field services.
- Special funding appropriations.

Sustainment commanders must continuously assess risk and develop plans to mitigate risk throughout operations. Sustainment professionals must seek to understand, balance and take risks rather than avoid risks to ensure sustainment of the operational force. Below is a sample list of risk considerations during operations to consolidate gains:

- Does the consolidation area have the correct mix of forces to address combat with by-passed forces while assisting with consolidation of gains?
- Are MCT, movement control battalions, and convoy commanders prepared to task organize and redirect enroute missions as priorities change?
- Are movement control personnel engaged in updating the status of all roads within the consolidation area?
- Are sufficient amounts of class V available to provide support to limited combat operations within the consolidation area, while maintaining support ongoing large-scale combat operations?
- Are sufficient amounts of class VIII medical supplies for the local population, prisoner of war camps and contract laborers on hand?
- Are sufficient class IX parts available to allow maintainers to make repairs where time permits to vehicles temporarily repaired through battle damage assessment and repair methods?
- Are sufficient EOD personnel available to rapidly clear improvised explosive devices discovered within the area of operations?
- Are contracts in place that can be used to assist with restoring essential services if host nation or local, national or U.S. agencies are not able to assist? Are sufficient combat personnel available to provide security for contractors or non-governmental aid workers?
- In preparation for stability operations are meals planned for civilian emergency rations in the correct quantity and of the proper type? For POW centers? For displaced people? Are contracted assets available for transporting the class III to the area of need? Can tents be utilized for temporary shelter for civilians, contractors and prisoners of war?

Intelligent use of medical assets and agreements identified during shaping operations can set the conditions for operations to consolidate gains. Medical formations take into consideration the major military and civilian events in development and their impact on medical operations throughout the theater. Displaced civilians, humanitarian crisis, injured enemy combatants and non-combatants, and detainee medical care are within the scope of factors to consider. If planned correctly, medical plans addressing these issues should offer a sequential, interorganizational or whole of government approach to enable the transition of medical formations from first contact to successful hand-off of responsibilities or continuation of AHS support theater operations across all strategic roles.

Planning considerations for the provision of AHS support to consolidate gains include the duration, size, scope, and complexity of the operation, population supported, host-nation medical capabilities and internal capacity for medical care of the local populace and displaced civilians. Replacement by and integration of Army Component 2 and 3 medical units, AHS support to retrograde operations, and planning for transitions should also be considered.

Sustainment estimates should account for humanitarian requirements and potential increases in the size, scope, and duration of these mission requirements. Minimum essential stability tasks leverage all available sources, including local government and commercial sources, other government agencies, and non¬governmental organizations. In previous conflicts, international aid organizations have provided especially effective support for humanitarian needs, but these agencies require a safe and secure environment in which to operate.

During consolidation of gains operations, humanitarian needs may include providing food, water, shelter, and medical support to local and dislocated persons. Along with host nation and other governmental and non-governmental organizations when available, sustainment forces anticipate these requirements and plan to provide support necessary to sustain the population until local civil services are restored. Although operations to consolidate gains are designed to transition control of these areas to appropriate authorities as soon as possible, past experience indicates the duration and complexity of these operations frequently exceed planning estimates. Sustainment commanders should be adaptive and prepare for the possibility of extended operations.

Explosive ordnance disposal commanders must provide support to sustaining operations during consolidation of gains by detecting, identifying, conducting on-site evaluation, rendering safe, exploitation, and disposal or disposition of explosive ordnance, including weapons of mass destruction. Considerations must be made to mitigate both explosive ordnance designed to function after a time delay as well as unexploded ordnance, as both pose significant risk to overall consolidation of gains. Considerations for class V resupply must be addressed with specific emphasis needed when contending with explosive ordnance filled with sensitive high explosives; such munitions require significantly greater amounts of class V to dispose of correctly. Counter-improvised explosive device considerations must be addressed based upon the identified threat of criminal and insurgent forces during this operation. Counterinsurgency operations will potentially occur simultaneously during this operation.

As security improves within the area of operations, the OCS process enables sustainment commanders to utilize contracted capabilities in support of force protection, reestablishment of basic services, and repairs to critical infrastructure. Planners must carefully consider the impact of OCS to avoid disrupting local economies or inadvertently funding enemy or criminal activities. As operations to consolidate gains shift towards stability tasks, commanders may increase their reliance on commercial support in order to shift organic forces to other critical missions.

Sustainment planning should include long-term sustainment for U.S. forces remaining in theater to prevent recurrence of hostilities. Sustainment planning should also be conducted for transfer to civil authority or other legitimate agency.

The consolidation area refers to an AO extending from higher echelon boundaries to the boundaries of forces in the close area. The consolidation area may or may not include support areas and focuses on supporting units engaged in on-going combat operations. The need to consolidate gains represents a separate and distinct mission, and planners should avoid assigning consolidation responsibilities to forces, including sustainment units, that are committed to close operations. Units assigned to consolidate gains require their own dedicated support, with sustainment forces task organized to meet that operation's unique mission requirements. Once operations to consolidate gains commence, the senior sustainment headquarters in support of that operation may establish one or more support areas within the consolidation area.

II. Sustainment Fundamentals

The TSC and ESC coordinate strategic sustainment support for the operational area to enable the corps, division, and brigade formations to conduct operations. The TSC and ESC anticipate consolidation of gains sustainment support requirements and are prepared to support with barrier materiel, explosive ordnance disposal support, and humanitarian assistance. The DSB and support battalions support BCTs and enabling units by providing critical supply, transportation, and humanitarian support typically required during consolidation of gains operations.

During operations to consolidate gains, sustainment priorities continue to focus on support to combat operations. Priorities include establishing security, restoring combat power, and preparing for continued operations to destroy remaining enemy forces. However, as the area becomes more stable and security increases, sustainment priorities include conducting stability tasks, providing medical treatment, and repairing LOCs.

Sustaining Operations to Consolidate Gains

- Sustain combat operations
- Restore combat power
- Stability tasks
- Medical treatment
- Lines of communication repair

The increased violence and lethality of large-scale combat operations create the potential for widespread devastation within areas occupied by U.S. forces. Commanders ensure the performance of minimum essential stability tasks as soon as possible after initiation of operations to consolidate gains. These tasks depend heavily on sustainment and include the provision of security, water, food, shelter, power generation, and medical treatment. The urgency of these requirements may dictate significant shifts within the commander's priorities of supply and support. The demand for potable water may be especially challenging.

Because combat operations may damage or destroy existing roads, bridges, canals, airfields, port facilities and rail systems, sustainment forces coordinate with engineer elements for repair or replacement of damaged LOCs. Additionally, sustainment forces coordinate with maneuver elements for route security. Enemy forces are likely to target LOCs to disrupt sustainment operations.

Successful stability operations require effective coordination between military and civilian efforts. Where possible, the Department of State coordinates U.S. stabilization and reconstruction tasks. Sustainment functions directly support these tasks. To achieve unity of effort, sustainment commanders synchronize efforts with U.S. and foreign government agencies, international agencies, nongovernmental organizations, other unified action partners, and contractors. As the security situation improves, Army forces may transfer responsibility for minimum-essential stability tasks to other forces or appropriate civilian organizations while transitioning to the performance of primary stability tasks.

Sustainment Execution

Ref: ADP 4-0, Sustainment (Jul '19), pp. 3-10 to 3-13.

Execution is putting a plan into action by applying combat power to accomplish the mission (ADP 5-0). It focuses on actions to seize, retain, and exploit the initiative.

Sustainment determines the depth and duration of Army operations. It is essential to retaining and exploiting the initiative and it provides the support necessary to maintain operations until mission accomplishment. Failure to provide sustainment could cause a pause or culmination of an operation resulting in the loss of the initiative. It is essential that sustainment planners and operation planners work closely to synchronize all of the warfighting functions, in particular sustainment, to allow commanders the maximum freedom of action.

I. Logistics

Logistics is planning and executing the movement and support of forces. It includes those aspects of military operations that deal with: design and development; acquisition, storage, movement, distribution, maintenance, and disposition of materiel; acquisition or construction, maintenance, operation, and disposition of facilities; and acquisition or furnishing of services.

See pp. 3-5 to 3-42 for further discussion.

II. Financial Management

Financial management leverages fiscal policy and economic power across the range of military operations. Financial management encompasses finance operations and resource management.

See pp. 3-43 to 3-44 for further discussion.

III. Personnel Services

Personnel services are those sustainment functions related to Soldiers' welfare, readiness, and quality of life. Personnel services complement logistics by planning for and coordinating efforts that provide and sustain personnel.

See pp. 3-45 to 3-46 for further discussion.

IV. Health Service Support

Health service support encompasses all support and services performed, provided, and arranged by the Army Medical Department to promote, improve, conserve, or restore the behavioral and physical well-being of Army personnel and as directed, unified action partners (UAPs).

See pp. 3-47 to 3-50 for further discussion.

Sustainment plays a key role in enabling the simultaneous **offensive, defensive, and stability or defense support of civil authorities tasks** that occur as part of unified land operations. For example, general engineering support provides construction support to protect key assets such as personnel, infrastructure, and bases. Horizontal and vertical construction enables assured mobility of transportation networks and survivability operations to alter or improve cover and concealment to ensure freedom of action, extend operational reach, and endurance of the force. Legal personnel supporting rule of law activities may find themselves working closely with HN judicial, law enforcement, and corrections systems personnel.

Sustainment Execution (Overview)

Ref: ADP 4-0, Sustainment (Jul '19), pp. 3-10 to 3-12.

Sustainment Execution

Sustaining Offensive Operations

An offensive operation is an operation to defeat and destroy enemy forces and gain control of terrain, resources, and population centers (ADP 3-0). The intent of an offensive operation is to impose the commander's will on the enemy. In the offense, the decisive action is a sudden, shattering action against an enemy weakness that capitalizes on speed, surprise, and shock. The sustainment of offensive tasks is high in intensity. Commanders and staffs plan for increased requirements and demands on sustainment. Sustainment planners work closely with other warfighting function staffs to determine the scope of the operations and develop estimates for quantity and types of support required They anticipate where the greatest need might occur and develop a priority of support. Sustainment planners may consider positioning sustainment units in close proximity to operations to reduce response times for critical support. They also consider alternative methods for delivering sustainment in emergency situations.

To maintain initiative and freedom of action, coordination between staff planners must be continuous. The most important commodities are typically fuel classes III(B), V and IX, along with the functions of maintenance, and medical evacuation. Based on planning assessments, sustainment commanders direct the movement of these and other support to meet anticipated requirements.

Another aspect of sustaining the offensive is the lengthening of line of communications. Widely dispersed forces, longer line of communications, and congested road networks increase stress on transportation systems. As a result, a combination of ground and aerial delivery may be planned to accommodate the distribution. Distribution managers and movement control units synchronize movement plans and priorities according to the commander's priority of support. Distribution must be closely coordinated and tracked to ensure delivery of essential support. The routing function of movement control becomes an essential process for coordinating and directing movements on main supply routes or alternate supply routes, and regulating movement on line of communications to prevent conflict and congestion.

Higher casualty rates associated with the offensive may increase the requirement for medical resources. Positioning medical support close to operations in order to facilitate treatment and evacuation. If increased casualty rates overwhelm medical resources, nonmedical transportation assets may be needed for evacuation. Additional combat and operational stress control teams may also be required to support combat and operational stress casualties following operations.

Higher casualty rates also increase the emphasis on personnel accountability, casualty reports, and replacement operations. G-1s and S-1s plan for accurate tracking of casualties and replacements through coordination with casualty liaison teams and the HRSC.

Religious support may become critical during the offensive. Religious support through counseling and appropriate worship can help reduce combat and operational stress, increase unit cohesion, and enhance performance.

Using contractors during the sustainment of offensive operations entails great risk and raises significant practical and legal considerations. However, when necessary, the force commander may be willing to accept this risk and use contractors in forward areas. Commanders should seek counsel from their judge advocates when considering the use of contractors during offensive operations.

Sustaining Defensive Operations

A defensive operation is an operation to defeat an enemy attack, gain time, economize forces, and develop conditions favorable for offensive or stability operations (ADP 3-0). Defensive operations can set conditions for a counteroffensive or counterattack that enables Army forces to regain the initiative.

Typically, the defensive operation requires more centralized control. Movements of materiel and troops within the AO have to be closely and continuously coordinated, controlled, and monitored. Distribution managers direct forecasted sustainment to designated units. Increases in class IV requirements, such as barrier and construction materials, are pushed to designated collection points for unit retrieval.

The task of medical units is to rapidly locate, acquire, triage casualties, treat, and return to duty or resuscitate and stabilize for evacuation to the next higher role of medical care or out of the theater of operations. Medical treatment facilities and other AHS support assets should be placed within supporting distance of maneuver forces but not close enough to impede ongoing combat operations.

Sustaining Stability Operations

Stability operations are operations conducted outside the United States in coordination with other instruments of national power to establish or maintain a secure environment, provide essential governmental services, emergency infrastructure reconstruction, and humanitarian relief. (See ADP 3-07, Stability.) Sustainment in operations with a dominant stability component often involves supporting U.S. forces, multinational forces, and other contributing partners in a wide range of missions and tasks. It will require interaction with other governmental agencies and nongovernmental organizations.

At the tactical or local level, sustainment, in support of stability operations, restores essential civil services as defined in terms of immediate humanitarian needs (such as providing food, water, shelter, and medical support) necessary to sustain the population until local civil services are restored. Efforts to restore essential services involve developing HN capacity to operate, maintain, and improve those services. Other sustainment tasks associated with restoration of services include support to dislocated civilians and demining operations.

Security and economic stability precede an effective and stable government. Operational contract support plays a significant role in supporting economic stabilization and infrastructure development, especially at the local level. Here the emphasis is on generating employment opportunities, infusing monetary resources into the local economy, stimulating market activity, fostering recovery through economic development, and supporting the restoration of physical infrastructure. Drawing on local goods, services, and lab[illegible] through contracting, presents the force an opportunity to infuse cash into [illegible] omy, which in turn stimulates market activity. A functioning econom[illegible] ment and reduces the dependence of the population on the [illegible] ities.

Sustaining Defense Support of Civi[illegible]

Defense Support of Civil Authorities is support provide[illegible] [f]ederal military forces, DOD civilians, DOD contract personnel, DOD Compone[illegible] [a]ssets, and National Guard forces (when the Secretary of Defense, in coordination with the Governors of the affected States, elects and requests to use those forces in title 32, U.S.C., status) in response to requests for assistance. These requests come from civil authorities for domestic emergencies, law enforcement support, and other domestic activities, or from qualifying entities for special events. This is also known as civil support.

The principal logistics functions of the Army during disaster response is to move, maintain and provide logistical capabilities necessary to respond to requests for assistance from civil authorities and also provide sustainment support to the responding DOD forces. Army units are expected to be self-sustaining during any civil support operations and units should deploy with sufficient basic sustainment commodities (food, water and fuel) to be self-sufficient for the initial phase of operations, usually 72 hours. Resupply is then coordinated through the higher headquarters tasked to coordinate, synchronize and execute the resupply mission. National Guard forces continue to be supported through their state's military infrastructure, primarily through the state's joint force headquarters.

Sustaining Operations to Consolidate Gains

Consolidate gains are activities to make enduring any temporary operational success and set the conditions for a stable environment allowing for a transition of control to legitimate authorities (ADP 3-0). Consolidation of gains is an integral and continuous part of armed conflict, and it is necessary for achieving success across the range of military operations. Army forces deliberately plan to consolidate gains throughout an operation as part of defeating the enemy in detail to achieve the overall political and strategic objectives.

During operations to consolidate gains, sustainment priorities continue to focus on support to combat operations. Priorities include establishing security, restoring combat power, and preparing for continued operations to destroy remaining enemy forces. However, as the area becomes more stable and security increases, sustainment priorities include conducting stability tasks, providing medical treatment, and repairing line of communications.

Sustaining Army Special Operations Forces

The sustainment of ARSOF is provided by sustainment forces on an area basis and they may draw from sustainment resources from a number of sustainment organizations supporting an area of operations. ARSOF forces may not be self-sufficient. They require external sustainment to support their operations.

ARSOF sustainment structures are lean and unable to provide all sustainment functions required to support their missions. ARSOF routinely arrive in an AO early, execute forced-entry operations, and operate independently in small teams. Because of these factors, logistics support to ARSOF must be tailored based upon mission variables. For more information on ARSOF see FM 3-05.

The integration of ARSOF support capabilities with those of the TSC facilitates synchronized and tailored support to specific ARSOF mission requirements. This relationship provides sufficient flexibility to respond to changing requirements. Additionally, the coordination and synchronization between the ARSOF and TSC leverages logistics resources and achieves greater operating efficiencies for the ARSOF.

The sustainment brigade (special operations) (airborne) plans, integrates, and assesses the need for sustainment to ARSOF. The brigade is designed to serve as an early entry element to provide mission command of one combat sustainment support battalion in support of a conventional force subordinate to the joint special operations task force. The brigade synchronizes and manages sustainment and distribution operations; determines and anticipates sustainment requirements; plans, coordinates, and synchronizes both current and future sustainment operations for deployed ARSOF units.

The most challenging support of ARSOF is during operations in austere environments. When an ARSOF unit deploys into an undeveloped theater of operations, it must bring sufficient resources to survive and operate until it establishes a bare-base support system or makes coordination for support through the TSC, HN, or MNF.

Chap 3

(Sustainment Execution) I. Logistics

Ref: ADP 4-0, Sustainment (Jul '19). pp. 1-5 to 1-9.

Logistics is planning and executing the movement and support of forces. It includes those aspects of military operations that deal with design and development; acquisition, storage, movement, distribution, maintenance, and disposition of materiel; acquisition or construction, maintenance, operation, and disposition of facilities; and acquisition or furnishing of services. For the sustainment warfighting function, explosive ordnance disposal tasks are discussed under protection and intelligence warfighting functions (ADP 3-37 and FM 2-0.) Army logistics include the following—

Sustainment Execution

Logisitics

- I **Maintenance** *(ATP 4-33)*
- II **Transportation** *(FM 4-01)*
- III **Supply** *(ATP 4-42)*
- IV **Field Services** *(ATP 4-42)*
- V **Distribution** *(ATP 4-0.1)*
- VI **Operational Contract Support** *(ATP 4-92)*
- VII **General Engineering Support** *(ATP 3-34.40)*

I. Maintenance

Ref: ATP 4-33 (w/Chg 1), Maintenance Operations (Jul '19).

Army maintenance primary purpose is to ensure unit readiness by maintaining weapon systems and equipment in a fully mission-capable status for immediate and continuous employment in complex and highly lethal environments. Army maintenance organizations are increasingly required to anticipate, analyze, adapt, and tailor available resources for effective and timely support of operations. Success in these types of operations and environments continues to be based on the bottom-line measurements of maintainability, reliability, and availability.

A. Two-Level Maintenance

Ref: ATP 4-33 (w/Chg 1), Maintenance Operations (Jul '19), pp. 1-7 to 1-12.

The Army utilizes a tiered, two-level maintenance system comprised of field and sustainment maintenance. Command teams, maintenance personnel, and planners must have a complete understanding of two-level maintenance fundamentals in order to properly plan and execute maintenance operations. Two-level maintenance provides the operating unit with more capabilities forward and the ability to respond rapidly.

- Soldiers perform field-level maintenance as far forward as possible with the equipment being retained by or returned to the owning unit. Crewmembers, equipment operators, and Ordnance-trained maintainers perform field maintenance.
- All Army modification table of organization and equipment (MTOE) maintenance units perform field maintenance.
- Sustainment maintenance is performed by U.S. Army Materiel Command (USAMC) elements normally comprised of civilians and contractors who return equipment to a national standard, after which the equipment is placed back into the Army's overall supply system. When a unit sends equipment to a sustainment maintenance organization the owning unit, in most cases, removes the equipment from its property book. Only in rare instances, such as unit reset, and watercraft maintenance, is the equipment returned to the owning unit.

The goal of the maintenance system is to reduce repair cycle times by repairing or replacing components, modules, and assemblies as far forward as possible, maximizing reliance on rapid repair parts distribution, and visibility. TM maintenance allocation charts provide repair time guidelines for both field and sustainment-level tasks.

Field Maintenance

Three distinct groups of Soldiers perform field maintenance: equipment operators, equipment crews,and Ordnance Corps trained maintainers. The owning unit or a supporting maintenance unit performs field maintenance utilizing its own tools and test equipment. The unit should retain and repair the item until it is ready to return to service. Maintainers perform field maintenance on all types of unserviceable items of equipment and weapon systems. Repairs include the replacement of an unserviceable line replaceable unit,component, module, or part. However, field maintenance is not limited to simply remove and replace actions. If the operator/crew or Ordnance Corps trained maintainers are authorized and possess the requisite skills,special tools, proper repair parts, references, and adequate time the item should remain on-site and not be evacuated for sustainment maintenance. This is especially relevant to BCTs. The expertise to fix major weapon systems (Abrams, Bradley, Paladin, and Stryker vehicles) resides only in the FSC. There are no maintenance units equipped or staffed to perform field-level maintenance repairs to these weapon systems outside the BCT. The brigade support battalion in a Stryker brigade combat team (SBCT) is one exception as the FMC includes mechanics to maintain the medical company's Stryker medical vehicles.

Operator and Crew Field Maintenance

Operators and crews perform field maintenance on their equipment as outlined in the operator's -10 TM. Some Soldiers receive formal training from their proponent on a specific system through advanced individual training or new equipment training. For example, the driver and gunner for an Abrams Main Battle Tank receive training that gives them an important skill set. Other Soldiers receive specialized training from their unit. Unit commanders are responsible for establishing, conducting, and maintaining a driver training program. Driver training programs emphasize the critical role that platform operators play in PMCS. The operator or crew are typically the first to observe a fault or identify the development of a fault. In many instances, they have the ability to repair the fault or minimize its impact using onboard spares enabling mission completion. Condition-based maintenance indicators or instrumentation can initiate operator/crew tasks. Typically, tasks consist of inspecting, servicing, lubricating, adjusting, and replacing minor components/assemblies as authorized by the maintenance allocation charts. TMs identify operator/crew tasks within a maintenance allocation chart. The maintenance allocation chart (MAC) identifies the proper basic issue item tools and onboard spares required to complete a given service.

Operators/crews are system specialists in those MOS that receive formal training from their proponent through advanced individual training or specialized functional courses on diagnosing specific system faults. These operator/crews include Patriot missile, signal, military intelligence, or a maneuver unit's master gunner. In some instances, these personnel receive special tools to perform maintenance on their assigned system. Operator/crew maintenance falls into the crew subcategory of field maintenance, detailed in the equipment operator's manual. Operators troubleshoot the entire system using the operator's -10 TM and simplified or embedded diagnostic equipment to identify, isolate, and trace problems.

Maintainer Field Maintenance

Maintainer maintenance is accomplished by Ordnance school trained maintainers utilizing the relevant TM on a component, accessory, assembly, subassembly, plug-in unit, shop replaceable unit within a line replaceable unit, or other portion either on the system or after it is removed by a trained maintainer. Depending on the system and MOS involved, the definition of a line replaceable unit or shop replaceable unit is flexible. The characterization of line or shop replaceable units for wheeled and tracked vehicles, radar, or the Warfighter Information Network-Tactical (WIN-T), shifts as the field maintenance troubleshooting increases in complexity.

Sustainment Maintenance

Sustainment maintenance consists of two subcategories: below depot-level sustainment maintenance and depot-level sustainment maintenance. Units utilize sustainment maintenance when crew, operator,operator-maintainer, or Ordnance Corps trained maintainers lack the requisite skills, special tools, proper repair parts, or references to complete repairs using field maintenance. Based on the extent of damage to the specific item, leaders must decide the best course of action based on operational and mission variables. There is no absolute checklist. Field maintenance is the preferred method of repair. The intent of sustainment-level maintenance is to perform commodity-oriented repairs to return items to a national standard, providing a consistent and measurable level of reliability. USAMC, through the Army Sustainment Command (ASC) and its subordinate Army field support brigades and Army field support battalions, execute sustainment maintenance missions. Sustainment maintenance supports both operational forces and the Army supply system. Unless prior planning occurs, USAMC subordinate elements typically enter after initial setting the theater tasks are completed.

Below Depot Sustainment Maintenance

Maintainers perform below depot-level sustainment maintenance on a component, accessory,assembly, subassembly, plug-in unit, or other portion generally after removal from the system. Sustainment maintenance performed in an operational environment will normally be below-depot sustainment maintenance.

Depot Sustainment Maintenance

Maintainers perform depot-level maintenance on end items, components, accessories, assemblies, subassemblies, or plug-in units either on the system or after it is removed.

B. Managing Battlefield Maintenance

Ref: ATP 4-33 (w/Chg 1), Maintenance Operations (Jul '19), pp. 3-1 to 3-6.

Maintenance planners in an operational headquarters generally do not drive the planning process but must be fully integrated throughout the process. Maintenance planners use the commander's intent, planning guidance, and MDMP to develop the maintenance concept of support. They derive the concept of support from running estimates developed using a variety of planning tools. These running estimates project casualty figures, maintenance requirements, software patches, and other sustainment requirements. Maintenance planners participate in all aspects of MDMP to ensure synchronization and unity of effort.

Effective maintenance plans balance three elements: identify requirements (the minimum number of available weapon systems required for mission success), identify available maintenance resources (what maintenance and repair parts capability is on hand to meet the mission), and manage the maintenance resources for maximum effect (establishing priorities, task organizing to weight the main effort, posturing class IX, anticipating shortfalls). This last requirement is aligned to the unit with maintenance priority, and the priority of work assigned to the key systems. Prioritization is used to identify how maintenance support is weighted for the mission.

When a shift or change in priorities could provide a greater overall effect, the maintenance manager takes appropriate action or makes recommendations through the chain of command. Figure 3-2 shows the basic concept for managing maintenance support. In a similar manner to the maintenance planning process above,the maintenance manager must identify the resources on-hand and those already committed to meet identified requirements. Staffs manage available resources within the established support framework to return the maximum number of items to a fully mission-capable status.

IDENTIFY REQUIERMENTS	IDENTIFY AVAILABLE RESOURCES	MANAGE FOR MAXIMUM RETURNS
• Compute Unit Needs. • Parts Required. • Battle Damage Assessment.	• Command • Personnel • Publications • Time • Facilities • Test Equipment • Tools • Parts • Skills	• Maintenance Support Forward. • Commander's Guidance. • Battle Damage Repair Controlled Exchange. • Cannibalization • Recovery • Evacuation

Ref: ATP 4-33 (w/Chg 1), fig. 3-2. Managing battlefield maintenance support.

Critical Information for Maintenance Support

Timely maintenance support is reliant on supported units providing critical information. This information includes unit locations, type of equipment requiring maintenance, type of fault, mobility status (can the equipment move on its own), parts required, number and status of supporting mechanics, and threat. Accurate reporting ensures commanders at all levels have a true picture of the capabilities of formations from the forward line of troops back. Units may utilize communications to rapidly pass information up the chain of command but must utilize a DA Form 5988-E (Equipment Maintenance and Inspection Worksheet) to maintain a permanent record.

The information flowing from maneuver units on vehicle and equipment status is complemented by the statuses entered into GCSS-Army. There is often a delay before a DA Form 5988-E is entered into GCSS-Army and vehicle and equipment statuses need to be synchronized on a regular basis to ensure the commander has an accurate maintenance picture.

This critical information can be incorporated into a unit's common operating picture, also known as a COP. Reports pulled from GCSS-Army can be utilized to help populate the common operating picture.

Time Limits

The BCT commander, with input from the BSB commander, may establish time factors for evacuation of battle-damaged or non-mission capable equipment. Time limits identify the amount of time a piece of non-mission capable equipment can be worked on before moving it to the rear. The decision on whether or not to implement time limits must include an analysis of time-distance factors. Moving a piece of equipment to another location for service may significantly increase the amount of time it takes to return the item to the unit in a combat ready status. Time limits, when adopted, are typically included within a unit's standard operating procedures (SOP), but should provide flexibility to adjust to local conditions and operational tempo.

The maneuver unit commander and BSB commander can array their maintenance personnel across the battlefield in a manner that they believe will most efficiently allow maintainers to repair disabled or battle-damaged equipment. Commanders must keep in mind if they elect to use time-factored evacuation that there is no maintenance support for key weapons systems above the FSC level. In order to have key weapons fixed in the BSA the commander must pre-position maintainers who possess the requisite skills, special tools, proper repair parts, references, and adequate time, otherwise key weapons evacuated to the BSA will not be fixed.

The FSC in the MCP must be aware of time limits established by unit SOP and manage workflow accordingly. This ensures that maintainers do not initiate repairs on equipment that cannot be completed before a planned move. The time between moves during offensive operations is typically shorter than in other operations. An accurate maintenance assessment of incoming equipment, similar to the triage process used by medical personnel, determines priorities of work. It may also be the basis for the MCP not accepting equipment for repair. Options for equipment that is not accepted include—

- The unit can self-tow the broken equipment during the move.
- The FSC can tow the equipment.
- Maintainers may utilize BDAR to enable the equipment to move under its own power.
- Units can cannibalize or sanitize and abandon the equipment.
- Evacuate it to a rear echeloned element of the FSC if echeloned. Typically, this will be the BSA,if the maintainers there have additional time, training, and tools to fix it.
- During tactical pauses, FMTs should work to bring as many systems back into operation as possible.

During large-scale combat operations units may reach a point at which the number of non-mission capable vehicles impacts the unit's freedom of maneuver and freedom of action. The unit must decide at that point whether or not to remove a non-mission capable item; when to remove the item; and to where and how far back it should be moved. The commander must also move maintainers and their tools along with the equipment in order to repair the vehicle or piece of equipment. The commander must also keep in mind that the further back a non-mission capable piece of equipment is evacuated the longer it will take to return it to the unit. This is especially significant during offensive operations where the unit can advance significant distances within a 24-hour period.

C. Battle Damage Assessment & Repair (BDAR)

Ref: ATP 4-31, Recovery and Battle Damage Assessment and Repair (Nov '20), chap 1.

ATP 4-31/MCRP 3-40E.1, Recovery and Battle Damage Assessment and Repair, provides techniques on how recovery and battle damage assessment and repair (BDAR) are employed during operations.

Battlefield recovery and battle damage assessment and repair (BDAR) are separate and distinct subsets of maintenance. Both are the owning unit's responsibilities and both have a fundamental purpose of returning combat assets to the battlefield as soon as possible. The purpose of recovery is to rapidly free mired equipment or remove disabled equipment from the battlefield. The purpose of BDAR is to apply expedient repairs allowing the equipment to self-recover and continue the mission.

Recovery

Recovery is actions taken to extricate damaged or disabled equipment for return to friendly control or repair at another location (JP 3-34). These actions typically involve extracting, towing, lifting, or winching. Towing is usually limited to moving equipment to a maintenance collection point (MCP). Crew, operator, or dedicated recovery personnel accomplish the recovery based on the mire level or the severity of battle damage.

Damaged and inoperable equipment on the battlefield can strain dedicated recovery resources. Dedicated recovery assets should be strategically placed within the area of operations to optimize battlefield recovery operations. Commanders must emphasize the use of self and like vehicle recovering methods to the greatest extent possible. These practices will minimize the use of dedicated recovery assets for routine recovery missions.

Recovery operations, both on and off the battlefield, can be extremely hazardous. Safety must remain a top priority for each recovery mission. Leaders must conduct a risk assessment, consider operational variables, and continually enforce safe practices. Proper maintenance of recovery vehicles and serviceability of authorized rigging and other equipment is essential to ensure safe recovery mission.

Expedient Repair

Expedient repairs are non-conventional repair procedures performed by operators and maintainers that rate as low to medium risk. TMs do not direct these repairs and the repairs may be regarded as either permanent or temporary. Soldiers perform expedient repairs not only on the battlefield but also during training exercises. Expedited repairs can also reduce vehicle and equipment downtime incurred waiting for the delivery of spare parts.

Battle Damage Assessment and Repair (BDAR)

Maintainers rapidly return disabled equipment to the operational commander using BDAR procedures and field-expedient repair of components. Battle damage assessment is the estimate of damage composed of physical and functional damage assessment, as well as target system assessment, resulting from the application of lethal or nonlethal military force (JP-3-0). Battle damage repair is an essential repair, which may be improvised, carried out rapidly in a hostile environment in order to return damaged or disabled equipment to temporary service (JP 4-09).

BDAR restores the minimum essential combat capabilities necessary to support a specific combat mission or to enable the equipment to self-recover. Commanders must authorize the use of BDAR. Soldiers accomplish BDAR by bypassing components or safety devices, exchanging parts from like or lower priority equipment, fabricating repair parts, jury-rigging, taking shortcuts to standard maintenance, and using substitute fluids, materials or components (complete controlled exchanges in accordance with AR 750-1). Depending on the repairs required and the amount of time available, repairs may or may not return the vehicle to a fully mission-capable status. Operators/crew, maintenance teams, or recovery teams may perform BDAR.

II. Transportation

Ref: FM 4-01, Army Transportation Operations (Apr '14).

The U.S. Army Transportation Corps provides an overwhelming capability for Army and joint forces in achieving operational reach, freedom of action and prolonged endurance. Army transportation, combined with strategic enablers, delivers to the Combatant Commander expeditionary capabilities. Introductory figure-1 depicts the strategic to tactical transportation system. It illustrates transportation agencies, organizations, and commands that deploy forces; distribute personnel and materiel; sustain forces for extended durations; and redeploy/retrograde forces and materiel upon mission completion.

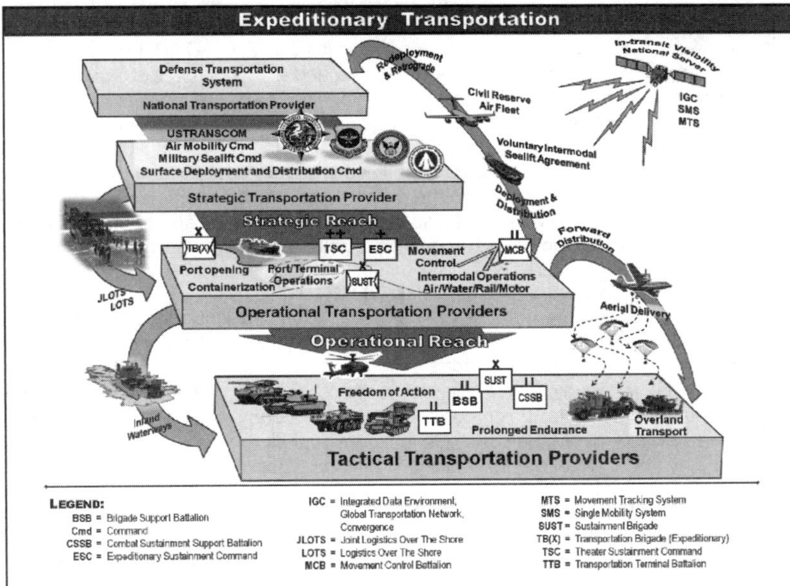

Ref: FM 4-01, Introductory figure-1. Army transportation operations overview.

Transportation operations are critical for theater opening. The Army's transportation expeditionary capabilities play an important role in early entry operations. The newly developed transportation brigade (expeditionary) provides a rapid deployment capability that quickly establishes ports operations. The Surface Deployment and Distribution Command provides essential port management for the duration of an operation. Army watercraft provides the capability to conduct joint-logistics-over-the-shore operations and the ability to maneuver small channels, rivers, or land on a bare beach.

Transportation is an integral part of the reception, staging of, onward movement, and integration (RSO&I) of forces. Movement control battalions (MCB) and movement control teams (MCT) regulate the movement forces along busy supply routes. Transportation units operate ports, terminals and intermodal sites. Transportation staffs within the support operations centers of the theater sustainment command (TSC) or expeditionary sustainment command (ESC), and sustainment brigades, manage segments of the theater distribution pipeline by aerial, airdrop, or surface capabilities. When required, transportation units may also play an advisory role in railway operations.

At the tactical level the sustainment brigade, combat sustainment support battalion and the brigade support battalion provide overland transportation assets such as heavy equipment transports; medium tactical vehicles; palletized load system; or other motor transport assets. Transportation capabilities assist a commander's maneuverability by positioning combat units for decisive action and delivering vital support for prolonged operational endurance.

Army transportation functions are divided into four categories; mode operations, intermodal operations, movement control, and theater distribution.

Note: See p. 3-38 for discussion of theater distribution.

A. Mode Operations

Mode operations are the execution of movements using various conveyances (truck, lighterage, railcar, aircraft) to transport cargo and PAX (ADRP 4-0). There are two transportation modes of operation, surface and air, available to support military operations. The surface mode includes motor, water and rail. The air mode consists of fixed-wing and rotary-wing aircraft. Although a pipeline is considered a surface mode of transportation this does not necessarily apply in this sense. However, it must be noted that a pipeline can be used to transport large quantities of bulk petroleum and water which would reduce the number of surface assets needed to transport this commodity. The Army does have the capability to lay and operate a petroleum pipeline which is known as the Inland Petroleum Distribution System.

Surface Modes of Transportation

There are three different types of surface modes of transportation; motor, waterway and rail. Each surface mode uses fundamentally different solutions in a separate environment. The highway mode uses ground transportation assets to travel across land. The waterway or sea mode of transport consists of assets such as ships, boats and barges which transport commodities over a body of water like a sea, ocean, lake, canal or river. The rail mode is a conveyance that moves cargo and passengers by way of a railway or railroad. *See facing page for further discussion.*

Air Modes of Transportation

The air mode of transport is comprised of fixed and rotary wing assets. Fixed and rotary wing airlift assets have definite cargo size and weight limitations that must be taken into account when planning for the movement of military personnel and equipment. Movement by air is the fastest but most expensive mode of transportation.

- **Fixed Wing.** The U.S. Air Force, through the AMC, provides a variety of fixed winged assets to fulfill strategic and theater airlift requirements. Air transport by fixed-wing aircraft is the most important transportation mode in terms of rapid strategic mobility. The most common fixed wing assets used to support strategic and theater movements are the U.S. Air Force's C-130, C-5, C-17 and the CRAF.
- **Rotary Wing**. Rotary wing aircraft are used mainly for short-range, tactical transport missions. They can transport essential military equipment and relief supplies directly to a forward area while avoiding rough terrain and/or damaged road or railway systems. These assets support the operational and tactical levels of war and can operate on a less improved support structure than fixed wing aircraft. Military aircraft used to transport personnel and cargo are classified either utility helicopters (UH), or cargo helicopters (CH). The most common of these are the UH-1, UH-60, CH-46, CH-47, and CH-53. All five can lift cargo from an external cargo hook. However the CH-46, CH-47, and Joint capabilities assigned to the common user pool are the only rotary wing assets with cargo compartments large enough to carry a significant amount of internal cargo. For sustainment operations, the Deputy Chief of Staff G3 Aviation section at the Army Service Component Command (ASCC), corps or division levels are responsible for consolidating, prioritizing and processing aviation maneuver sustainment requests.

Surface Modes of Transportation

Ref: FM 4-01, Army Transportation Operations (Apr '14), pp. 1-7 to 1-8.

There are three different types of surface modes of transportation; motor, waterway and rail.

Motor Transport Operations

Motor transport is a ground support transportation function that includes moving and transferring units, personnel, equipment and supplies by vehicle. Army motor transportation units are the single largest provider of land surface movement within joint forces and include organic and contracted assets. Motor transport is also the most flexible mode of transportation. This mode can incorporate multi-stop schedules as well as making rerouting or diverting assets less difficult than other modes. Motor transportation provides essential distribution capabilities to sustain forces, prolong endurance, and extend operational reach. Motor transport is an integral aspect of the Army's support and force sustainment.

Refer to ATP 4-11, Army Motor Transport Operations, for additional discussion.

Watercraft Operations

Watercraft consists of Navy and Army sealift assets. The Navy assets are managed by the MSC and consist of the active and reduced operational status fleet and those assets acquired from the ready reserve fleet. Commercial shipping organizations may provide assets at the request of USTRANSCOM. These assets primarily support the DTS at the strategic level.

The Army's contribution to the water mode of transport consists of a variety of lighterage and floating utility watercraft to perform a wide range of waterway operations.

- **Lighterage** - is the process in which small craft are used to transport cargo or personnel from ship-to-shore using amphibians, landing craft, discharge lighters, causeways, and barges (JP 4¬01.6) Lighterage can also be used to transfer equipment and personnel between ships or for intratheater transport. Lighterage are further classified into conventional displacement (landing craft), or modular causeway systems (powered ferry).
- **Floating Utility** - perform operations incidental to water terminal operations, except lighterage service. Watercraft in this category are harbor and ocean going tugs, pusher tugs, floating cranes, barges, floating causeway and roll-on/roll-off discharge facilities.

Rail Operations

Rail operations are limited to fixed routes that will traverse areas outside the non-contiguous operational environment protected by combat forces. Establishing rail in a theater is based on broad, initial planning that includes the assessment of a HN's rail capabilities and infrastructure. The Army's expeditionary railway center can provide this assessment and provide a feasibility study on the HN rail capability. The expeditionary railway center can also advise on the employment of those rail capabilities. Once operations are established, there are several methods of operation, which can be used depending on the theater's environment and circumstances. Rail, when available, is the most cost effective and expeditious means of moving large quantities of cargo.

For additional information on rail and the expeditionary railway center, refer to ATP 4-14, Expeditionary Railway Center Operations.

B. Intermodal Operations

Ref: FM 4-01, Army Transportation Operations (Apr '14), pp. 1-8 to 1-10.

The term intermodal refers to the type of international freight system that permits transshipping among sea, highway, rail, and air modes of transportation through the use of American National Standards Institute and the International Organization for Standardization containers, line haul assets, and handling equipment (JP 4-09). Intermodal operations is the process of using multiple modes (air, sea, highway, rail) and conveyances (i.e. truck, barge, containers, pallets) to move troops, supplies and equipment through expeditionary entry points and the network of specialized transportation nodes to sustain land forces (ADRP 4-0). A node refers to a location in a mobility system where a movement requirement is originated, processed for onward movement, or terminated (JP 3-17). Intermodal operations involve the transportation of cargo to or from different nodes in a standardized container or vehicle using multiple modes of transport without the intermediate handling of the cargo itself. The results of intermodal operations are increased distribution effectiveness and efficiency, reduced delivery times and decreased operating costs.

The two key components in intermodal operations are port/terminal operations and containerization.

Port / Terminal Operations

Port/terminal operations encompass the reception, processing, and staging of passengers; the receipt, transit storage, and marshalling of cargo; the loading and unloading of modes of transport conveyances; the manifesting and forwarding of cargo and passengers to a destination.

There are two types of ports, aerial ports and water ports. An aerial port is an airfield that has been designated for the sustained air movement of personnel and materiel as well as an authorized port for entrance into or departure from the country where located (JP 3-17). Aerial ports provide the most expeditious method for rapid force deployment and normally serve as a link to theater land transportation systems in the theater. Water ports have three categories; fixed, unimproved and bare beach and are categorized by their ability to accommodate various vessel sizes and the port's cargo handling capacity. The SDDC is designated the port manager at seaports and provide resources to manage, operate and clear those ports.

A terminal is a facility designed to transfer cargo from one means of conveyance to another (JP 4¬01.6). Terminals are essential nodes to the total distribution network that support the commander's concept of operation. When linked by modes of transport (air, highway, rail and water), terminals define the transportation structure for the operation. Force projection is the ability to project the military instrument of national power from the United States or another theater, in response to requirements for military operations (JP 3-0). Force projection requirements necessitate the early identification and establishment of terminals. They may be among the initial major objectives seized by U.S. forces during a forcible entry, or the focal point of activities for Foreign Humanitarian Assistance (FHA) operations. Terminal selection is important because deployment and sustainment of the military operation will hinge heavily on the terminal's effectiveness. Without adequate terminals, a geographic combatant commander's mobilization, deployment, employment, sustainment, and redeployment concepts may become unsupportable (JP 4-01.5).

Determining the cargo capacity of terminals and ports is a critical factor in the planning process of transportation operations.

Additional information on terminal operations can be found in ATP 4-13, Army Expeditionary Intermodal Operations.

Containerization

Containerization is a system of cargo transport based on a range of intermodal International Organization for Standardization (ISO) containers. These containers can be loaded and unloaded, stacked, transported efficiently over long distances, and transferred from one mode of transport to another without the intermediate handling of the container's contents. The Army has the largest requirement for strategic lift (both air and sea) and has historically relied on commercial sealift to move 85-90% of its cargo during contingency operations. The U.S. and world merchant fleet are dominated by large, fast containerships and have the necessary infrastructure support. This is a huge advantage over break bulk ships in terms of the amount of cargo that can be moved, time and cost. Containerization increases the types of ships available to support strategic movements as well as increasing the cargo capacity of other available ships. Other benefits of containerization are increased protection against shipping damage and safeguards against pilferage. Containerization also increases interoperability between Service components and commercial industry.

Container management is the process of establishing and maintaining visibility and accountability of all cargo containers moving within the Defense Transportation System (ADRP 4-0). It incorporates the planning, organizing, directing, controlling, and executing of all functions and responsibilities required to provide for positive and effective use of DOD/Service-owned, leased or controlled ISO containers. This includes, but is not limited to, the functions and responsibilities of life cycle asset and operational management supporting the full spectrum of operations.

Global Container Management is designated by USTRANSCOM as the global container manager. It serves as the DOD's single manager and authority for the control of DOD containers moving through the DTS. It is composed of the SDDC's Container Management Division, located at Scott Air Force Base, and the Army Intermodal and Distribution Platform Management Office (AIDPMO), located at Tobyhanna Army Depot. AIDPMO serves as the Army's single manager for Army owned or leased ISO containers, flatracks, and other distribution platforms. The primary focus of AIDPMO is to maintain control and readiness condition of Army-owned/leased intermodal assets. They monitor the inventory, management, and accountability of all ISO intermodal containers for the Army. AIDPMO also provides container management support to Army units and activities worldwide. Container management at the installation level consists of establishing roles and responsibilities for both Installation activities and assigned units on the installation or supported by the installation. The responsibilities associated with this level of container management are; expeditious return of container assets back to the DTS, conducting scheduled inventories, maintenance, inspections and the continuous reporting of container status.

Refer to ATP 4-12, Army Container Operations, for additional information on container management.

C. Movement Control

Ref: FM 4-01, Army Transportation Operations (Apr '14), p. 1-10 and ATP 4-16, Movement Control (Apr '13), chap. 4.

Movement control is the dual process of committing allocated transportation assets and regulating movements according to command priorities to synchronize the distribution flow over lines of communications to sustain land forces (ADRP 4-0). Movement control, as a transportation subordinate function, provides commanders a mechanism to synchronize movements for deployment, redeployment, and distribution operations to support unified land operations and provides them situational understanding to effectively influence the movement in their area of responsibility. Movement control is inherently joint allowing other Services access to Army common user transportation and is a medium for the Army to utilize joint transportation to enhance military operations. Not vested in a singular unit, movement control responsibilities are embedded in a network that relies on coordination for the planning and execution while ensuring transportation assets are utilized effectively, efficiently and that LOCs are de-conflicted to support decisive action.

Movement control enablers are organizations, units or sections within an organization that allocate transportation assets, plan, regulate and synchronize movements. These components should be among the early units deployed in the theater opening force modules. Early deployment will allow for the timely establishment of a transportation system with the capability to receive and program the onward movement of the deploying force and manage its growth. For theater distribution, these facilitators are found at the TSC and/or ESC Support Operations (SPO) Mobility sections and are responsible for movement control planning. The MCBs and MCTs are the executers of movement control in their assigned areas of operation.

Movement Control Battalion (MCB)

The MCB will also coordinate with host nation authorities for cargo transfer locations, road clearances, border clearances, escort support, and transportation support. The MCB will have as many subordinate MCT's as needed to operate in its area of operations, based on the number of customers, air terminals, rail terminals, seaports, and MSR's it must support. The MCB will provide logistics support to the MCT's under its C2. However, MCT's operating away from their HQ will require logistics support from other units. The MCB will—

- Continue to provide command, control, and technical guidance to 4 to 10 MCT's
- Provide asset visibility and maintain ITV of tactical and non-tactical moves within its assigned geographical area (including unit moves and convoys)
- Assist in planning and executing plans and operations
- Apply and meet movement priorities provided by the TSC and sustainment brigade
- Support the entire spectrum of distribution

Movement Control Team (MCT)

The MCT is the basic and most critical level in the movement control process. MCT's are the common point of contact for mode operators and users of transportation. The MCT is a 21-Soldier team created with the capability to perform every type of movement control mission. It is designed to provide maximum flexibility in its employment. Each team has a headquarters section and four identical subunits (or sections). The MCT can operate as a single team or separately at up to four different locations. For example, a single MCT can be deployed initially to provide movement control functions at an airfield while simultaneously providing cargo documentation. As the mission expands, the team can deploy a section onto the MSR's to conduct movement control operations. As the operation matures, that same MCT can operate at a second airfield or seaport. The operational use

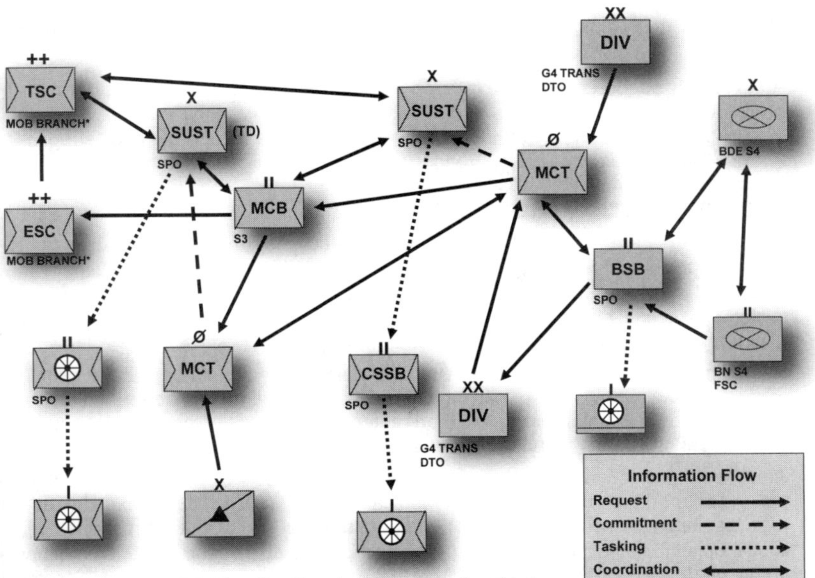

* The mobility branch within the TSC/ESC DMC provides guidance, plans and policies and staff supervision for transportation operations.

of the MCT can be specifically tailored to the mission and operational environment. The standardization of MCT's increases the number of teams available for deployment, since each unit is tailorable in the truest sense of that term. If METT–TC factors dictate that the MCT needs to be split into four sections in different locations, each section can be properly equipped with the vehicles, communications equipment, STAMIS, and generators it needs to operate independently.

MCT's process movement requests and arrange transport for moving personnel, equipment, and sustainment supplies. They process convoy clearance requests and special hauling permits. MCT's coordinate with the MCB for the optimal mode (air, rail, inland waterway, or highway) for unprogrammed moves and commit the mode operators from the sustainment brigade, LOGCAP, multinational elements, and the host nation. They also assist in carrying out the movement program.

MCT's are arrayed on the battlefield at various nodes to best support the TSC concept of support. In general, a MCT collocates with a Sustainment Brigade to provide movement control support on an area basis to EAB units or units not assigned/attached to a Division. An additional MCT can work directly for the Division G-4 Transportation Officer (DTO) to assist in the planning and execution of movement control in the division area of operations. MCT's can operate at an APOD to facilitate the movement of ground cargo (primarily palletized), assist in passenger manifesting, and provide air load planning capability. MCT's operating at SPOD's facilitate the ground movement of containerized cargo and assist in the marshalling/staging of equipment for onward movement. Elements of a MCT can locate at other critical nodes (for example Central Receiving and Shipping Point (CRSP), Convoy Support Center (CSC), TDC, and container yards, as needed) to provide cargo documentation capability or for the added visibility and management of distribution operations. MCT's are also used to enforce the highway regulation plan established by the TSC and will deploy its sections along Main Supply Routes (MSR) to augment convoy tracking operations and provide on the ground traffic de-confliction.

III. Supply

Ref: ATP 4-42, Materiel Management, Supply, and Field Services Operations (Nov '20).

Army supply support is critical to all operations. It is necessary to ensure personnel, equipment, weapon systems, and facilities are sustained and operational. Although the importance and priority of any one commodity varies by operation, all supplies are necessary for successful operations

A. Supply Fundamentals

Commanders determine the composition, frequency, and method of resupply to best support operations. Supply operations are time sensitive, preplanned pauses in the battle rhythm that allow combat forces to resupply. Supply operations are inseparable from tactical operations. The most brilliantly conceived and executed tactical plan will fail unless reinforced by a sound supply plan.

Supply personnel request, receive, issue, prepare and maintain detailed accounting records. A unit submits a request, which results in an obligation and eventual disbursement of unit funds for the required materiel. Unit leaders, materiel managers, and budget analysts supervise disbursements, cash, and credit card purchases.

Army health service support and force health protection depend upon specialized materiel and support services managed by medical logistics (MEDLOG) personnel. Medical logistics encompasses medical supply support, medical equipment maintenance and repair, optical fabrication and repair, blood storage and distribution, patient movement items, medical contracting, regulated medical waste, medical gases, and medical facilities and infrastructure. This publication addresses the Army's medical services only to the extent that it clarifies the medical supply support process and the Army Health System (AHS) support to class I, management of class VIII, and to field services.

Routine resupply operations include regular resupply of items in classes I, III, V, VIII, and IX, and of any other items requested by units. Battalions plan routine resupply and it usually takes place at every opportunity. The logistics package (LOGPAC) contains company and battalion assets that transport supplies to the company. For example, a company supply sergeant assembles the LOGPAC in the brigade support area under the supervision of the headquarters and headquarters company commander or the battalion supply officer. Likewise, medical brigades plan routine resupply for field hospitals and other echelon above brigade medical formations. Class VIII resupply is usually integrated into LOGPACs provided cold chain storage requirements can be maintained. Figure 1-1 depicts a notional field site of forward support elements in a logistics resupply point.

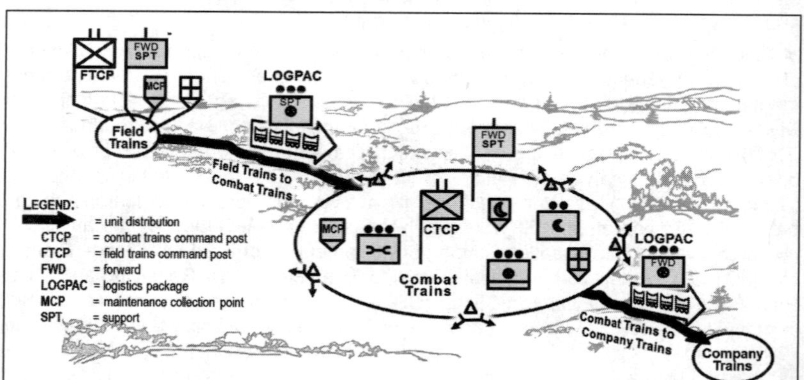

Ref: ATP 4-42, fig. 1-1. Resupply operations.

Sustainment Execution

B. Classes of Supply

Ref: ADP 4-0, Sustainment (Jul '19). table 1-1, pp. 1-7 to 1-8.

The Army divides supply into ten classes for administrative and management purposes.

Class	Symbol	Description
Class I		Subsistence, including health and welfare items.
Class II		Clothing, individual equipment, tentage, organizational tool sets and kits, hand tools, administrative and housekeeping supplies and equipment (including maps).
Class III		POL, petroleum and solid fuels, including bulk and packaged fuels, lubricating oils and lubricants, petroleum specialty products; solid fuels, coal, and related products.
Class IV	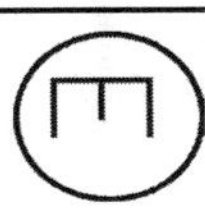	Construction materials, to include installed equipment and all fortification/barrier materials.
Class V		Ammunition of all types (including chemical, radiological, and special weapons), bombs, explosives, mines, fuses, detonators, pyrotechnics, missiles, rockets, propellants, associated items.
Class VI		Personal demand items (nonmilitary sales items).
Class VII		Major items: A final combination of end products which is ready for its intended use.
Class VIII		Medical material, including medical peculiar repair parts.
Class IX	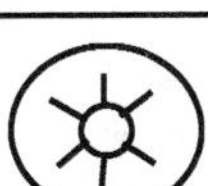	Repair parts and components, including kits, assemblies and subassemblies, reparable and non repairable, required for maintenance support of all equipment.
Class X		Material to support nonmilitary programs; such as, agricultural and economic development, not included in Class I through Class IX.

Emergency resupply, usually involving class III (bulk), class V, and class VIII is executed when the platoon has such an urgent need for resupply that it cannot wait for the routine LOGPAC. For example,emergency resupply procedures start with immediate redistribution of ammunition in individual vehicles,followed by cross leveling ammunition within the platoon while awaiting ammunition to be brought forward. Cross-leveling ammunition refers to distributing equal amounts and kinds of ammunition throughout the platoon.

There are two standard methods of supply conducted in an operational environment: supply point distribution and unit distribution:

Supply Point Distribution

Supply point distribution requires the supported unit to move to a supply point to pick up supplies. The supply point issues materiel to the supported unit that transports their supplies back to the unit with organic transportation. Supply points include supply support activities (SSA) for all supply classes. A supply support activity is an activity assigned a Department of Defense activity address code and that has a supply support mission (JP 4-09).

Unit Distribution

When unit distribution is used, the supported unit receives supplies in its area. The logistics release point, established by the supported unit, may be any place on the ground where unit vehicles pickup supplies and then take them forward to their unit.

Refer to ATP 3-20.98, Scout Platoon, for information on how to establish unit distribution operations.

Throughput distribution bypasses one or more intermediate supply echelons in the supply system to avoid multiple handling. It leverages configured loads and containerization to deliver supplies directly to units in forward areas. For example, engineer barrier materials may be shipped directly to the emplacing units, instead of their supporting sustainment units. Throughput is not automatic. It must be specified in appropriate plans and coordinated.

Materiel managers push supplies to supported units or supported units pull supplies from the supply system:

Push Supply Flow

A process where supplies are pushed forward based on running estimates of the supported unit's consumption and anticipated requirements. The flow of supplies occurs without a request from the supported unit. Typically during the early stages of an operation, sustainment units, via unit distribution, push certain classes of supplies (I, III (bulk), VIII, and V) to subordinate sustainment units and to supported units. Commander priorities and other planning considerations dictate when and where supplies are pushed.

Pull Supply Flow

A process where the flow of supplies occurs after a request from the supported unit and based on the real consumption of the supported unit. As distribution capabilities expand,sustainment units implement the pull system where supported units and maintenance elements request supplies either manually or through automation.

Scheduled supplies are those for predicted requirements. Normally, scheduled supplies do not require a request for replenishment. Requirements are based on troop strength, equipment density, forecasts, and daily usage factors. Scheduled supplies are pushed to supported units based on planned distribution arrangements derived from the logistics status report (LOGSTAT). Subsistence, water, bulk fuel, and munitions are normally treated as scheduled supplies.

Demand supplies are those for which a unit submits a request. Items in supply classes II, III (packaged),IV, VII, VIII, and IX are typically demand supplies.

Command regulated supplies are those a commander has decided to closely control because of scarcity,high cost, or mission need. Commanders normally control supply classes II, III (bulk), IV, V, VII, and VIII. Only the commander who designated it for control can release regulated items for issue.

To sustain tactical operations for specific periods, all units, starting at the company level, determine their ammunition requirements and submit a required supply rate report to the next higher headquarters.

The authorization and allocation of ammunition within an area of operations (AO) is determined by using the required supply rate and the controlled supply rate:

gets

Controlled Supply Rate

The controlled supply rate is the rate of ammunition consumption that can be supported, considering availability, facilities, and transportation. It is expressed in rounds per unit, individual, or vehicle per day. (FM 4-30)

want

Required Supply Rate

The required supply rate is an estimated amount of ammunition needed to sustain tactical operations, without ammunition expenditure restrictions, over a specified time. The required supply rate is expressed as rounds per weapon per day, or as a bulk allotment per day or based on mission. (FM 4-30)

Aerial delivery is used for routine and urgent resupply to units in remote or forward locations where terrain limits access. Aerial delivery reduces the need for route clearance of ground lines of communications as it bypasses enemy activity. In order for aerial delivery to be effective, friendly forces must control airspace throughout the AO and enemy ground-based air defenses must be neutralized.

Refuel on the move is most often used to support extended moves to or from a tactical assembly area. When vehicles enter a refuel on the move site for refueling, they receive a predetermined amount of fuel(usually timed) and they move out to return to their convoy or formation.

Controlled exchange is the removal of serviceable components, assemblies, and subassemblies from unserviceable, uneconomically reparable equipment for reuse in restoring equipment to a serviceable condition. Controlled exchanges allow units to maintain combat readiness by repairing non-mission capable equipment with the use of serviceable parts from another non-mission capable vehicle. This process must be strictly controlled and documented or it becomes unauthorized cannibalization.

Consumable items are the expendable property that a unit needs to perform its mission. Consumable items, such as ammunition, fuel, cleaning and preserving materials, surgical dressings, drugs, medicines, repair parts, building materials, and office supplies lose their separate identity in use. Expendable items require no formal accountability after issue from a stock record account. Shop stock, bench stock, authorized stockage lists, and unit supply lists contain consumable items. These items require proper use and safeguarding, but do not require formal accounting after issue to the user.

Nonexpendable property retains its original identity throughout its life cycle. Nonexpendable properties are material planning objects that are assigned a material planning objects number that includes the item's line item number (LIN). Nonexpendable property requires formal property book and stock record accountability throughout its life cycle.

Durable property is an item of Army property that does not require property book accountability after issue from the stock record account. For example, hand tools, used over a relatively long period without being depleted or consumed, are typically controlled through hand receipts.

C. Issue Materiel

Ref: ATP 4-42, Materiel Management, Supply, and Field Services Operations (Nov '20), pp. 5-24 to 5-26.

This section discusses establishing a supply point because it influences the effectiveness of the issue process. Every supply point, all supply classes, must adapt to changing locations. It is important to note that every deployment is different and every site is different. When deployed, the supply point will issue supplies to the supported unit from either a developed or an undeveloped location. A developed location will have an infrastructure (roads, buildings) and undeveloped location will lack that infrastructure. Every tactical supply point, regardless of supply class, will use the same basic principles when developing the field layout plan.

Spend time and effort planning day-to-day operations while developing the storage layout plan. The layout of a supply point can adversely affect the ability to process receipts and issues, but a well-planned layout can reduce handling of items, thereby improving processing times. Sketch the supply point area to show the use of the space (receiving, shipping, hazardous materials storage, yard open storage, office space),and the materiel stored (bulk fuel, bulk water, subsistence, medical supply, repair parts, construction materials, end items). A sketch is a rough drawing that allows the testing of several ideas to zero in on the most likely layouts for the supply point. Include latrines and offices in the sketch. See figure 5-7 for an example of a supply point field site sketch. In this instance the supply is placed in an area with some trees and no roads other than the main road.

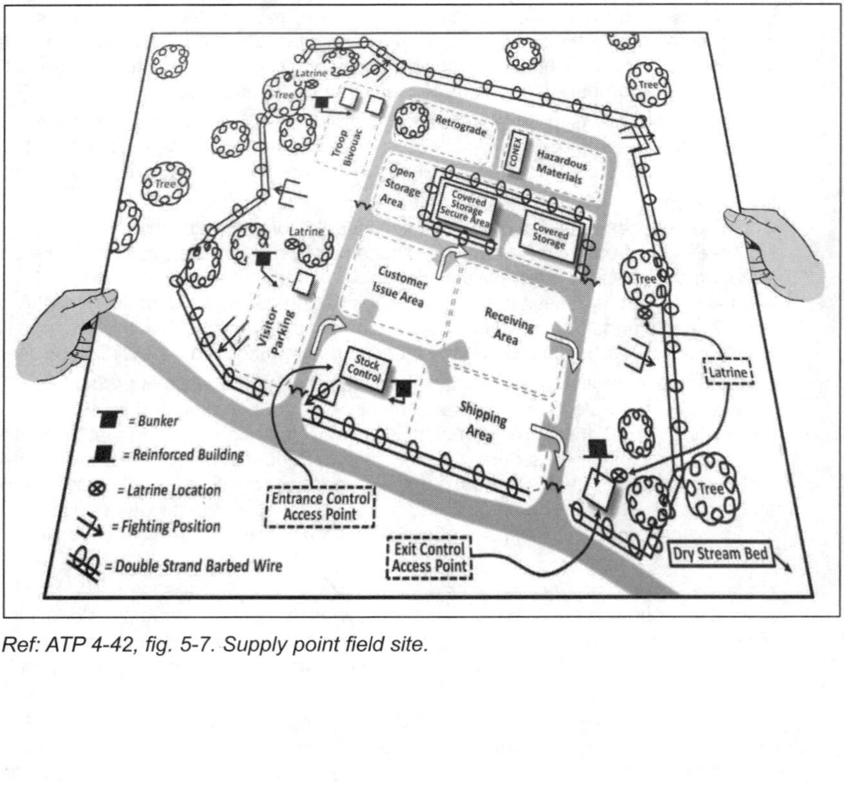

Ref: ATP 4-42, fig. 5-7. Supply point field site.

Supply point storage layouts depend upon the supply class and where the supply point is placed on battlefield. However, the following planning considerations apply to every supply point:

- Establishing a secure perimeter.
- Designating, marking and enforcing the traffic flow route. Normally, traffic flow will be in one direction to control access to the supply point storage locations.
- Establishing controls at the entrance and exits to control the flow of traffic. The entrance and exit point may be the same depending on the terrain or other circumstances associated with the supply point site.
- Placing the parking area near the entrance to control access to the supply point storage locations.

While each supply class has its own set of requirements, the basic principles for issue are similar. Issue sections must carry out the issue activities efficiently if supported units are to receive supplies on time,in the quantity requested, and in a usable condition. If supply point distribution is used, the number of times a unit picks up supplies during a week depends on the distance the unit has to travel and how often it can arrange for transportation. Some nearby units may pick up supplies two or three times a day. If unit distribution is used, coordination between the supply point and the supported unit is crucial to ensure that the materiel arrives at the logistics release point on time and with the correct supplies in the correct amount.

Unlike other supply classes, planners must establish the bulk water supply point at the raw material source. Raw water, at its ground or surface source, may be fresh, brackish, or salt water. Prior to issuing to supported units, water support companies purify raw water to make it safe for human consumption. Planning considerations for a water supply point include the raw water source, a water intake point, a treatment area,a storage area, and an issue point.

Bulk fuel supply points range from strategic and operational level storage to refueling a tactical unit engaged in offensive operations. Bulk fuel supply points may consist of ocean tanker loading and unloading facilities, storage terminals, above ground or underground steel tanks, pump stations, pipelines, hose lines,tank vehicles, rail tank cars, fabric tanks, and tactical systems. Each tactical unit refueling operation is unique depending on the refueling system used, number of vehicles, the distance the unit is traveling, and how many times the unit needs refueling.

Refer to ATP 4-42.2, Supply Support Activity Operations, ATP 4-41, Army Field Feeding and Class I Operations, ATP 4-43, Petroleum Operations, ATP 4-44, Water Support Operations and ATP 4-35,Munitions Operations and Distribution Techniques, ATP 4-02.1, Army Medical Logistics, AR 40-61,Medical Logistics Policies for detailed information on issue procedures for each commodity.

Retrograde of Materiel

Retrograde of materiel is as important as its forward distribution. All classes of supply require retrograde procedures because all units accumulate serviceable, unserviceable, and uneconomically repairable items requiring cross levelling, recycle, return to the supply system, or disposal. Depending on an item's condition, materiel managers for a commodity will direct recycle to original use, recycle for alternate use, redistribution, destruction, or demilitarization based on policy for that commodity.

Refer to ATP 4-42.2, Supply Support Activity Operations, ATP 4-41, Army Field Feeding and Class I Operations, ATP 4-43, Petroleum Operations, ATP 4-44, Water Support Operations, and ATP 4-35,Munitions Operations and Distribution Techniques, ATP 4-02.1, Army Medical Logistics, AR 40-61,Medical Logistics Policies for detailed information on retrograde procedures and requirements.

Sustainment Execution

D. Supply Support

Supply support includes all processes from acquisition until the materiel is disposed of or consumed.

Success during large-scale combat operations depends on the Army's ability to feed and clothe its Soldiers, provide medical care, fuel its combat vehicles, fortify its positions, and replace its weapons systems. Materiel management is the means for providing supply support necessary to accomplish these tasks. Theater materiel managers develop a distribution plan to deliver supplies from the theater ports of entry to the supply points and to the supported unit to execute mission requirements. Sustainment planners establish supply points throughout the theater to support the readiness of supported units. Tactical supply personnel execute unit level procedures that ensure food, water, clothing, ammunition, petroleum products, barrier material,medical materiel, and major end items are available when and where needed.

Supply support depends upon the flow of information and materiel from the supply source to the user, back to the supply system, and disposal. This flow includes the last tactical mile (unit formations that come directly into contact with enemy forces) and the retrograde of materiel. Speed alone is not the desired result. Effective supply support requires the reliable, predictive, rapid, and precise delivery of materiel when and where required. Success is measured when the commander's validated materiel requirements are delivered.

Figure 5-1 on the following page (p. 3-27) and figure 5-2 (on p. 3-39) contain the distribution management functions that influence effective and efficient supply support. These graphics illustrate the coordination between the players throughout the distribution management process and the high-level tasks associated with each of the functions. These tasks are not all inclusive and each task contains too many details to list in this publication.

Requirements Determination

Unit leaders and supply personnel at every echelon forecast requirements for all classes of supply. Requirements by supply class depends upon the type of unit, unit mission, equipment density, current operations, future operations, and other factors.

After a unit or maintenance facility determines there is a requirement and the requirement validated through unit channels and procedures, the unit or maintenance facility generates a demand on the supply system to deliver materiel. A using unit may satisfy materiel requirements through several different avenues. Options include—

- Submit a request either manually or through an automated system.
- Purchase supplies locally with a government purchase card or through a contract.
- Initiate a demand on the supply system through timely and accurate LOGSTAT reporting.
- Submit an operational needs statement to initiate a materiel solution to an operational requirement.
- Submit an equipment sourcing document for filling equipment shortages that are validated and authorized by DA.

Note: The Army supply system is the supply process of choice. Units should try to meet requirements within the system before using other options, such as local purchase.

Loads

Ref: ATP 4-42, Materiel Management, Supply, and Field Services Operations (Nov '20), p. 1-8.

During the initial phases of a deployment, supplies have to be at required levels for each individual unit. Deploying units must be self-sustaining until linking up with sustainment elements in theater. In the event of a contingency operation, units deploy with prescribed amounts of all classes of supplies. Loads combine durable and expendable supplies kept by units to sustain operations. Assault and follow-on echelons carry these loads. During the initial phase of deployment, this is the only source of supply. Emergency resupply will likely be limited to class I, class III (bulk), class V, and class VIII.

Basic Loads

Basic loads are designated quantities of supplies that allow units to initiate combat operations. The quantity of each item of supply in a basic load depends on the number of days in combat the unit might expect to fight without resupply. The components of required medical equipment sets comprise the basic load for medical units. Items from the basic load will be used during peacetime only when no operational loads are available. Units must replace basic load items that are subject to deterioration or have a limited shelf life as required. Medical equipment sets and medical materiel sets at Role 3 comprise the basic load for medical units. Medical units usually do not have the capacity to carry more than the required medical equipment sets;therefore unit sustainment beyond the medical equipment sets must be built into the sustainment plan.

Operational Loads

Operational loads are quantities of classes I through V (to include maps) and class VIII supplies that sustain peacetime operations for a specific time. Operational loads of class VIII are rare because Army medical units do not treat patients in peacetime beyond exercises or for range support. Typically the medical units uses the supplies in their medical equipment sets in support of these exercises because it is the only way to rotate potency and dated medical supplies.

Prescribed Loads

Prescribed loads are quantities of maintenance significant class II and IV, class VIII repair parts, and IX organizational repair parts kept to support a unit's maintenance program and are based on command-designated items and demand history data.

DODAAC management is one of the most important functions for materiel managers (all supply classes) during requirements determination activities. Every organization that requests, requisitions, contracts for, receives, has custody of, issues, or ships DOD assets, or pays bills for materiel and services is identified by six-position alphanumeric, the DODAAC. Units may have more than one DODAAC, for example, a DODAAC for bulk fuel and a DODAAC for other materiel.

The property book office, located in the G-4 or S-4, maintains property book records for subordinate elements and any separate elements operating in its area of responsibility. It accounts for nonexpendable property assigned to organizations and activities. It manages nonexpendable material through record adjustments, issues, turn-ins, property losses, and status reports. Property book offices request, issue, receive,and store materiel.

The G-4 or the S-4 manages supply operations. They analyze battlefield characteristics for effects on the type of support required and for the manner in which support will be delivered. The G-4 or S-4 ensure a continuous flow of all classes of supply to their subordinate units. They coordinate delivery schedules and distribution methods between subordinate and supporting units. The G-4 or S-4 coordinate logistics requirements with external activities and internally monitors field feeding, water purification, mortuary affairs, aerial delivery, laundry, and shower for organic and attached elements.

G-4 or S-4 determine potable water consumption rates. To complete their missions, deployed personnel must drink sufficient quantities of potable water to maintain adequate physical and mental health. The daily water requirements depend on a number of factors including the weather, geographical area, the tactical situation, and the operational tempo. Planners must also consider the following activities that also require potable water: ice making, food preparation, medical treatment, personal hygiene (brushing teeth, shaving, and showering), laundry, mortuary affairs operations, chemical, biological, radiological, and nuclear decontamination.

The SPO fuel and water section is responsible for bulk water distribution and materiel management. Comprehensive and detailed requirement forecasts and running estimates depend upon supported unit daily consumption, forecasted requirements, on-hand bulk water quantity, and available storage capacity. The SPO fuel and water section synchronizes water production capability with water support company storage capacity because of the finite amount of storage capacity at each level. Bulk water planners consider supported unit storage capacity and on-hand stockage to avoid:

- A unit unable to accept the shipment because its storage capacity is at or near full.
- Delivering a quantity of bulk water that does not meet the supported unit requirements.

Request

AR 710-2 defines a request as a supply request initiated by a supported unit. A requisition, defined in AR 710-2, is a supply request initiated by the supply point in a Military Standard Requisitioning and Issue Procedures format or a unit supply request converted to a Military Standard Requisitioning and Issue Procedures format by the supply point for submission to the next higher source of supply. Units do not request bulk water and fuels through traditional, request forms such as the DA Form 2765–1 (Request for Issue or Turn-in) or electronically through an automated system such as GCSS-Army. Petroleum and bulk water materiel managers, at each echelon, manually consolidate consumption data forwarded from subordinate and supported company-level units. They prepare comprehensive and detailed requirement forecasts and running estimates of water and fuel to ensure there are adequate stocks available to meet daily requirements.

Distribution Management Process

Ref: ATP 4-42, Materiel Management, Supply, and Field Services Operations (Nov '20), fig. 5-1. Distribution management framework.

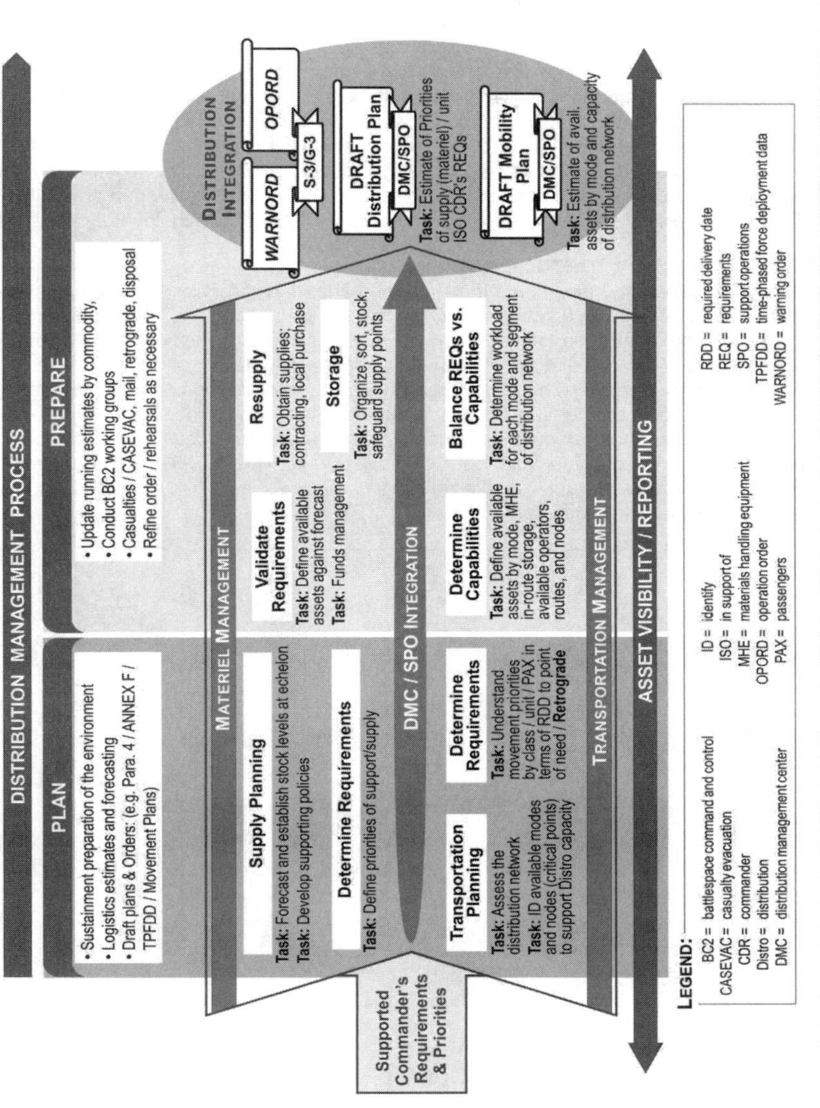

Sustainment Execution

Materiel Management (Select Classes)

Ref: ATP 4-42, Materiel Management, Supply, and Field Services Operations (Nov '20), pp. 5-5 to 5-6.

The following section describes select materiel management roles by position and by commodity. Personnel roles and responsibilities are similar regardless of echelon or size of the organization. Duty titles vary and, in some instances, one individual may perform the responsibilities embodied in two or more of the materiel management roles described in this section of the publication. Support operations at all echelons perform the same basic distribution management functions, but their focus changes based on the scope of responsibilities. For example, the TSC DMC includes a DIB that focuses on the entire theater, but the CSSB has a SPO that focuses only on that CSSB's support area. This ATP does not provide echelon specific role descriptions for multi-echelon duty positions.

Field Feeding Operations

The food advisor is the food operations subject matter expert for field feeding and class I. Food advisors provide technical supervision over internal food service activities and advises the commander on feeding operations, available feeding options, and the service capability. The food advisor coordinates with DL A, the Joint Culinary Center of Excellence, veterinary personnel, and contracting officers to acquire required menu components and helps the theater subsistence officer determine requirements, schedule issue and turn-in times, and decides the best methods for ration break down and distribution. Food advisors ensure available resources such as equipment, facilities, and personnel are adequate to receive, store, and issue class I supplies. Figure 3-1 depicts a notional battlefield field feeding operation.

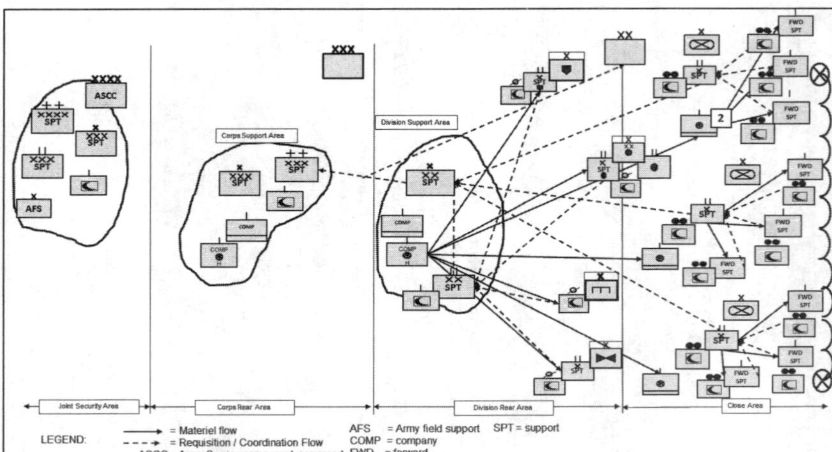

Figure 3-19. Field feeding operations.

Class V Materiel Management

The ammunition NCO advises the command and supported units on all ammunition issues. The ammunition NCO coordinates and controls use of class V supplies, recommends controlled supply rates to the logistics staff officer. The ammunition NCO monitors class V to ensure adherence, within operational considerations, to the established controlled supply rate and that resupply at rates that exceed the controlled supply rate are coordinated according to SOP. Figure 3-20 depicts class V materiel management.

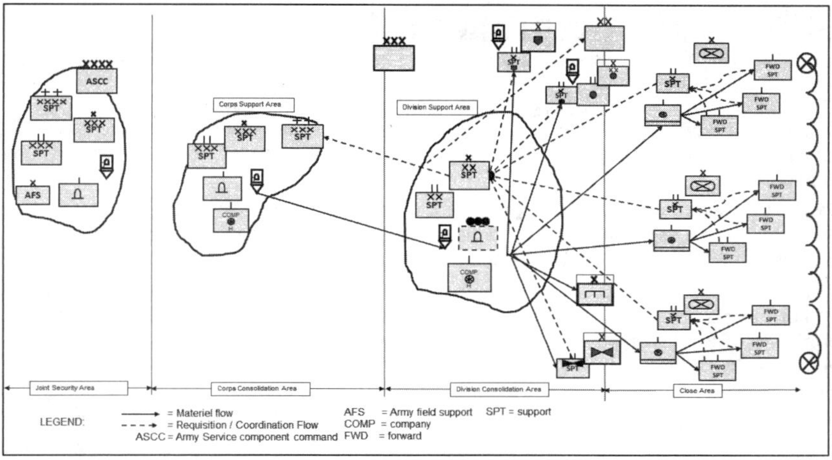

Figure 3-20. Class V materiel management.

Multi-Class Supply Support Activity

The following section describes the roles and responsibilities for supply personnel staffing multi-class SSAs. Figure 3-21 depicts multi-class commodity distribution management and materiel management. All supply classes entering the theater of operations through an air or seaport of debarkation entail theater-level SSA and central receiving point personnel sorting and packaging materiel for transport to the forward points of need. Transportation managers assign the most efficient and effective mode of transportation by the priority of the requesting unit's place within the priority of support, the priority of the item, and the type of cargo being moved. Often, materiel passes through specified areas, such as, trailer transfer points or central receiving points for sorting according to SSA for supply point distribution or supported unit for unit distribution.

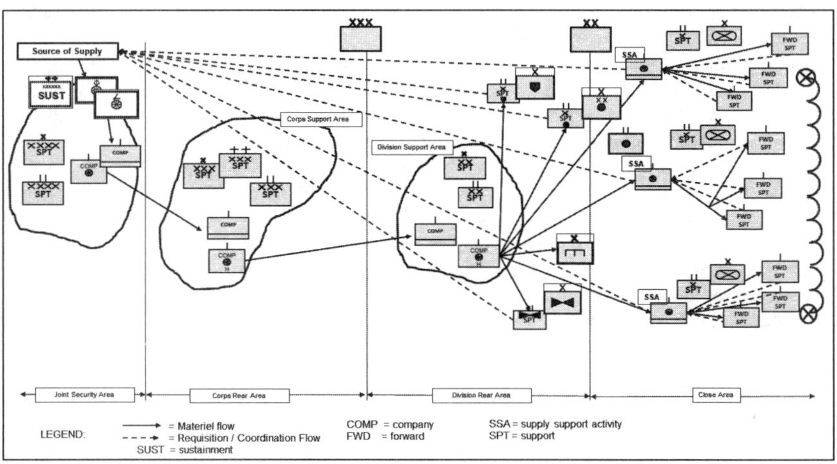

Figure 3-21. Multi-class distribution and materiel management

Class I supply support operations are included in this discussion because class I operations are a separate section of the current composite supply company's supply platoon.

Bulk Water Request Process

Ref: ATP 4-42, Materiel Management, Supply, and Field Services Operations (Nov '20), pp. 5-5 to 5-6.

Bulk water requirements determination and request process includes the same basic process as all classes of supply, but is managed manually through coordination and communication.

Figure 5-3 (facing page) represents the requirements determination process for bulk water in a BSB. However, please note that this process works the same for all sustainment units at all echelons. It is the focus that changes.

Circle 1. The BSB provides bulk water support to its parent brigade with organic storage and distribution capability. The BSB is dependent upon the DSB and DSSB for bulk water production. The BSB's water platoon does not have the capability to produce bulk water and the FSCs do not have the capability to store and distribute bulk water.

Circle 2. Concurrently, the DSB SPO fuel and water section maintains awareness of division current and future operations to forecast bulk water requirements. The DSB SPO fuel and water section anticipates the division's bulk water requirements for the next 96-120 hours and plans bulk water production and resupply based on the division commander's priorities.

Circle 3. The BSB SPO supply and field services section receives daily LOGSTAT reports showing bulk water consumption and requirements from their supported FSCs. The BSB supply and field services section analyzes the FSC's on hand quantities, expected consumption and capacity. The BSB supply and field services section anticipates the brigade's bulk water requirements for the next 96 hours to plan bulk water resupply based on brigade priorities. The BSB SPO supply and field services section sends current consumption information and forecasted bulk water requirements to support future brigade operations to the DSB SPO fuel and water section for resupply.

Circle 4. The DSB SPO fuel and water section receives and analyzes the BSBs' daily LOGSTAT showing on hand quantities, expected consumption and expected capacity for the next 96 hours. The DSB SPO fuel and water section validates their forecasts and production throughput against the BSBs' bulk water requirements to verify proper amounts and identify discrepancies with estimates.

Circle 5. The DSB SPO fuel and water section provides the DSB DIB with the division's current bulk water on-hand quantities by unit, bulk water requirements by unit and storage capacity for the next 96 hours. The DSB SPO fuel and water section also provides bulk water production, storage and distribution assets at the division support area. The DSB SPO DIB synchronizes and plans bulk water distribution to all supported units with other commodity requirements in accordance with division priorities and within the constraints of division movement corridors.

Circle 6. The DSB SPO fuel and water section consolidates the division's daily bulk water consumption and daily requirements, and provides the information to the ESC DMC fuel and water section. The information also provides the ESC DMC awareness of any DSB production, storage or distribution capability shortfalls. The DSB SPO fuel and water section also provides a copy of the bulk water information to the division G-4 for visibility.

The DSSB composite truck company transports water distribution assets belonging to the DSSB CSC and resupplies the BSB Distribution Company water distribution assets. The BSB Distribution Company resupplies FSC and Maneuver Battalion water buffaloes.

The DSSB composite truck company transports water distribution assets belonging to the DSSB CSC and resupplies the BSB Distribution Company water distribution assets. The DSSB composite truck company throughputs bulk water to FSC and maneuver battalion water buffaloes.

Bulk Water Request Process

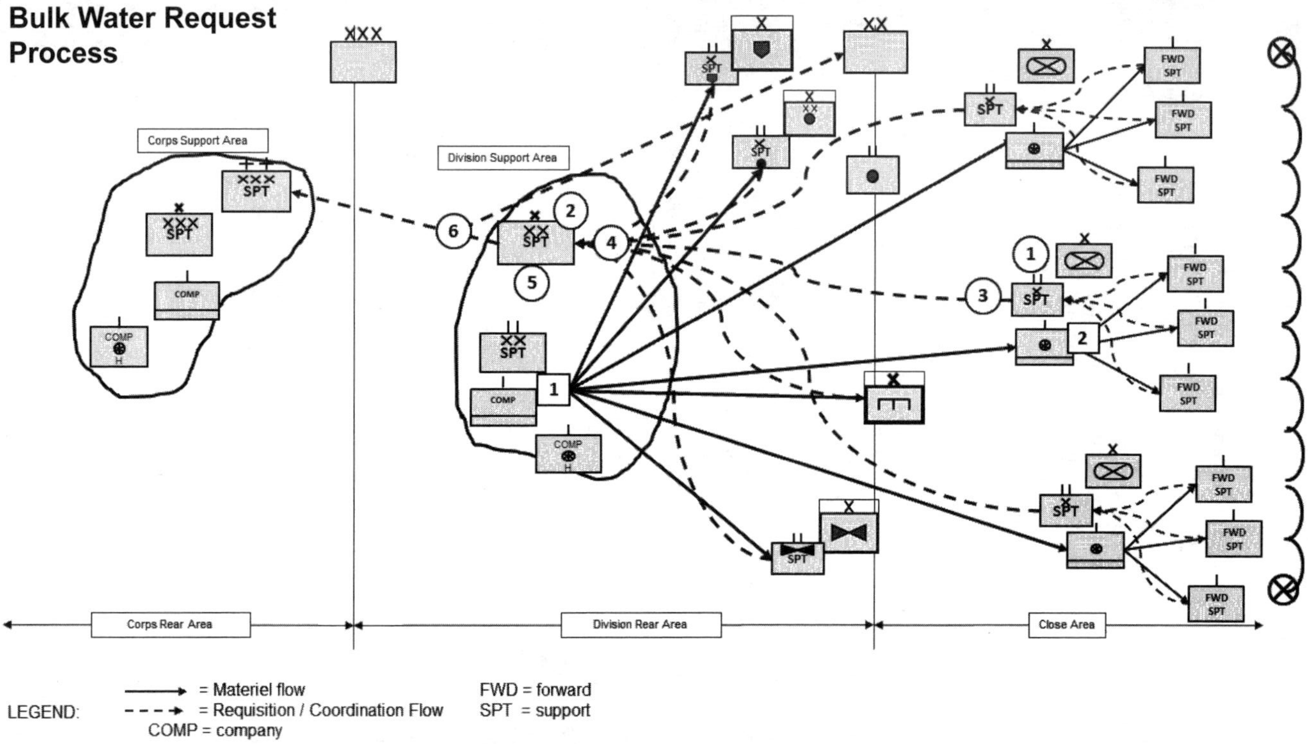

Ref: ATP 4-42, fig. 5-3. Each of the numbered points represent a node where actions take place and are discussed in subsequent paragraphs. The circle points represent requisition and coordination actions between organizations. The square points represent distribution actions. The dotted lines represent transmission of information. The solid lines represent the physical movement of materiel.

Bulk Fuel Request Process

Ref: ATP 4-42, Materiel Management, Supply, and Field Services Operations (Nov '20), pp. 5-6 to 5-8.

Petroleum materiel managers communicate along command channels and technical channels to ensure that supported units receive on-specification bulk petroleum. The command channel involves the petroleum units within the Army while the technical channel includes joint and Army organizations. Command and technical channels incorporate a two-way flow of information to ensure that the petroleum materiel managers receive the most current guidance and to inform the command or suppliers of real or potential problems in the supply chain. Army petroleum materiel managers work closely with DLA-Energy to ensure that petroleum units can deliver fuel to supported units in a timely manner.

Figure 5-4 (facing page) depicts the class III (bulk) request process.

Circle 1. The BSB provides support to its parent brigade with organic storage and distribution capability. The BSB SPO supply and services section forecasts brigade fuel requirements based on current and future operations.

Circle 2. Concurrently, the DSB SPO fuel and water section maintains awareness of division current and future operations to forecast fuel requirements. The DSB SPO fuel and water section anticipates the division's fuel requirements for the next 96-120 hours and plans fuel resupply based on the division commander's priorities.

Circle 3. The BSB SPO supply and service section receives daily LOGSTAT showing fuel consumption and requirements from the FSCs. The BSB SPO supply and service section analyzes the FSC's on hand quantities, expected consumption and capacity. The BSB SPO supply and service section anticipates the brigade's fuel requirements for the next 96 hours to plan fuel resupply based on brigade priorities. The BSB SPO supply and service section sends current consumption information and forecasted fuel requirements to support future brigade operations to the DSB SPO Fuel and water section for resupply.

Circle 4. The DSB SPO Fuel and water section receives and analyzes each support battalion's daily LOGSTAT showing on hand quantities, expected consumption and expected capacity for the next 96 hours. The DSB SPO fuel and water section validates their forecasts against the BSB's fuel requirements to verify proper amounts and identify discrepancies with estimates.

Circle 5. The DSB SPO fuel and water section provides the DSB SPO DIB with the division's current on-hand quantities by unit, fuel requirements by unit and storage capacity for the next 96 hours, and fuel distribution assets at the division support area. The DSB SPO DIB synchronizes and plans fuel distribution to all supported units with other commodity requirements in accordance with division priorities and within the constraints of division movement corridors.

Circle 6. The DSB SPO Fuel and water section consolidates the division's fuel consumption and requirements and prepares the bulk petroleum contingency report. The DSB Fuel and water section provides this report to the corps ESC in order to provide the ESC SPO visibility on the division's anticipated requirements for the next 120 hours. The DSB SPO Fuel and water section also provides a copy of the bulk petroleum contingency report to the division G-4 for visibility.

Square 1 and 2. During division logistics synchronization meetings, the DSB SPO DIB coordinates with the support battalion SPOs to synchronize DSB resupply operations to each supported unit. The DSB tasks the DSSB with the resupply mission. The possible courses of action for resupply, based on mission variables, includes but are not limited to:

- The DSSB composite supply company resupplies the BSB fuel assets in the brigade support area after the BSB's Distribution Company completed resupply to the FSCs.
- The DSSB composite throughputs fuel directly to the FSC location or logistics release points.

Bulk Fuel Request Process

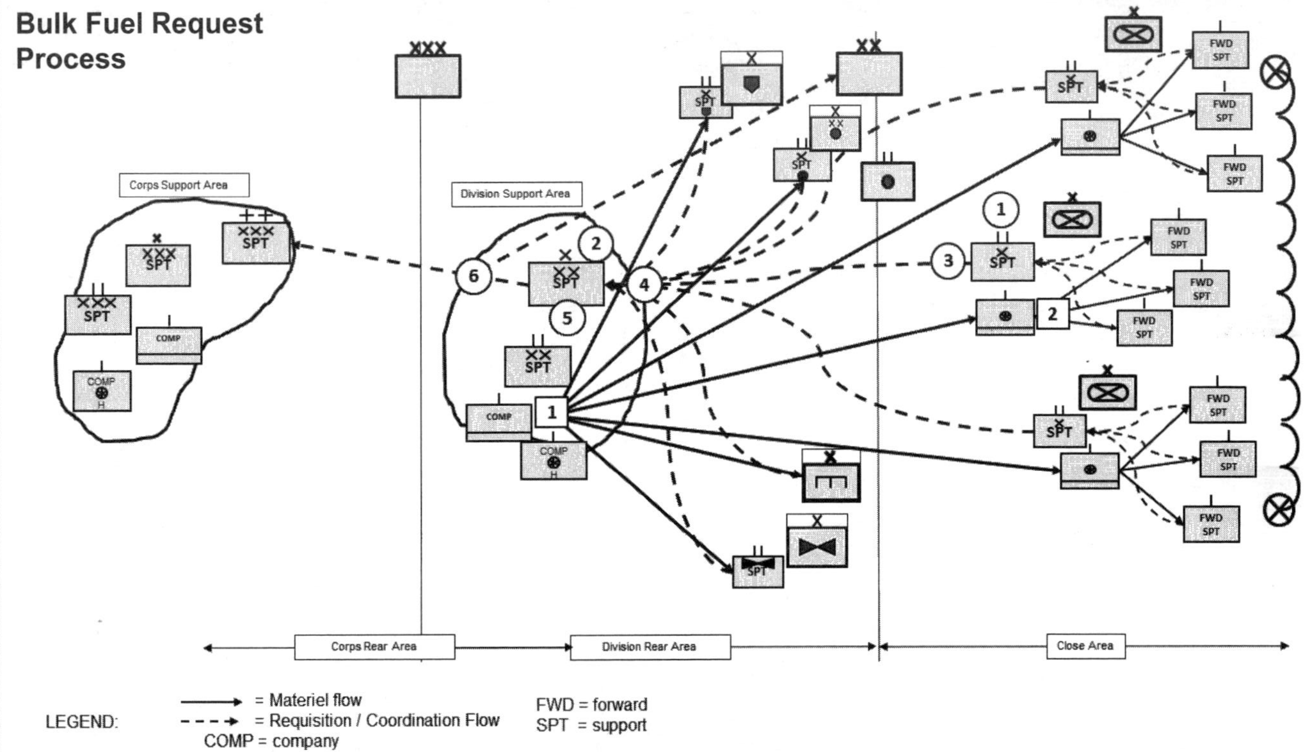

Ref: ATP 4-42, fig. 5-4. Each of the numbered points represent a node where actions take place and are discussed in subsequent paragraphs. The circle points represent requisition and coordination actions between organizations. The square points represent distribution actions. The dotted lines represent transmission of information. The solid lines represent the physical movement of materiel.

Class VII Evacuation and Replacement

Ref: ATP 4-42, Materiel Management, Supply, and Field Services Operations (Nov '20), pp. 5-28 to 5-29.

As units suffer equipment losses during operations, the class VII section supports the division in evacuating unrepairable equipment from the division AO, maintaining asset visibility, and recommending replacement solutions within the division.

Figure 5-8 depicts class VII evacuation and replacement. For simplicity, the process depicted and subsequent narrative focuses on a mechanized infantry company team and one unrepairable class VII item. However, this process applies to all units when field-level maintenance cannot repair destroyed equipment.

Square 1. A mechanized infantry company team recently involved in an engagement suffered battle damage to one of their M1A2 SEP main battle tanks. The supporting field maintenance team conducted a battle damage assessment and determined repairs exceed the maintenance allocation chart tasks at the field maintenance level. The FSC coordinates with BSB SPO transportation section to transport the tank to the brigade support area.

Circle 1. The infantry company supply sergeant completes the turn-in transaction in GCSS-Army,to include maintenance and code-out documentation, and turns in the equipment to the BSB's SSA. At this point, the company commander is relieved of accountability. The BCT PBO forwards the requirement for are placement to the division G-4.

Square 2. The BCT SPO transportation section coordinates with the DSB SPO transportation operations branch and creates the requirement to back haul the equipment from the BSA. The DSB SPO transportation operations branch conforms available assets within the DSB and coordinates with the DSB SPO DIB to add the backhaul requirement to the logistics synchronization matrix and the DIB allocates transportation assets to the requirement.

Square 3. The DSB tasks the DSSB to backhaul the tank. The DSSB's composite truck company transports the equipment from the BSA to the DSA to await corps transportation assets and further transport out of the division AO.

Circle 3. If a replacement M1A2SEP is available within the division, the division G-4 initiates a transfer directive and the division G-3 directs the losing and gaining unit to conduct the transfer through orders.

Square 4. The losing unit prepares the M1A2SEP for transfer and the BSB SPO coordinates with the DSB SPO transportation operations branch for transportation support. The DSB SPO transportation operations branch receives the transportation support requirement from the losing unit and reviews assets available. The DSB SPO transportation operations branch coordinates with the DSB DIB to add the transportation requirement to the logistics synchronization matrix and the DSB DIB plans and allocates transportation assets.

Circle 4. Upon receipt of the replacement tank, the gaining unit supply sergeant signs the transfer document and returns it to the losing unit supply sergeant to finalize the supply transaction.

Square 5. The DSB SPO transportation branch coordinates with the corps ESC DMC transportation operations branch to backhaul the damaged M1A2SEP from the DSA to the classification and identification company . The ESC DMC transportation operations branch reviews assets available within the sustainment brigade and coordinates with the ESC DMC DIB to add the backhaul requirement to the ESC's logistics synchronization matrix.

Circle 5. If a replacement M1A2SEP cannot be sourced from within the division, the division G-4 coordinates with the corps G-4 and requests replacement from within the corps. The G-4, supported by the ESC DMC, reviews property books and determines if a replacement can be sourced from another division within the corps.

Square 6. A CSSB CTC is tasked to backhaul the damaged M1A2SEP from the DSA to the classification and identification company to be sorted, classified and inspected in accordance with theater directives and priorities.

Class VII Evacuation and Replacement

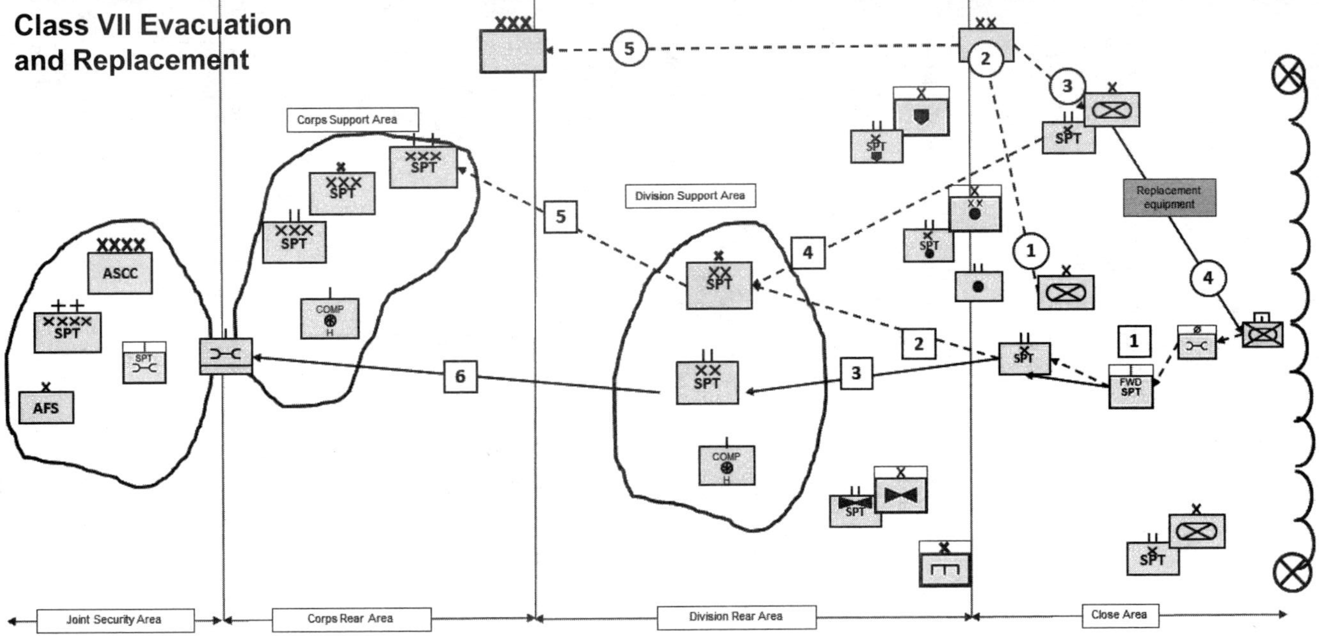

LEGEND:

→ = Materiel flow
- - → =Requisition / Coordination Flow
ASCC=Army Service component command
AFS=Army field support
COMP = company
FWD = forward
SPT = support

Ref: ATP 4-42, fig. 5-8. Each of the numbered points represent a node where actions take place and are discussed in subsequent paragraphs. The circle points represent requisition and coordination actions between organizations. The square points represent distribution actions. The dotted lines represent transmission of information. The solid lines represent the physical movement of materiel.

Sustainment Execution

IV. Field Services

Ref: ATP 4-42, Materiel Management, Supply, and Field Services Operations (Nov '20), pp. 6-1 to 6-17.

Field services are essential services that enhance a Soldier's quality of life during operations. They consist of aerial delivery, shower and laundry, mortuary affairs, water purification, and field feeding. The type and level of field services provided depends upon a supported unit's requirements and the existing infrastructure in a theater, theater of operations, or joint operations area.

Planners should consider military and political objectives while planning for the field services mission because the consequences of a failed field services mission can have a negative impact on morale, national public support, coalition support, and local national support. For example, prior to offensive and defensive operations, planners should analyze the mission, threat, fatality estimates, duration of the operation, delivery modes, availability of lift assets, and available AD support assets.

A. Aerial Delivery

Aerial delivery provides operational reach for commanders. The parachute riggers that pack personnel parachutes, cargo parachutes, and rig cargo for airdrop are the element that make aerial delivery a field service. Parachute rigger responsibilities include airborne assault support, parachute packing, aerial delivery equipment repair, and the rigging of supplies and equipment. Aerial delivery is the capability in the distribution system that provides supply to the force when land based lines of communication have been disrupted or the terrain is too hostile.

Personnel parachute packer's planning considerations are described in ATP 4-48, Aerial Delivery, AR 56-4, Distribution of Materiel and Distribution Platform Management, AR59-4, Joint Airdrop Inspection Records, Malfunction/Incident Investigations, And Activity Reporting, and AR750-32, Airdrop, Parachute Recovery, and Aircraft Personnel Escape Systems.

Planners should analyze which units complement each other when planning whom, when, and whereto deploy units. Planners should consider how to deploy rigger units in a timeframe that ensures the theater has AD support. During shape and prevent operations, sustainment planners should consider AD courses of action that range from the optimal time to deploy the rigger units to where to place these units on the battlefield. Since AD cannot be contracted, courses of action should contain planning considerations associated with each strategic role. For example, during large-scale combat offensive operations AD units contain elements that can be deployed separately as far forward as the BSB. During consolidation of gains,AD planning considerations may include deploying reserve component AD support elements early enough to accommodate transitioning of the resupply cycle from deployment accompanying items to replenishment items.

B. Mortuary Affairs

Mortuary affairs units assist commanders with the initial care of deceased personnel by evacuating human remains, with personal effects, to mortuary affairs facilities. Figure 6-2 on page 6-4 depicts mortuary affairs operations on a notional battlefield. Planners should consider military and political objectives for the mortuary affairs mission because the consequences of a failed mortuary affairs mission or disrespectful handling of human remains can have a negative impact on troop morale within the losing unit, national public support, coalition support, local national support, and enemy treatment of human remains or prisoners. For example, prior to offensive and defensive operations, planners should analyze the mission, threat, fatality estimates, duration of the operation, human remains evacuation modes, availability of lift assets, and available mortuary affairs assets.

For additional information, refer to JP 4-0, Joint Logistics, the DODD 1300.22,

Mortuary Affairs Policy, AR 638-2, Army Mortuary Affairs Program and ATP 4-46, Contingency Fatality Operations.

C. Water and Field Services

Commanders fight DNBI by promoting the principles of good personal hygiene and sanitation, proper preparation and storage of foods, and the importance of safe and clean drinking water. Planners at each echelon should consider DNBI prevention while considering courses of action for deploying and placing field services elements on the battlefield. Historical accounts of wars, battles, and military training consistently show that the majority of losses to armies are not the result of combat injuries but rather DNBI.

Water Purification

Impure water may serve as a means for the transmission of various diseases. Water from streams,shallow wells, ponds, swamps, and lakes is especially likely to carry disease organisms. All sources of water must be considered unsafe for drinking until properly treated. Therefore, all water supplies should be treated. Commanders should never permit the drinking of untreated water or not approved by preventive medicine personnel.

The water purification field service plays a crucial role in prevention of DNBI. Planning for water support begins with determining the amount and quality of water required. This depends on mission guidance from the tactical commander, mission scope, mission duration, stage of operation, operational environment,enemy chemical, biological, radiological and nuclear capabilities, and size of the force. The desirability of a water supply depends upon the quantity available, accessibility of source, and type (surface or underground).Water purification units require the following—

- Drainage to eliminate wastewater from treatment systems.
- Large enough storage facilities to meet daily supported unit requirements.
- Road networks for the accessibility of vehicles and personnel.

D. Food Service

Food service depends upon purified water for cooking and sanitation. Planning considerations should include potentially deploying reserve component water purification elements early enough to accommodate transitioning of the ration cycle from packaged meals to fresher ingredients.

E. Shower and Laundry

Warm showers, clean clothes, and potable water increase morale and help prevent DNBI. The reserve component has the majority of the shower and laundry capability, causing planners to consider deployment priority through time phased force and deployment data. Planning for mobilization of units must take into account the operational requirements, training requirements, equipment status, and the readiness of the units to be mobilized. Historically, these units often lack all authorized equipment and supplies. They may be deficient in prescribed load lists or have a shortage of authorized equipment and supplies. Planners should analyze which units complement each other when planning whom, when, and where to deploy units that provide field services. For example, water purification units support the quartermaster field service companies that provide shower and laundry services.

Planners should consider deploying quartermaster field service companies in a timeframe that ensures they have water purification support. Planning considerations include transportation priorities, available common-servicing and cross-servicing command agreements, formalized bilateral and multilateral support agreements, personnel attrition factors, and ports of debarkation for planning strategic lift of forces and sustainment.

V. Distribution

Distribution is defined as the operational process of synchronizing all elements of the logistics system to deliver the right things to the right place and right time to support the CCDR. It is a diverse process incorporating distribution management and asset visibility.

A. Theater Distribution

Theater distribution is the flow of personnel, equipment, and materiel within a theater to meet the geographic combatant commander's missions (JP 4-09). Theater distribution is a process of synchronizing all elements of the logistics system to deliver the right things to the right place at the right time to support the geographic combatant commander. Theater distribution planning is both art and science and can be defined as the process that coordinates and synchronizes the fulfillment of joint force requirements from point of origin to point of employment. Theater distribution planning involves the operations, logistics, and acquisition communities. It is an iterative process that includes detailed analysis and evaluation of the distribution networks and functions supporting the end-to-end distribution process. Distribution planning is a collaborative effort spanning the strategic, operational, and tactical levels of war.

Sustainment Execution

See pp. 7-29 to 7-30 for related discussion of distribution from ATP 4-94. Refer to ATP 4-0.1, Theater Distribution (Oct '14) for more information.

Theater distribution begins at the port of debarkation (POD) and ends at the unit. PAX and materiel enter the theater by surface (land or water) or by air and move through the various transportation modes/hubs/nodes to their destination or point of use. Essential to theater distribution is preparing and executing a theater movement program. The theater movement program serves as an authority to commit transportation assets, sanctions movements, directs mode operators to furnish assets and alerts receiving units to accept programmed movements.

Distribution is vital for operational reach. Operational reach is enabled by an Army theater distribution system that provides forces with a continuous flow of sustainment. A theater distribution system is a distribution system comprised of four independent and mutually supported networks within a theater to meet the geographic combatant commander's requirements; the physical network, the financial network, the information network, and the communications network (JP 4-01).

Two important aspects of theater distribution are in-transit visibility and visibility of assets:

In-Transit Visibility (ITV)

In-transit visibility is the ability to track the identity, status, and location of Department of Defense units, and non-unit cargo (excluding bulk petroleum, oils, and lubricants) and passengers; patients; and personal property from origin to consignee or destination across the range of military operations (JP 4-01.2). This includes force tracking and visibility of convoys, unit cargo/equipment and containers/pallets. ITV enables sustainment commanders and the logistics community to anticipate and manage the flow and the ability to take initiative and make changes as the operational environment permits.

Asset Visibility

Asset visibility provides users with information on the location, movement, status, and identity of units, personnel, equipment, and supplies which facilitates the capability to act upon that information to improve overall performance of the Department of Defense's logistics practices (JP 3-35). Maintaining asset visibility at the strategic, operational and tactical levels is an important standard of Army transportation. In order to have a successful transportation system, it is vital to know what assets are available, which assets are on mission and when assets will become available. This includes common-user, coalition and supporting Service assets as well as organic assets.

B. Distribution Management Framework

Ref: ATP 4-42, Materiel Management, Supply, and Field Services Operations (Nov '20), fig. 5-2. Distribution management framework.

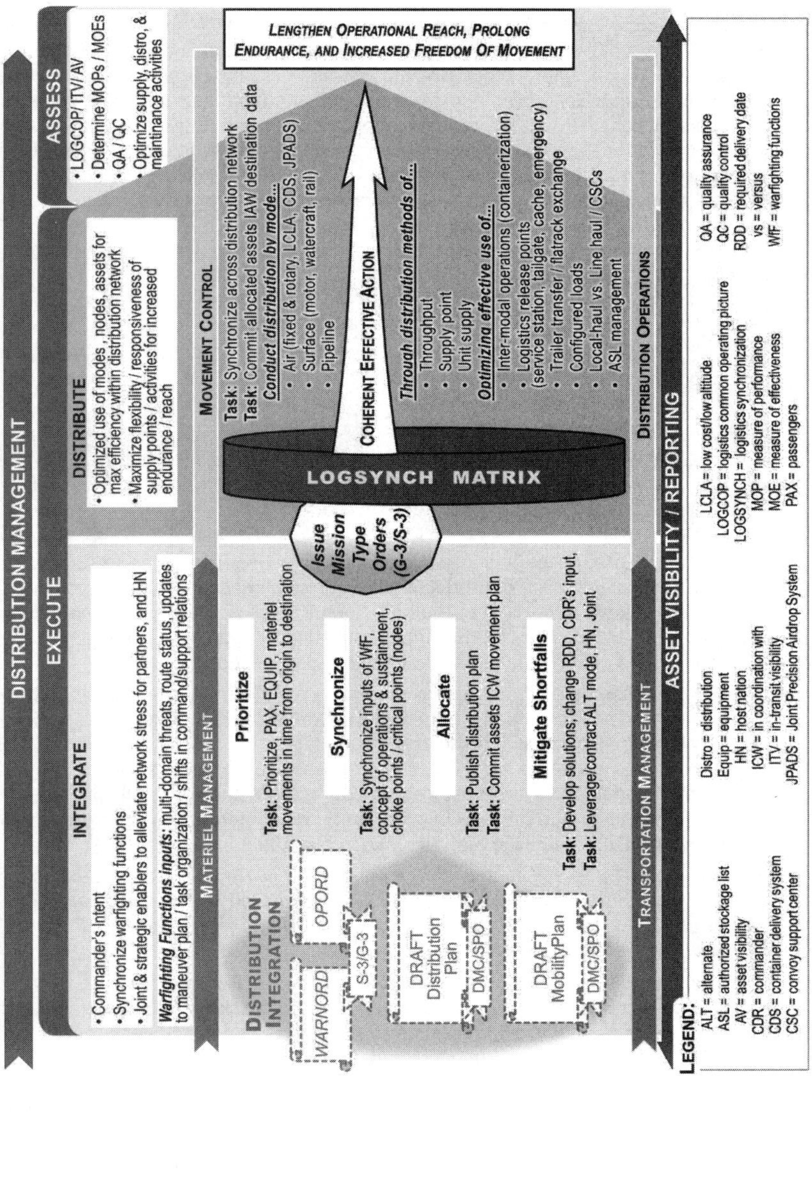

VI. Operational Contract Support

Operational contract support is the process of planning for and obtaining supplies, services, and construction from commercial sources in support of operations along with the associated contractor management functions.

Ref: ATP 4-42, Materiel Management, Supply, and Field Services Operations (Nov '20), pp. 5-18 to 5-10. See p. 1-8 for related discussion from ADP 4-0 (Jul '19).

Contracted commercial support plays an important role in the sustainment warfighting function. Contracts provide deployed units with technical and surge capabilities that either do not exist or exceed existing capabilities. Employment of contracts as a sustainment capability occurs across all echelons and across supply and some field services functions. OCS is the process of planning for and obtaining supplies and services from commercial sources within a theater of operations. OCS becomes a component of the supply support process because it augments existing capabilities, bridges gaps in the deployed force structure,leverages assets, and reduces dependence on CONUS based logistics.

Local Purchase

Materiel may be procured locally from outside of the Army supply system using government purchase cards and operational contracting:

- Units purchase nonstandard or commercial items under $3,000 with the government purchase card. Commanders follow local policy for use of the government purchase card.
- Typically, commands use contracts to request nonstandard or commercial items over $3,000. To acquire contract support, the unit should coordinate with the local contracting agency to develop and submit a requirements package. Only warranted contracting officers may award contracts.

Leaders and sustainment planners should know and understand the OCS process to avoid lapses in services or product delivery. This process includes pinpointing the location of key partner agencies and responsible organizations. During planning Army sustainment planners, consider contract support capabilities, limitations, and how to integrate contractors into the overall concept of support.

Materiel managers think through complex problems to identify specific support requirements to be included in a performance work statement. After the requiring activity determines and validates its requirements and the funding is approved, the request package goes into the contracting process, which ultimately leads to vendor selection. Materiel managers must also be cognizant of special considerations, requirements to manage materiel in accordance with emergency use approvals for non-Food and Drug Administration approved class VIII items.

Refer to ATP 4-10/MCRP 4-11H/NTTP 4-09.1/AFMAN 10-409-O, Multi-Service Tactics, Techniques, and Procedures for Operational Contract Support, and ATP 4-92, Contracting Support to Unified Land Operations, and AR 700-137, Logistics Civil Augmentation Program, for additional information.

The requiring activity develops an acquisition-ready requirement package for submission to the supporting contracting activity. The package includes a detailed performance work statement for service requirements or detailed item descriptions for a commodity requirement. In addition to the performance work statement, these packages include an independent cost estimate and DA Form 3953 (Purchase Request and Commitment). Finally, the requiring activity must be prepared to support, and possibly lead, an acquisition review board to approve and set priorities on high demand, special command interest contract support actions.

The requiring activity staff assists the contracting agency by nominating a COR. CORs are required for every service contract and task order. CORs ensure that contractors provide the contracted service or item in a safe and effective manner.

The requiring activity completes receiving reports. After acceptance, the receiving official must certify receipt of supplies for payment purposes either in wide area work-flow or by completing a Service specific inspection and receiving statement, for example, DD Form 250, (Material Inspection and Receiving Report).Requiring activities should ensure the receiving official has the skills to perform the duties as receiving official and is made available as needed to perform these duties

A. Contractor Functions on the Battlefield

Contractors can provide the full range of supply and field services functions during military operations,including item management, stockage, and delivery of all classes of supply. Contractor support may also provide field services (such as, laundry, shower, water purification, waste disposal, and portable latrine support) within specific parameters. The availability and type of contract support on the battlefield depends on mission, enemy, terrain, troops, time, and civilian considerations.

To meet the challenges of large-scale combat operations, the Army is rebalancing its force structure to improve the velocity, precision, and survivability of future sustainment operations. Force structure improvements will provide commanders with more organic sustainment capability, especially in the forward area. However, gaps in the Army's fuel distribution, unit mobility, and maintenance capabilities will require several years to reduce, and the recalibration of capabilities between active and reserve components will take time. In the short term, these force structure improvements will not eliminate the need for commercial support.

Contractors will play an especially critical role in shaping operations, when Army service component commands depend heavily on the LOGCAP and other external support contracts to open ports, establish intermediate staging bases, and execute the reception, staging, and onward movement of arriving units. As operations develop, most uniformed sustainment capabilities will deploy forward to support the corps headquarters, divisions, and brigades engaged in direct combat, leaving an enormous joint security area dependent on contractors to operate the theater distribution network and provide communications, security,and base life support.

Army formations conduct operations to consolidate gains in those areas where large scale combat operations have ceased. The transition to consolidation of gains, however, reflects a change in the scale of combat operations, not necessarily a change in their lethality.

Refer to JP 4-10, Operational Contract Support, and ATP 4-10/MCRP 4-11H/NTTP 4-09.1/AFMAN 10-409-O, Multi-Service Tactics, Techniques, and Procedures for Operational Contract Support for more information on the contracting process.

B. Working with Contractors

Contractors provide additional sources of support for required supplies and services. Contractor support does not directly correlate to military support. Keep in mind that contractor support involves payment to vendors for goods and services. The type and quantity of support a contractor provides is similar to that provided by a military support unit, when considered from a supported unit perspective; however, commanders and staffs must remain aware of some fundamental differences. Contractors perform only the tasks specified in the contract. Contractors and their employees are not combatants, but rather civilians accompanying the force. Commanders do not have command of contractor employees and contractor personnel are not government employees meaning contractors manage and supervise their employees. Commanders manage contractors through the contracting officer and COR in accordance with the terms and conditions of the contract.

Refer to ATP 4-10/MCRP 4-11H/NTTP 4-09.1/AFMAN 10-409-O, Multi-Service Tactics, Techniques, and Procedures for Operational Contract Support. For in-theater contractor and contract management, see AR 715-9, Operational Contract Support Planning and Management.

VII. General Engineering (GE)

Ref: ATP 3-34.40, General Engineering (Feb '15).

GE is a U.S. Army engineer discipline (see FM 3-34) and a joint engineer function (see JP 3-34). GE complements and supports combat engineering (mobility, countermobility, and survivability) and geospatial engineering and encompasses those engineer tasks that establish and maintain the infrastructure required to conduct and sustain military operations. Such tasks are conducted with unified action partners and are integrated into the force commander's plan (ADRP 3-0). This force may be led by any one of the Services, and GE support may come from any or all Service engineers, contractors, HN capabilities, or the engineers of other nations (ADRP 4-0). GE occurs throughout the area of operations (AO) and across the range of military operations.

Engineer Framework

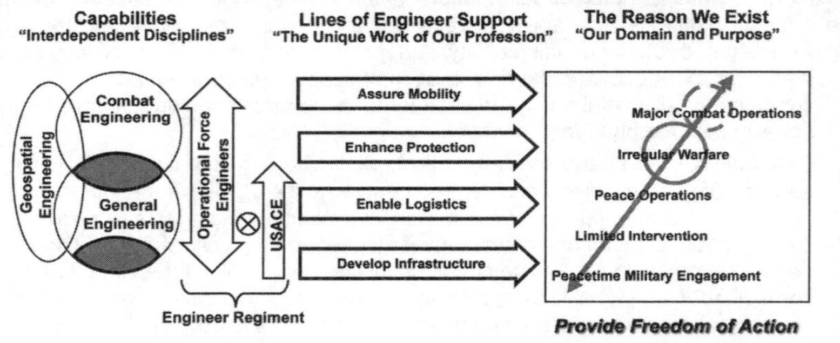

Legend:
USACE – United States Army Corps of Engineers

General engineering consists of those engineering capabilities and activities, other than combat engineering, that modify, maintain, or protect the physical environment (JP 3-34). GE encompasses the engineer tasks that establish and maintain the infrastructure required to conduct and sustain military operations. Examples include the construction, repair, maintenance, and operation of infrastructure, facilities, line of communications (LOC), bases, terrain modification and repair, and selected explosive hazard activities.

Combat Engineering

GE supports combat engineering (including mobility operations and countermobility operations), enhances protection (including survivability), enables force projection and logistics, builds partner capacities, and develops infrastructure. Countermobility operations is the construction of obstacles and emplacement of minefields to delay, disrupt, and destroy the enemy by reinforcement of the terrain (JP 3-34). GE may support combat engineers, but GE normally supports countermobility with earthwork and timber construction in the support or sustainment area for operational area security and survivability.

Refer to AODS6-1 (w/SMARTupdate 1):The Army Operations & Doctrine SMARTbook, 6th Rev. Ed. (Guide to Army Operations and the Six Warfighting Functions) for complete discussion of the engineer disciplines, lines of engineer support, and "mobility and countermobility" as part of the movement and maneuver warfighting function.

(Sustainment Execution) II. Financial Management

Ref: ADP 4-0, Sustainment (Jul '19), pp. 1-10 to 1-11.

I. Financial Management

Financial management is guided by the following principles: stewardship, synchronization, anticipation, improvisation, simplicity, and consistency. These principles are described in appendix A of FM 4-0. Financial management consists of finance operations and resource management.

A. Finance Operations

The finance operations mission is to support the sustainment of Army, joint, and multinational operations through the execution of key finance operations tasks. These key finance operations tasks are to provide timely commercial vendor services and contractual payments, various pay and disbursing services, and oversight and management of the Army's Banking Program. Finance operations also implement financial management policies and guidance prescribed by the Office of the Under Secretary of Defense (Comptroller) and national providers (such as the U.S. Treasury, Defense Finance and Accounting Service, and Federal Reserve Bank).

B. Resource Management

The resource management mission, accomplished by a battalion or brigade finance officer (S-8)/assistant chief of staff, finance/finance directorate of joint staff personnel, is to analyze resource requirements and ensure commanders are aware of existing resource implications. This allows commanders to make resource informed decisions, and obtain the necessary funding that allows them to accomplish their mission. Resource management is the critical capability within the financial management competency that matches legal and appropriate sources of funds with thoroughly vetted and validated requirements. Key resource management tasks are providing advice and recommendations to the commander, identifying sources of funds, forecasting, capturing, analyzing and managing costs; acquiring funds, distributing and controlling funds; tracking costs and obligations; establishing and managing reimbursement processes; and establishing and managing the Army's Managers' Internal Control Program.

II. Financial Management Core Competencies

The core competencies of financial management are fund the force, banking and disbursing, pay support, accounting support and cost management, and audit readiness.

Fund the Force

Fund the force is a critical capability, which matches legal and appropriate sources of funds with thoroughly vetted and valid unit requirements. Fund the force provides flexibility through monetary integration to augment, and in some cases, leads the effort in obtaining the effects the commander is trying to achieve on the battlefield. Key tasks include identifying, acquiring, distributing and certifying funds and obligating funds to units, providing payments in cash or via electronic funds transfers in support to contracting operations, and commercial vendor services.

Banking and Disbursing

Banking support encompasses financial management activities ranging from currency support (local or U.S.) of U.S. military operations to liaising with HN banking officials and strengthening local financial institutions. Other activities within banking support include establishing e-commerce and limited depository capabilities, as well as coordination with U.S. embassies, U.S. Army Financial Management Command, Defense Finance and Accounting Service, and Department of the Treasury entities to support theater banking initiatives.

Disbursing is the arm within financial management that ensures all payments are made in accordance with DOD regulations. It is strongly recommended that all elements of the fiscal triad, financial management, contracting and legal counsel, be co-located to facilitate fiscal communication, accuracy of documentation, and timely payment of goods and services. Disbursing is the paying of public funds to entities to which the

U.S. government is indebted; the collection and deposit of monies; the safeguarding of public funds; and the documenting, recording, and reporting of such transactions (FM 1-06).

Pay Support

This support function provides for full U.S. pay (including civilian pay where not supported by the Defense Finance and Accounting Service); travel support; local and partial payments; check-cashing and currency exchange to Soldiers, civilians and U.S. contractors; and non-U.S. pay support (such as detainee(s), HN employees, day laborers). Pay support also includes support to noncombatant evacuation operations in the form of travel advances. Finance offices ensure all Soldiers, regardless of component, receive timely and accurate pay in accordance with existing statutes and regulations.

Accounting Support and Cost Readiness

Accounting support entails the accurate and complete recording of financial transactions within the Army financial management support information systems. Additional accounting support involves review and reconciliation of these financial transactions to ensure the proper expenditure of entrusted funds.

Cost management transforms accounting data into valuable and accurate cost information that enables the commander's decision-making process. It collects and links financial (cost) data with non-financial (output and performance) data and presents information in a way directly related to the major mission objectives of the unit or organization. The process is the accurate measurement and thorough understanding of the full cost of a unit/organization's processes, products and services to support leader's decision-making and fiscal stewardship, thereby maximizing the effectiveness and efficiency of the organization's operations.

Audit Readiness

Audit readiness ensures that organizations produce timely, accurate, and relevant information that is consistently auditable. With the Chief Financial Officers Act of 1990, Congress mandated that all federal agencies produce auditable financial statements. The Office of the Under Secretary of Defense Comptroller outlines the DOD priorities for improving financial information and achieving financial statement audit readiness and directs DOD components to improve budgetary information, as this is the information most useful to the Department's decision makers. Specifically, DOD components focus on improving budgetary processes and information to obtain an unqualified audit opinion on the Statement of Budgetary Resources. Transition to enterprise resource planning systems will help in this area.

(Sustainment Execution)
III. Personnel Services

Ref: ADRP 4-0, Sustainment (Jul '19), pp. 1-11 to 1-18.

Personnel services are sustainment functions that man the force, maintain Soldier and Family readiness, promote the moral and ethical values of the nation, and enable the fighting qualities of the Army. Personnel Services provide support to the force utilizing the following principles: synchronization, timeliness, accuracy, and consistency. These principles are described in appendix A of FM 4-0. Personnel services involve planning and coordinating efforts that sustain personnel. Personnel services relate to personnel welfare (for example readiness, quality of life). Personnel services include the following—

Personnel Services

- I **Human Resources (HR) Support** *(FM 1-0)*
- II **Legal Support** *(FM 1-04)*
- III **Religious Support** *(FM 1-05)*
- IV **Army Band Support** *(ATP 1-19)*

I. Human Resources Support

Personnel services includes human resources (HR) support. The functions of human resources support are described below under their respective human resources core competencies. The core competencies include: man the force, provide human resources services, coordinate personnel support, and conduct human resources plans and operations. All of these functions and tasks affect the personnel aspects of building combat power within an organization.

- **Man the force** consists of personnel readiness management, personnel accountability, strength reporting, and personnel information management.
- **HR services** are those functions conducted by HR professionals specifically impacting Soldiers and organizations and include essential personnel services, postal operations, and casualty operations. Essential personnel services functions are performed by the assistant chief of staff, personnel (G-1) and the S-1. Postal operations are performed by HR personnel in G-1s, postal organizations, military mail terminal teams, human resources sustainment centers (HRSCs), HR companies, and monitored within the human resources operation branches. Casualty operations are performed by S-1s and HR unit personnel

(for example, HRSC; human resources operations branches; HR company; and casualty liaison teams and monitored within the human resources operations branches.

- **Personnel support functions** are normally coordinated by the G-1s and S-1s and generally fall under their responsibility. These functions include morale, welfare and recreation; command interest programs; and Army band operations.
- **HR planning and operations** are the means by which HR leaders envision a desired HR end state in support of the operational commander's mission requirements through the use of the military decision-making process, rapid decision and synchronization process, and Army design methodology. It communicates to subordinate HR professionals the intent, expected requirements, and desired outcomes in the form of an operation plan and operation order. It also communicates the process of tracking execution of planned HR support to ensure effective support to the operational commander.

Refer to FM 1-0.

II. Legal Support

The Judge Advocate General's Corp is the Army proponent for legal support matters. The Judge Advocate General has overall responsibility for legal support based on statutory, regulatory, delegated, and general authority, as well as by designation and assignment. The Judge Advocate General directs the delivery of Army legal support across a broad spectrum of legal functions.

Legal support to operations encompasses all legal services provided by members of the Judge Advocate Legal Service in support of units, commanders, and Soldiers. The Judge Advocate Legal Service provides legal services to the Army and joint force under two core competencies: legal support to the Army and legal support to Soldiers and Family members.

Refer to FM 1-04.

III. Religious Support

The religious support mission of the Chaplain Corps is to assist commanders in the responsibility to provide for the free exercise of religion, and to provide religious, moral, and ethical leadership to sustain a ready force of resilient and ethical Soldiers and leaders. Unit ministry teams and chaplain sections, composed of at least one chaplain and one religious affairs specialist, possess three core competencies: nurture the living, care for the wounded, and honor the dead. The religious support mission is executed through two required capabilities—providing support and advising the command.

For additional information on religious support, refer to FM 1-05.

IV. Army Band Support

Band support is an element of personnel services and is aligned under the sustainment warfighting function. Army bands support the Army through the provision of tailored music that enhances Warrior morale, supports Army recruiting efforts, provides comfort to recovering Soldiers, reinforces relations with HN populations, and maintains a connection with the American public and Army Families. Army band operations is a subordinate key function of the HR Support core competency, coordinate personnel support. (FM 1-0)

Army Bands promote the Army and our national interests, enable commanders to shape the environment to accomplish their mission, and set the conditions that lead to trust and confidence in America's Army and its readiness to conduct operations in peacetime, conflict, and war.

Refer to ATP 1-19 for more details.

(Sustainment Execution) IV. Health Service Support

Ref: ADP 4-0, Sustainment (Jul '19), pp. 1-18 to 1-20 and FM 4-02, Army Health System (Nov '20).

The Army Health System is composed of both health service support and force health protection. Health service support is part of the sustainment warfighting function, and force health protection is part of the protection warfighting function. The principles of the Army Health System are conformity, proximity, flexibility, mobility, continuity, and control. These principles are defined in appendix A of FM 4-0. Health service support encompasses all support and services performed, provided, and arranged by the AHS to promote, improve, conserve, or restore the behavioral and physical well-being of Army personnel and as directed, UAPs.

Health Service Support

Health service support includes the following—

- **Casualty care, which encompasses a number of medical functions, to include:**
  - Medical treatment (organic and area medical support).
  - Hospitalization.
  - Dental care (treatment aspects).
  - Behavioral health/neuropsychiatric treatment.
  - Clinical laboratory services.
  - Treatment of chemical, biological, radiological, and nuclear patients).
- **Medical evacuation**
- **Medical logistics**

Medical Planning Tools

The Joint Medical Planning Tool, the Medical Planners' Toolkit, and the Medical Contingency Requirements Workflow are planning tools approved for calculation of medical requirements. The Joint Medical Planning Tool is a computer-based simulation tool developed by the Naval Health Research Center that models patient flow from the point of injury through more definitive care. The Medical Planners' Toolkit is a suite of tools developed to support the joint medical planning community. The Medical Planners' Toolkit combines the Patient Conditions Occurrence Frequency tool, the Casualty Rate Estimation Tool, and the Expeditionary Medicine Requirements Estimator into a single desktop application. The Medical Contingency Requirements Workflow is the authoritative source for estimating medical materiel contingency requirements by time and role of care.

See pp. 4-27 to 4-30 for discussion of Army Health Service (AHS) planning.

I. Casualty Care

Ref: ADP 4-0, Sustainment (Jul '19), pp. 1-18 to 1-19.

Casualty care encompasses all issues pertaining to the provision of clinical services for the treatment of Soldiers from the point of injury to successive roles of care. Casualty care includes the following sub-functions: medical treatment (organic and area medical support), hospitalization, the treatment aspects of dental care and behavioral health/ neuropsychiatric treatment, clinical laboratory services, treatment of chemical, biological, radiological and nuclear patients.

Casualty Care (Sub-Functions)

1. Medical Treatment
2. Hospitalization
3. Dental Care
4. Behavioral Health
5. Clinical Laboratory Services
6. Treatment of CBRNE Patients

Medical Treatment (Organic and Area Medical Support)

The medical treatment function encompasses Roles 1 and 2 medical treatment support. Role 1 medical treatment is provided by the combat medic or by the physician, the physician assistant, or the health care specialist in the battalion aid station/Role 1 medical treatment facility. Role 2 medical care provides greater resuscitative capability than is available at Role 1 and is rendered by the medical company in the BSB or by the medical company (area support), which is an echelon above brigade asset. These roles of care are provided by organic assets or on an area support basis from supporting medical companies or detachments. The area support function encompasses emergency medical treatment, advanced trauma management, routine sick call, emergency dental care, preventive medicine, and combat and operational stress control support.

Hospitalization

The Army's hospitalization capability consists of Role 3 combat support hospitals and hospital centers purposely positioned to provide support in the AO. At Role 3, the combat support hospital and hospital centers expand the support provided at Role 2 and are staffed and equipped to provide care for all categories of patients, to include resuscitation, initial wound surgery, damage control surgery, and postoperative treatment. Hospitalization capabilities deploy as modules or multiple individual capabilities that provide incrementally increased medical services in a progressively more robust AO. The hospitalization capability in the AO offers essential care to either return the patient to duty (within the theater patient movement policy) and/or stabilization to ensure the patient can tolerate evacuation to a definitive care facility outside the area of operations (this support is key to early identification and treatment of mild traumatic brain injuries).

Dental Care

Dental care provided as part of health service support includes far forward dental treatment, treatment of oral and dental disease, and early treatment of severe oral and maxillofacial injuries. Dental personnel may also be used to augment medical personnel (as necessary) during mass casualty operations.

Behavioral Health

The primary focus of behavioral health/neuropsychiatric treatment is to screen and evaluate Soldiers with maladaptive behaviors. The purpose of this function is to provide diagnosis, treatment, and disposition for Soldiers with neuropsychiatric/behavioral health-related issues.

Clinical Laboratory Services

Clinical laboratory services provide basic support within the theater, to include procedures in hematology, urinalysis, microbiology, and serology. Role 2 medical company (area support) and brigade support medical companies, when operating with or collocated with a forward resuscitative and surgical team or other resuscitative and surgical team, may augment the clinical laboratory and blood storage capabilities of that element. The combat support hospital and hospital center perform procedures in biochemistry, hematology, urinalysis, microbiology, and serology in support of clinical activities. The hospital also provides blood-banking services.

Treatment of Chemical, Biological, Radiological, and Nuclear Patients

Health service support operations in a chemical, biological, radiological and nuclear environment are complex. Medical personnel may be required to treat chemical, biological, radiological and nuclear injured and contaminated casualties in large numbers. Medical treatment is provided in protected environments and protective clothing must be worn. Movement of chemical, biological, radiological and nuclear casualties can spread contamination to clean areas. All casualties are decontaminated as far forward as the situation permits and are decontaminated before they are admitted into a clean medical treatment facility. The admission of one contaminated casualty into a clean medical treatment facility may contaminate the facility, thereby reducing treatment capabilities in the facility. *Refer to ATP 4-02.7 has additional information.*

Tactical Combat Casualty Care (TCCC)

Tactical combat casualty care (TCCC) is divided into the three phases:

- Care under fire
- Tactical field care
- Tactical evacuation care

Tactical combat casualty care occurs during a combat mission and is the military counterpart to prehospital emergency medical treatment. Tactical combat casualty care in the military is most commonly provided by enlisted personnel and includes self-aid and buddy aid (first aid), combat lifesaver (enhanced first aid), and enlisted combat medics and critical care flight paramedics in the Army, corpsmen in the United States Navy (USN), United States Marine Corps, and United States Coast Guard, and both medics and pararescue in the United States Air Force (USAF). Tactical combat casualty care focuses on the most likely threats, injuries, and conditions encountered in combat and on a strictly limited range of interventions directed at the most serious of these threats and conditions.

The TCCC initiative originated with the Naval Special Warfare Command and later continued by the United States Special Operations Command. Special operations forces do not have a dedicated, designed, and equipped MEDEVAC capability. Therefore, they use nonmedical platforms augmented with medical personnel to perform the evacuation function. The conventional force doctrinal categories of MEDEVAC and CASEVAC as defined in Army doctrine on MEDEVAC are not changed, however, during this phase of TCCC both types of evacuation occur depending upon the availability of assets and the time window available to execute the evacuation process.

Ref: FM 4-02, Army Health System (Nov '20), p. 1-6.

II. Medical Evacuation (Including Medical Regulating)

Medical evacuation is the timely and effective, movement of the wounded, injured, or ill to and between medical treatment facilities on dedicated and properly marked medical platforms with en route care provided by medical personnel. The provision of en route care on medically equipped vehicles or aircraft enhances the patient's potential for survival and recovery and may reduce long-term disability. Medical regulating is the coordination and control of moving patients to medical treatment facilities that are best able to provide the required specialty care.

Refer to ATP 4-02.2 for additional information.

III. Medical Logistics

Medical logistics encompasses planning and executing all class VIII supply support to include management of the following functions: medical materiel, medical equipment maintenance and repair, optical fabrication and repair, patient-movement items, medical gases, blood storage and distribution, regulated medical waste (including hazardous material), medical facilities and infrastructure, and medical contracting. The system is anticipatory with select units capable of operating in a split-based mode.

Refer to ATP 4-02.1 for more information.

* Force Health Protection

The mission sets of the Army Medicine (which historically had been shown under the combat service support battlefield operating system) are under two warfighting functions-the protection and sustainment warfighting functions. This change more closely aligned the Army Medicine mission sets with the overall warfighting functions of the Army.

Force health protection are measures that promote, improve, or conserve the behavioral and physical well-being of Soldiers comprised of preventive and treatment aspects of medical functions that include: combat and operational stress control, dental services, veterinary services, operational public health, and laboratory services. Enabling a healthy and fit force, prevent injury and illness, and protect the force from health hazards. Although nutrition plays a significant role in maintaining a healthy and fit force, nutrition is discussed as an integral part of the hospitalization function under the HSS mission.

Refer to FM 4-02, Army Health System (Nov '20), for further discussion. The FHP mission is discussed in Part Two of FM 4-02, while the HSS mission is discussed in Part Three. Although Parts Two and Three discuss the mission sets as separate entities, the medical personnel and staffs that plan, coordinate, and synchronize these operations are responsible for the execution of both mission sets. These interrelated and interdependent medical functions are complex in nature and require medical command and control for synchronization and integration. This ensures the interrelationships and interoperability of all medical assets and optimizes the effective functioning of the entire AHS system.

Chap 4

Planning Sustainment Operations

Ref: ADP 4-0, Sustainment (Jul '19), (and adaptations from previous references).

Planning sustainment support of an operation is vital to mission success. Sustainment commanders and their planning staffs must coordinate and synchronize every stage of the planning process with the operational staff. They must also coordinate, synchronize and integrate the sustainment plan with joint and multinational partners to ensure a continuous linkage with strategic level providers.

Sustainment planning begins with the operational commander's intent and concept of operations. This single, unifying idea provides direction for the entire operation. Based on a specific idea of how to accomplish the mission, commanders refine the concept of operations during planning. They adjust it throughout the operation as subordinates develop the situation or conditions change.

See chap. 8, Joint Logistics, pp. 8-27 to 8-38 for discussion of planning joint logistics.

I. Sustainment Preparation of the Operational Environment (OE)

Sustainment preparation of the operational environment is the analysis to determine infrastructure, environmental, or resources in the operational environment that will optimize or adversely impact friendly forces means for supporting and sustaining the commander's operations plan. The sustainment preparations of the operational environment assist planning staffs to refine the sustainment estimate and concept of support. It identifies friendly resources (HNS, contractable, or accessible assets) or environmental factors (endemic diseases, climate) that impact sustainment.

Some of the factors considered (not all inclusive) are as follows:

- **Geography**. Information on climate, terrain, and endemic diseases in the AO to determine when and what types of equipment are needed. For example, water information determines the need for such things as early deployment of well-digging assets and water production and distribution units.
- **Supplies and Services**. Information on the availability of supplies and services readily available in the AO. Supplies (such as subsistence items, bulk petroleum, and barrier materials) are the most common. Common services consist of bath and laundry, sanitation services, and water purification.
- **Facilities**. Information on the availability of warehousing, cold-storage facilities, production and manufacturing plants, reservoirs, administrative facilities, hospitals, sanitation capabilities, and hotels.
- **Transportation**. Information on road and rail networks, inland waterways, airfields, truck availability, bridges, ports, cargo handlers, petroleum pipelines, materials handling equipment (MHE), traffic flow, choke points, and control problems.
- **Maintenance**. Availability of host nation maintenance capabilities.
- **General Skills**. Information on the general skills such as translators and skilled and unskilled laborers.

See pp. 1-29 and 2-12 for discussion of sustainment preparation of the battlefield from ADP 4-0. See pp. 4-5 to 4-10 for related discussion of logistics preparation of the battlefield.

II. Sustainment Planning Overview

Ref: ADRP 4-0, Sustainment (Jul '19), pp. 3-2 to 3-4.

Sustainment planning begins with the operational commander's intent and concept of operations. This single, unifying idea provides direction for the entire operation. Based on a specific idea of how to accomplish the mission, commanders refine the concept of operations during planning. They adjust it throughout the operation as subordinates develop the situation or conditions change.

Sustainment planning continues with an analysis of the conditions in the operational environment with emphasis on the enemy and operational variables mission, enemy, terrain and weather, troops and support available, time available, and civil considerations. It involves understanding and framing the problem and envisioning the set of conditions that represent the desired end state.

Planning sustainment support of an operation is vital to mission success. Sustainment commanders and their planning staffs must coordinate and synchronize every stage of the planning process with the operational staff. They must also coordinate, synchronize and integrate the sustainment plan with joint and multinational partners to ensure a continuous linkage with strategic level providers.

The sustainment staff's role in synchronizing sustainment planning with operations is necessary to assist operational commanders and their staffs establish the conditions for what is possible. To ensure maximum freedom of action sustainment planners must understand the commander's intent, be able to visualize the operation and articulate the operational risks. Limitations like, insufficient infrastructure or the availability of a key class of supply or replacement weapon systems has bearing on the commander's ability to execute the mission. Sustainment commanders and staffs must present credible courses of action commensurate with sustainment capabilities to allow as much freedom of action as possible to accomplish the operational end state. While sustainment should not be an impediment to an operation, poor planning, lack of coordination, and understanding could severely influence the success of the operation.

Doctrinal Linkage

Sustainment commanders must ensure their staffs are conversant in operational and sustainment doctrine as well as joint and multinational doctrine. Doctrine is a guide and establishes a basic framework for the conduct of operations that facilitates the planning process. Sustainment is a complex operation with many branches and sequels. Doctrine, based on past experiences and knowledge, helps leaders understand the broad context of what is possible, and then allow commanders and staffs to use judgment in the application of doctrine to adjust to the unique circumstances facing them.

To conduct sustainment in unified land operations, sustainment planners must fully understand the operations process—planning, preparing, executing, and continuously assessing the operation; operational art and mission command. Additionally, sustainment planners must understand the mechanics of joint operations and the necessary links that ensure strategic level support.

A firm doctrinal grasp enables sustainment staffs to use and apply the planning tools of the operations process. In addition to a firm foundation in planning, operational, and sustainment doctrine, planners must understand maneuver doctrine in order to arrange sustainment actions in a manner to effectively support the operation.

Refer to BSS6: The Battle Staff SMARTbook, 6th Ed. for further discussion for complete discussion of the operations process—planning, preparing, executing, and continuously assessing the operation; operational art and mission command.

The Operations Process (Underlying Logic)

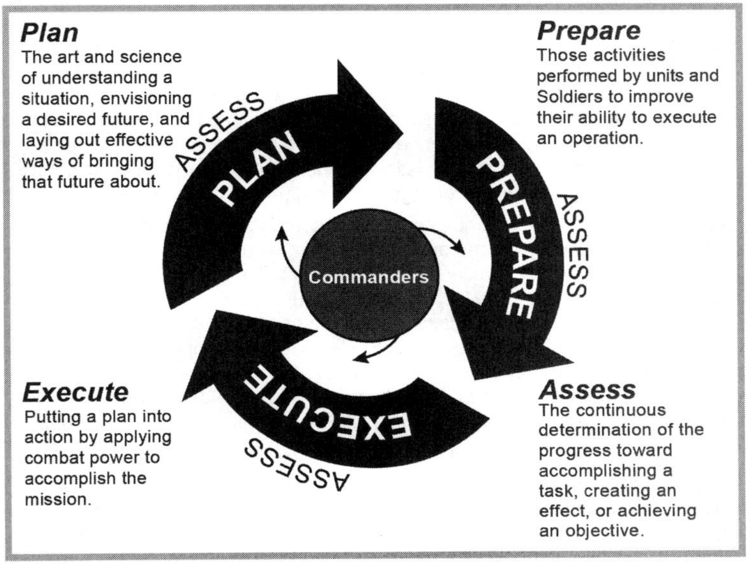

Ref: ADP 5-0, The Operations Process (Jul "19), intro fig. 1.

Planning Considerations

Sustainment staffs create viable plans that are well coordinated and synchronized, facilitate operational tempo and support the commander's priorities before, during, and after operations. Sustainment planners in an operational headquarters generally do not drive the planning process, but must be fully integrated throughout planning to help understand, visualize, and describe solutions. Sustainment planners use the commander's intent, planning guidance, and the military decision making process to develop the sustainment concept of support.

The concept of support is derived from running estimates developed using a variety of planning tools. These running estimates project consumption rates for key classes of supply, casualty figures, maintenance requirements, and other sustainment requirements. Sustainment planners participate in all aspects of the military decision making process to ensure synchronization and unity of effort.

Planning in a sustainment headquarters requires planners to take an active role in the maneuver planning process. They assist in the development of the commander's understanding of the operational environment, identify shortfalls, articulate risks, and articulate the sustainment commander's vision. Sustainment planners must have the most current products from the organizations they support as well as planning products from their higher headquarters to ensure proper nesting and synchronization. Developing effective plans facilitates well-synchronized transitions between operational phases.

A comprehensive analysis of HN capabilities and plans incorporating these resources provides sustainment commanders with an array of options. For example, the availability of reliable contracting resources could reduce the burden on military resources and an already strained distribution system. Contracted resources could enable military resources to be focused on high priority operations that are unsuitable for civilian personnel. The use of contractors and HNS are often directly tied to the level of violence and threat in the operational environment.

III. Sustainment Planning Tools

Ref: ATP 4-42, Materiel Management, Supply, and Field Services Operations (Nov '20), pp. 4-8 to 4-9.

Using planning guidelines and planning factors, materiel managers determine the quantities of supplies and services needed to support an operation. *AR 700-8, Logistics Planning Factors and Data Management*,dictates policy and responsibilities for managing Army logistics planning data. Logistics planning data includes a variety of information, such as consumption rates, reference data, and planning factors. Strategic, operational, and tactical leaders use Army logistics planning data and factors to estimate the amount and type of efforts required for a given operation.

There are many strategic, operational, and tactical sources (manual and automated) for information that provide logistics and operational information that can be used as planning data. For example, at the tactical level GCSS-Army, collects consumption information in real-time and near real-time for all classes of supply managed in that system. Other information, such as bulk fuel consumption,water purification, and water consumption, comes from LOGSTAT reporting. At the strategic level and operational level DLA and USAMC collects consumption information in real-time and near real-time with ERP systems. Operations Logistics (OPLOG) Planner is the most widely used source for logistics data because it contains Army and joint planning factors for all classes of supply.

See p. 4-10 for further discussion of the Operations Logistics (OPLOG) Planner.

Units forecast each commodity using logistics planning factors. A planning factor (rate, ratio, length of time) is a multiplier used to estimate the amount and type of effort involved in an operation. Population based planning factors have three variables: weight of the commodity for a given period, population supported during the same period, and estimated number of days for the operation. Equipment based planning factors consider equipment usage profiles for each fuel burning LIN in each organization.

Class VIII planning is an integral function of the multi-functional Army Health System. Planning considerations include patient bed occupancy, projected patient movement and operating room utilization,status of blood and other medical stocks, and the commander's health service support priorities. Class VIII requirements are challenging to forecast because consumption is dependent on many operational and geographic variables and demands are not generated until forces are actually deployed (peacetime demands do not reflect casualty care). Medical planners use modelling tools such as The Joint Medical Planning Tool and the Medical Contingency Requirements Workflow to models patient flow and estimate medical materiel requirements based on patient conditions and associated medical treatment data.

Logistics Synchronization

Daily logistics synchronization meetings help to ensure successful supply support execution. Effective logistics synchronization meetings have appropriate participation for validating logistic status reports,synchronize resupply operations, and create shared understanding among all of the sustainment planners. Leaders use the logistics synchronization meeting to update a preformatted and practiced synchronization matrix.

The synchronization matrix is a tool for planning and synchronizing operations between maneuver and supporting units during an operation. Using the synchronization matrix throughout the planning process enables planners to think through the many details that ensure mission success. Planners spell out who is getting what (commodities and amount), when (time window), where (grid coordinates), and how (supply point, logistics release point, forward logistics element) on the synchronization matrix. It includes key activities of adjacent units, and may incorporate unit and commodity management data.

I. Logistics Preparation of the Battlefield (LPB)

Ref: ATP 4-93, Sustainment Brigade (Aug '13), app. A. See pp. 8-27 to 8-38 for additional information on planning joint and theater Logistics.

Logistics preparation of the battlefield (LPB) is a key conceptual tool available to personnel in building a flexible strategic/operational support plan. Logistics preparation of the theater of operations consists of the actions taken by logisticians at all echelons to optimize means (force structure, resources, and strategic lift) of supporting the joint force commander's plan. These actions include identifying and preparing ISBs and forward operating bases; selecting and improving LOC; projecting and preparing forward logistics bases; and forecasting and building operational stock assets forward and afloat. LPB focuses on identifying the resources currently available in the theater of operations for use by friendly forces and ensuring access to those resources. A detailed estimate of requirements, tempered with logistics preparation of the theater of operations, allows support personnel to advise the JTF/ASCC/ARFOR commander of the most effective method of providing responsive support.

See pp. 8-36 to 8-37 for related discussion of key joint logistics planning process outputs to include the theater logistics analysis (TLA), theater logistics overview (TLO), logistics estimate, and concept of logistic support (COLS).

I. Intelligence in Support of Logistics

The logistician uses intelligence to develop and execute the logistics support plan. Logistics intelligence is critical to the planning effort. Some of the areas that should be included in the intelligence analysis are listed below:

- Intent to engage in multinational operations and the extent of logistics support to be provided to non-DOD agencies and allies
- Available resources in the AO
- Conditions that alter consumption factors, such as severe climate changes or a requirement to provide support to allies
- Capabilities of local facilities to support reception and staging operations
- Foreign military logistics structure, national infrastructure capabilities, and political inclination to facilitate joint forces support
- Environmental, geographical, climatological, and topographical factors that may affect support operations
- Analysis of the capabilities of the host nation's and region's LOC's and capabilities to support the operation

Intelligence is equally critical for war and stability operations. Logisticians must have a complete logistics database or file to develop a solid plan for the LPB.

LPB is those actions (force structure, resources, and strategic lift) taken to reduce the cost of logistically supporting an OPLAN or a contingency plan. LPB minimizes or eliminates potential problems at the outbreak of hostilities, during deployment, and throughout the operations. It is a systematic tool used by logisticians and commanders to complete their mission. It becomes the basis for deciding where, when, and how to deploy limited resources (supplies, equipment, people, and money).

The ASCC of a combatant command will prepare supporting Army plans with logistics planners concentrating on the logistics plans. Once logistics planners know the contingency country or geographic region, they can begin to build a logistics information database. This applies even if the command has a small chance of being deployed to a particular area. Once completed, the information data base file can be used to develop a comprehensive plan for LPB. The relative priority given to this effort will depend on the concept of operations and other command priorities. The key point is that the logisticians cannot afford to wait until maneuver units deploy to begin the LPB. It is a complex and time-consuming function.

Any actions that can reduce the cost of moving supplies, equipment, and people into an objective or contingency area are candidates for inclusion in the LPB plan. Planning must provide for the timely arrival of sustainment assets that are balanced according to the mission. Strategic lift assets are extremely limited.

II. Relevant Logistics Information

The following paragraphs contain types of relevant logistics information. These can be added to or taken away as individual missions dictate.

1. Geography

Collect information on climate and terrain in the AO. Determine if current maps are available. Use this information to determine when various types of supplies, equipment, and field services will be needed. For example, use water information to determine the need for early deployment of well-digging assets and water production and distribution units.

2. Supply

Collect information on supply items that are readily available in the AO. Determine which of these can be used in support of US forces. Subsistence items, bulk petroleum, and barrier materials are often available in country. Collect information on the supply system of the armed forces of the supported country; determine if it is compatible with the US system. Has the host nation bought, through foreign military sales, repair parts supporting current US systems? Can contingency contracting provide resources from HNS sources or third country sources until Army capabilities arrive in the AOR? Answers to these types of questions will aid in analyzing whether HNS negotiations are possible.

3. Facilities

Collect information on warehousing and cold storage facilities, production and manufacturing plants, reservoirs, administrative facilities, sanitation capabilities, and hotels. Their availability could reduce the requirement for deployment of similar capacity.

4. Transportation

Collect information on road nets, truck availability, rail nets, bridges, ports, cargo handlers (longshoremen), petroleum pipelines, and MHE. Also collect information on traffic flow, choke points, and control problems.

5. Maintenance

Collect information on maintenance facilities that could support US or coalition equipment. Examine the supported country's armed forces. Could they supplement our capability? Is there a commonality in equipment and repair parts? Does the country have adequate machine works for possible use in the fabrication of repair parts?

6. General Skills

Collect information on the general population of the supported country. Is English commonly spoken? Are personnel available for interpreter/translator duties? Will a general labor pool be available? What skills are available that can be translated to joint forces? Will drivers, clerks, MHE operators, food service personnel, guards, mechanics, and longshoremen be available?

7. Miscellaneous

Include any other information that could prove useful. Set up other categories as needed.

III. Sources of Information

Ref: ATP 4-93, Sustainment Brigade (Aug '13), p. A-3.

Collecting intelligence is not as difficult as it first may appear. A lot of information is routinely collected. There are several sources that can aid the logistician in building the LOGEEI data file. The below are some possible sources of information.

Intelligence Preparation of the Battlefield (IPB) Related Data

The weather and terrain data bases in the intelligence preparation of the battlefield, with its overlays, can provide excellent current information. This can be used in preselecting LOCs and sites for sustainment facilities. The event analysis matrix and template in the intelligence preparation of the battlefield *(ATP 2-01.3)* can also be used to assess the need for road, rail, airfield, port, and bridge improvements.

Intelligence preparation of the battlefield (IPB) is a systematic, continuous process of analyzing the threat and other aspects of an operational environment within a specific geographic area. Led by the intelligence officer, the entire staff participates in IPB to develop and sustain an understanding of the enemy, terrain and weather, and civil considerations. IPB helps identify options available to friendly and threat forces. IPB consists of four steps. Each step is performed or assessed and refined to ensure that IPB products remain complete and relevant. The four IPB steps are—

- Define the Operational Environment
- Describe Environmental Effects on Operations
- Evaluate the Threat
- Determine Threat Courses of Action

Civil Affairs Data

Civil affairs (CA) units CMO can provide information and knowledge of the operational environment to US and other coalition forces as directed. CA with their unique interaction and continuous dialogue with indigenous populations, institutions, intergovernmental organizations, NGOs, and US agencies and organizations can provide valuable information in support of the sustainment brigade mission and GCC's overall plan.

Embassy Staffs

The State Department has a worldwide network of embassies. These embassies are excellent sources of detailed information on a particular country. Embassy staffs routinely do country studies that, when current, can provide the bulk of the information you may need. A good library is another excellent source of information.

See pp. 8-36 to 8-37 for related discussion of key joint logistics planning process outputs to include the theater logistics analysis (TLA), theater logistics overview (TLO), logistics estimate, and concept of logistic support (COLS).

Refer to BSS6: The Battle Staff SMARTbook, 6th Ed. for further discussion of IPB. BSS6 covers the operations process; commander's activities; Army planning methodologies; the military decisionmaking process and troop leading procedures; integrating processes (IPB, information collection, targeting, risk management, and knowledge management); plans and orders; mission command, C2 warfighting function tasks, command posts, liaison; rehearsals & after action reviews; and operational terms and military symbols.

Logistics Prep of the Battlefield (LPB)

Ref: Adapted from FM 4-93.52 Tactics, Techniques, and Procedures for the DISCOM (Digitized), chap. 5.

Logistics preparation of the battlefield is the process of gathering data against pertinent battlefield components, analyzing their impact on sustainment, and integrating them into tactical planning so that support actions are synchronized with maneuver. It is a conscious effort to identify and assess those factors, which facilitate, inhibit, or deny support to combat forces. Just as intelligence preparation of the battlefield is important to the conduct of actual combat operations, logistics preparation of the battlefield is equally important to sustaining the combat power of the force.

The process requires tacticians to understand the data needed by logisticians to plan and provide timely, effective support. It requires TF logisticians to understand the mission, the tactical plan, and the battlefield's time and space implications for support. It is a coordinated effort to prepare the battlefield logistically. The basic steps in systematizing the process are:

- Determine battlefield data pertinent to support actions
- Determine sources from which raw data can be derived
- Gather pertinent data
- Analyze collected data elements and translate them into decision information by assessing their impact on the mission and the competing courses of action
- Integrate decision information into tactical planning by incorporating it in sustainment estimates and TF plans and orders

When determining what battlefield data are relevant to sustainment, it's helpful to break down CSS operations into certain key elements against which data can be collected for study and analysis. These data elements are called the components of tactical sustainment. The following descriptions of the components of tactical sustainment are not intended to be all-inclusive. They are offered here, however, to stimulate thought and to facilitate an understanding of those factors which impact on tactical sustainment:

- Logistics resources are the wherewithal to effect support, including CSS organizational structures, command and control, task organizing for support, communications, information automation systems, medical facilities, and materiel such as transportation assets and supply, maintenance and field services equipment.
- Logistics capabilities include soldier and leader skills and the personnel staffing which, collectively, activate sustainment resources and bring to life the required support.
- Logistics capacities include reception and clearance capacities, carrying capacities of transportation assets, volumes of storage facilities, maintenance production output rates, and supply route characteristics such as surface composition, tunnels, overhead obstructions, bridge weight limits and traffic circulation rates.
- Materiel stocks include the quantity and status of weapon systems, ancillary equipment, ammunition, repair parts and consumable supplies required or available to sustain or reconstitute combat power of deployed units. Also included are sustainment status reports and known or projected shortfalls.
- Consumption and attrition rates include experienced or expected usages of consumable supplies and weapon systems, which must be considered to anticipate support requirements.
- Time and space factors are those requirements and restrictions of the battlefield, which influence whether logistic support is provided to deployed forces at the right place and time. Included here are plans, orders, rehearsals, priority of support,

positioning for support, tempo of support (intensity of demand), security, risk assessment, the effects of terrain, weather, contaminated areas, minefields, night time enemy threat on sustainment operations, and the battlefield signatures of logistic resources. Time and space factors, especially, impact on the synchronization and integration of sustainment on the battlefield.

Sources from which relevant battlefield data are derived include:

- Higher headquarters briefs, plans and orders
- The commander's planning guidance
- The commander's intent (or concept)
- Operations and intelligence briefings and overlays
- Modified table of equipment (MTOE) of task force units
- Sustainment status reports
- Scouts
- Engineer route reconnaissance overlays
- Traffic circulation and highway regulating plans
- Personal reconnaissance

Logisticians should treat the components of tactical sustainment as essential factors that should be assessed for each plan. By doing so, they bring a professional approach to the contributions they make in the planning process. The components are variables. Some are dynamic and change with METT-TC so they should be validated daily, even hourly, if necessary. Commanders should appreciate the unique contributions their logisticians make in the planning process and when they've done a thorough job of collecting and analyzing pertinent battlefield data. Commanders must not accept less.

The commander and staff conduct LPB. Successful LPB contributes immeasurably to the favorable outcome of battle. Logistics preparation of the battlefield is an on-going process by which logisticians analyze:

- Tactical commander's plan/concept of operation
- Tactical commander's intents
- Supported force sustainment requirements
- Available sustainment resources
- Combat service support shortfalls
- The enemy (intentions, capabilities, weaknesses, doctrine)
- Terrain and weather
- Intelligence preparation of the battlefield (IPB) products
- Transportation infrastructure
- Host nation support available
- Time/distance factors

Logistics preparation of the battlefield (LPB) products are:

- A logistics estimate
- A visualization of the pending battle and logistics activity required by phase
- Anticipated logistics challenges and shortfalls
- Solutions to logistics challenges and shortfalls
- How, when, and where to position logistics units to best support the tactical commander's plan
- A synchronized tactical and logistical effort

Operations Logistics (OPLOG) Planner

Ref. https://www.cascom.army.mil/private/CDI/FDD/Multi/PDB/OPLOGPlanner.HTM
** Requires Army Knowledge Online (AKO) log-in and password.*

The OPLOG Planner is a web-based interactive tool that assists commanders and staff from strategic through operational levels in developing a logistics estimate. It is designed to support operations typically associated with multi-phase operations plans and orders. The OPLOG Planner enables staffs to develop estimated mission requirements for supply Class I, Class II, Class III(P), Class IV, Class VI, Class VII, Class X including water, ice, and mail.

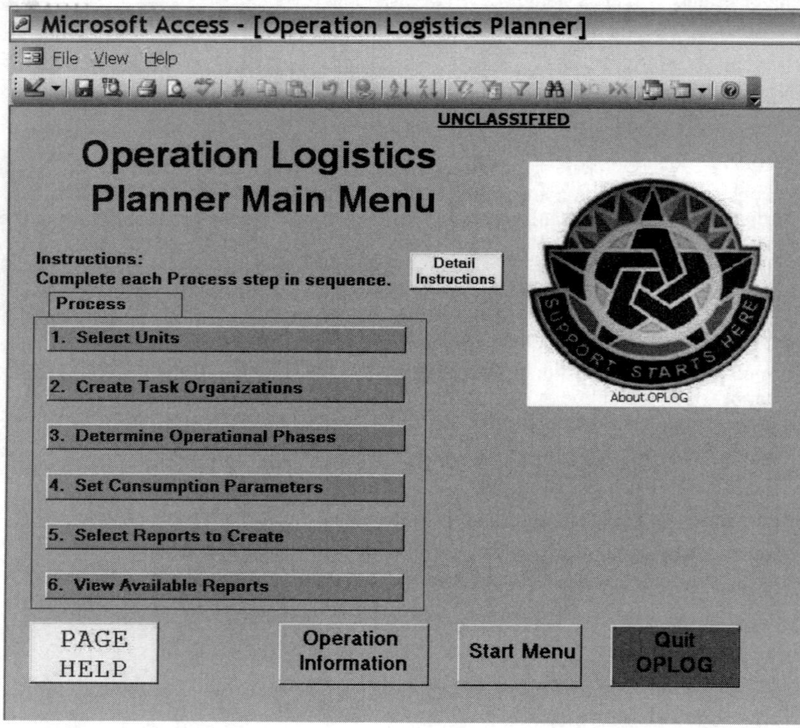

OPLOG Planner allows planners to build multiple task organizations from a preloaded list of units and equipment or from custom built units that are generated or imported. Each task organization is assigned a consumption parameter set that establishes the rates, climate, Joint Phases (Deter, Seize Initiative, and Dominate), and Army full spectrum operations (Offense, Defense, and Stability). The planners have the option of using predefined default planning rates or customizing rates based on what a unit is experiencing.

OPLOG Planner generates the logistics supply requirements which can be viewed by the Entire Operation, each Phase of the Operation, each Task Organization, each Unit, or each Unit's Equipment. Logistics requirement reports can be printed or exported to automated spread sheets for further analysis or saved for recall to be used in course of action analysis.

II. The Military Decision Making Process (MDMP)

Ref: FM 6-0 (C2), Commander and Staff Organization and Operations (Apr '16), chap. 9.

The military decisionmaking process is an iterative planning methodology to understand the situation and mission develop a course of action, and produce an operation plan or order (ADP 5-0).

The Military Decisionmaking Process (MDMP)

Key Inputs	Steps	Key Outputs
▪ Higher headquarter's plan or order or a new mission anticipated by the cdr	**I Receipt of Mission** (WARNORD)	▪ Commander's intial guidance ▪ Initial allocation of time
▪ Commander's initial guidance ▪ Higher HQs plan/order ▪ Higher HQs' knowledge and intelligence products ▪ Knowledge products from other organizations ▪ Army design methodology products	**II Mission Analysis** (WARNORD)	▪ Problem statement ▪ Mission statement ▪ Initial commander's intent ▪ Initial planning guidance ▪ Initial CCIRs, and EEFIs ▪ Updated IPB & running estimates ▪ Assumptions ▪ Evaluation criteria for COAs
▪ Mission statement ▪ Initial cdr's intent, planning guidance, CCIRs, and EEFIs ▪ Updated IPB, running estimates ▪ Assumptions ▪ Evaluation criteria for COAs	**III COA Development**	▪ COA statements and sketches -Tenative task organization -Broad concept of operations ▪ Revised planning guidance ▪ Updated assumptions
▪ Update running estimates ▪ Revised planning guidance ▪ COA statements/sketches ▪ Updated assumptions	**IV COA Analysis (War Game)**	▪ Refined COAs ▪ Potential decision points ▪ War-game results ▪ Initial assessment measures ▪ Updated assumptions
▪ Update running estimates ▪ Refined COAs ▪ Evaluation criteria ▪ War-game results ▪ Updated assumptions	**V COA Comparison**	▪ Evaluated COAs ▪ Recommended COAs ▪ Updated running estimates ▪ Updated assumptions
▪ Updated running estimates ▪ Evaluated COAs ▪ Recommended COA ▪ Updated assumptions	**VI COA Approval** (WARNORD)	▪ Commander approved COA and any modifications ▪ Refined commander's intent, CCIRs, and EEFIs ▪ Updated assumptions
▪ Commander-selected COA and any modifications ▪ Refined commander's intent, CCIRs, and EEFIs ▪ Updated assumptions	**VII Orders Production, Dissemination, and Transition**	▪ Approved operation plan or order ▪ Subordinates understand the plan or order

Refer to BSS6: The Battle Staff SMARTbook, 6th Ed. for discussion of the military decision making process (MDMP). Related topics include the operations process, integrating processes, plans & orders, mission command, rehearsals, after action reviews (AARs), and operational terms and symbols.

MDMP - Sustainment Considerations

Ref: Adapted from FM 4-93.52 TTP for the DISCOM (Digitized), chap. 4.

Refer to The Battle Staff SMARTbook for complete discussion of the military decisionmaking process (MDMP).

I. Mission Analysis

See pp. 4-15 to 4-22 for discussion of mission analysis to include personnel, logistics, and casualty estimates.

The MDMP begins when a mission is received from higher headquarters. Very rarely will this be in the form of a complete operations order. More likely it will begin after a verbal or written warning order is received. The commander, upon receiving a mission, should provide his staff with guidance as to how they should proceed with their analysis, and a warning order, in five-paragraph field order format, should be issued to subordinate units to allow them to begin to prepare for a new mission.

The staff begins mission analysis by developing their initial staff estimates based upon the higher headquarters order and their commander's guidance. Mission analysis also determines what the mission of higher headquarters is, what this equates to as a mission for their unit, and the situation/circumstances that may impact upon their unit's ability to execute a particular course of action that will be proposed to accomplish the mission. Each staff officer produces an estimate in his or her area. The results of mission analysis should include completed staff estimates, including an initial intelligence preparation of the battlefield (IPB) by the S-2. The staff will also produce a proposed restated mission for their unit. These products will be presented to the commander and he will provide additional planning guidance to include: number of courses of action he wants the staff to develop, initial commander's critical information requirements (CCIR), timeline, risk guidance, and rehearsals to be conducted. Additionally, the commander will provide his initial intent that will include the method and end state for the operation. A second warning order should be issued to subordinate units no later than the end of mission analysis.

The Logistics Estimate

The logistician's input during mission analysis primarily comes from the logistics estimate. The logistics estimate is a continuous process that begins during mission analysis and is continually refined and updated through mission completion.

II. COA Development

During COA development, the logistics planners must refine the logistics estimate they developed during mission analysis. Facts and assumptions developed during mission analysis must be verified and updated. Sustainment planners must identify any significant Sustainment considerations and requirements that have a major impact on mission accomplishment. Additionally, the sustainment planner must develop a draft concept of support during this phase of the MDMP.

COA Sustainment Feasibility

For each course of action, the logistics planner must access its sustainment feasibility. The sustainment feasibility is determined by whether or not the unit possesses the required resources to sustain the unit throughout the tactical operation. Tailoring your logistics estimate for each course of action can help make this determination. If requirements do not exceed capabilities, the sustainment of the course of action will generally be feasible. If any requirements do exceed capabilities you must again determine its significance and potential impact upon the mission. If the shortfall is a "WAR-STOPPER", and there are no workable solutions to the problem, then sustainment of the COA is not feasible. Ensure you have exhausted all possible means to solve the problem, to include support from higher headquarters, before you deem the COA not feasible.

Synchronization Requirements

The synchronization of sustainment during COA analysis is critical to ensure continuous support during the operation. During the war game, the logistical planner will determine, based on the scheme of maneuver, what supplies and services must be where at a given time. This will generate critical Sustainment actions that must be accomplished to sustain the mission. He must consider time-distance factors and determine which support activity will be available to provide the required support. This is where the logistical planner begins to directly link the actions of task force logistics assets with the support battalion sustainment activities and division/corps resupply activities. He must ensure that all critical Sustainment activities are included in the synchronization matrix to successfully synchronize all levels of support.

Critical Sustainment Requirements

During this phase of MDMP, the courses of action are compared using the synchronization matrices and notes taken for each evaluation criteria used. A decision matrix with the evaluation criteria and some type of weighting factor (e.g., numbers, +/-, etc.) should be used to record the results of the course of action comparison. A decision matrix can be used as an aid to obtain a decision from the commander as to what course of action will be selected.

III. COA Comparison

In order to compare COAs and determine which is more supportable, logistical planners must calculate estimated attrition rates, project battle losses for critical weapons systems, and project personnel battle losses. The RSR for each COA must be refined and compared to any CSR that may be in effect. Quantities of supplies required, demands on transportation assets, and reconstitution requirements must be compared to determine which COA stresses the units' logistical system the most. An analysis of the risks to Sustainment assets and operations must be compared and considered.

IV. Orders Production

See pp. 4-23 to 4-26 for further discussion of the concept of support (para. 4a).

The brigade logistic planners are responsible for paragraph 4 of the OPORD as well as the logistics annex (annex I). These products must be complete, concise and synchronize all levels of logistics support from top to bottom with the tactical plan.

Final Concept Of Support

The paragraph four consists of the final approved concept of support. Remember that this paragraph is written primarily to inform the commanders how they will be supported throughout the tactical operation, so do not include details on how the support elements are to execute the plan. Save all those details for the logistics annex. The concept of support should include a brief synopsis of the support command mission, support area locations to include the locations of the next higher logistics bases, the next higher levels support priorities, the commander's priority of support, significant or critical Sustainment activities, any significant risks, and the major support requirements in each tactical logistics function. If the tactical concept of operation is phased, the concept of support should also be phased to facilitate changes of priorities and logistics focus during each phase.

Complete/Concise

It is essential that the OPORD be complete, concise, and include all critical tasks that must be accomplished to support the tactical mission. Ensure you consider the command and support relationships of all units within your area of operation and ensure all elements receive support. Ensure you address all of the tactical logistics functions and properly phase the support concept if the tactical concept of operation is phased.

Synchronized Top-to-Bottom

To ensure proper synchronization, include all critical tasks and coordination requirements that were developed during the war gaming phase. Consider developing a logistics synchronization matrix, if possible. Ensure that all priorities of support are in agreement with the scheme of maneuver and weight the main effort. Coordinate with the other battlefield operating system elements to ensure that there are no inconsistencies in logistics information within the maneuver, engineer, artillery, and Sustainment annexes. As subordinate OPORDs are developed, you must ensure that their support plans are consistent and executable within your support framework. Synchronization of resupply operations from the corps/division, to the support battalion, to the task force level is critical.

Materiel Management Planning Considerations

Ref: ATP 4-42, Materiel Management, Supply, and Field Services Operations (Nov '20), p. 4-8.

Materiel managers and sustainment planners collect information on materiel that is readily available within the AO. Subsistence items, bulk petroleum, and barrier materials are the most commonly available commodities. Materiel managers and sustainment planners seek answers to the following questions—

- What materiel can be purchased locally?
- What supply systems are coalition partners using? Are the systems compatible?
- What classes of common user items are Army sustainment units responsible for providing joint and/or coalition forces? (*Refer to JP 4-02, Joint Health Services, for discussion on common user logistics.*) Does a commonality exist in equipment and repair parts?
- Does the host nation have repair parts that could support current systems?
- Are theater support contracts available?

A successful concept of operations is logistically supportable throughout the duration of the battle. National level materiel managers must know when to buy, how much to buy, where to distribute, how best to support complex logistics operations, and how to forecast and plan for ongoing operations. Operational level materiel managers plan for what to stock, how much to stock, and where to stock in order to maintain a continuous pipeline of supplies. Tactical level materiel managers plan for operational readiness of supported units.

Friction points and resource shortfalls can be avoided through early identification of customs and clearance requirements; contract support requirements; reception, staging, and onward movement; and closeout processes. It is vital that sustainment planners leverage strategic partners to identify all requirements during initial, mid and final planning conferences to ensure that the execution unit is prepared. Moreover,sustainment planners create a shared understanding from knowledge gained during planning conferences by creating a logistics playbook to use for synchronizing the sustainment warfighting function during execution.

By using known requirements, capabilities, and consumption rates for all classes of supply, materiel managers produce a logistics estimate with a logistics task organization that mitigates shortfalls and backhaul. The following list provides some examples of supply actions that range from strategic to tactical level activities—

- Analyze LOGSTATs to anticipate supply support requirements.
- Consider placement of supply points in the support areas.
- Establish supply stockage and safety levels.
- Monitor daily battle losses.
- Identify source, location, and requirements for water.
- Recommend policies, priorities, allocations, and criteria for controlled items.
- Establish and distribute materiel evacuation instructions, condition standards, and policies for the collection and evacuation of materiel.
- Track performance measures such as speed, accuracy, and reliability that indicate supply pipeline responsiveness.

III. Running Estimates & Mission Analysis

Ref: Adapted from FM 6-0 (C2), Commander and Staff Organization and Operations (Apr '16) and ADP/ADRP 4-0, Sustainment (Aug '12).

Running Estimates

A running estimate is a staff section's continuous assessment of current and future operations to determine if the current operation is proceeding according to the commander's intent and if future operations are supportable (FM 3-0). Building and maintaining running estimates is a primary task of each staff section. The running estimate helps the staff provide recommendations to commanders on the best course of action to accomplish their mission. Running estimates represent the analysis and expert opinion of each staff section by functional area.

See pp. 2-20 to 2-21 for related discussion of refining plans and estimates.

Mission Analysis

Mission analysis helps commanders to understand the situation to include their mission. This enables commanders to issue the appropriate guidance that drives the rest of the planning process. Commanders—supported by their staffs—gather, analyze, and synthesize information to orient themselves on current conditions in the AO. Such orientation helps commanders to better understand the relationships among the operational and mission variables. Mission analysis helps commanders understand the problem they have been called upon to resolve and how their units fit into the higher headquarters' plan.

During mission analysis, the staff conducts intelligence preparation of the battlefield and updates running estimates in relation to the new mission. The commander and staff analyze the higher headquarters' order to completely understand the higher headquarters commander's intent, mission, and concept of operations. They develop facts and assumptions about the upcoming operations and determine specified, implied, and essential tasks. They identify forces available for the mission, resource shortfalls, and any constraints placed on them from the higher command.

The logistician's input during mission analysis primarily comes from the logistics estimate. The logistics estimate is a continuous process that begins during mission analysis and is continually refined and updated through mission completion. The logistics estimate does not have a doctrinal format at the brigade level.

Mission analysis considerations feed information into the estimate process. The estimates are as thorough as time permits. Personnel/logistics estimates are kept current. As factors that influence operations change, new facts are developed and assumptions become facts or become invalid.

The duration of the mission analysis briefing may vary. It may be with only a few staff briefing the commander, or it may take be several days in the form of a conference that includes commanders, subordinate commanders, staff, and other partners.

See following pages (pp. 4-16 to 4-17) for a listing of suggested sustainment mission analysis considerations by element.

Refer to BSS6: The Battle Staff SMARTbook, 6th Ed. (Plan, Prepare, Execute, & Assess Military Operations) for discussion of mission analysis and running estimates. Additional related topics include the operations process, the three Army planning methodologies, integrating processes, plans and orders, mission command, rehearsals and after action reviews, and operational terms and military symbols.

Mission Analysis Considerations

Ref: Adapted from ADP/ADRP 5-0 and ADP/ADRP 4-0.

A comprehensive mission analysis briefing helps the commander, staff, subordinates, and other partners develop a shared understanding of the requirements of the upcoming operation. Time permitting, the staff briefs the commander on its mission analysis using the following outline:

- Mission and commander's intent of the headquarters two levels up
- Mission, commander's intent, and concept of operations of the HQs one level up
- Review of the commander's initial guidance
- Initial IPB products, including modified combined obstacle overlays and situation templates
- Civil considerations that impact the conduct of operations
- Specified, implied, and essential tasks (to include minimum essential stability tasks)
- Pertinent facts and assumptions
- Constraints
- Forces available and resource shortfalls
- Initial risk assessment
- Recommended initial information themes and messages
- Recommended initial critical commander's critical information requirements (CCIRs) and essential elements of friendly information (EEFIs)
- Initial intelligence, surveillance, and reconnaissance (ISR) plan
- Recommended timelines
- Recommended collaborative planning sessions
- Recommended restated mission

Sustainment Elements

I. Logistics

A. Supply

1. General Supply

a. Facts

- Classes I, II, III(p), IV, VI, VII, X, and water status
- Distribution system
- Critical shortages

b. Assumptions

- Resupply rates
- Host nation/multinational support
- Other

c. Conclusions

- Projected supply levels and status
- Shortfalls and critical sustainment risks/events

2. Class III (B)

a. Facts

- Class III(b) status
- Restrictions
- Distribution Systems
- Critical shortages

b. Assumptions

- Resupply rates
- Host nation/multinational support
- Other

c. Conclusions

- Projected supply status
- Projected distribution system
- Shortfalls and critical risks/events

3. Class V

a. Facts

- Class V status
- Distribution system
- Restrictions
- Critical shortages

b. Assumptions

- Resupply rates
- Host nation/multinational support
- Other

c. Conclusions

- Projected supply status
- Projected distribution status
- Shortfalls and critical risks/events

B. Field Services

Life support functions, including field laundry, showers, light textile repair, food preparation, mortuary affairs, aerial delivery support, food services, billeting, and sanitation.

a. Facts

b. Assumptions

c. Conclusions

C. Maintenance

Field and sustainment maintenance.

a. Facts

- Maintenance status (equip readiness)
- Class IX status
- Repair times, evacuation policy, assets
- Critical shortages

b. Assumptions

- Host nation/multinational support
- Other

c. Conclusions

- Projected maintenance status
- Shortfalls and critical CSS risks/events
- Recommendations

D. Transportation

Movement control, mode operations, terminal operations, container management.

a. Facts

- Status of transportation assets
- Critical LOC and MSR status
- Critical shortages

b. Assumptions

- Host nation/multinational support
- Other

c. Conclusions

- Projected status of transportation assets
- Projected status of LOCs and MSRs
- Shortfalls and critical risks/events
- Recommendations

E. Operational Contract Support

Operational contract support is the process of planning for and obtaining supplies, services, and construction from commercial sources along with associated contract management functions.

A. Facts

B. Assumptions

C. Conclusions

II. Personnel Services

A. Human Resource (HR) Support

a. Facts

- Personnel strengths and morale
- Replacements and medical RTD
- Critical shortages

b. Assumptions

- Replacements
- Host nation/multinational support
- Other

c. Conclusions

- Projected personnel strength
- Projected critical MOS status
- Shortfalls and critical risks/events

B/C/D/E. Finance/Legal/Religious/Band

a. Facts

b. Assumptions

c. Conclusions

III. Health Services Support (HSS)

See pp. 4-31 to 4-36 for discussion of considerations for HSS in support of joint and multinational operations. Include medical logistics considerations (class VIII supply).

a. Facts

b. Assumptions

c. Conclusions

IV. Other Sustainment-Related Functions

As a result of the movement from battlefield operating systems to the Warfighting Functions (WFF) construct, some tasks are realigned. Two of those tasks include:

A. Explosive Ordnance Disposal (EOD)

From a WFF perspective, EOD falls under the Protection WFF

a. Facts

b. Assumptions

c. Conclusions

B. Internment/Resettlement (I/R) Opns

Internment/Resettlement (I/R) operations are included under the Sustainment WFF (FM 3.0). While not a major sub-function of the sustainment WFF; I/R are supported by logistics, personnel services, and HSS.

Recommendations

- Sustainment unit and system capabilities, limitations, and employment
- Risk identification and mitigation
- Organization for operations, allocations to subordinate units, command and support relationships among subordinate units
- Resource allocation and employment synchronization of organic and supporting units (including other joint assets)
- General locations and movements of units

I. The Logistics Estimate

Ref: Adapted from ADP/ADRP 5-0, ADP/ADRP 4-0 and FM 4-93.50, app. A.

A running estimate is a staff section's continuous assessment of current and future operations to determine if the current operation is proceeding according to the commander's intent and if future operations are supportable (FM 3-0). Building and maintaining running estimates is a primary task of each staff section. The running estimate helps the staff provide recommendations to commanders on the best course of action to accomplish their mission. Running estimates represent the analysis and expert opinion of each staff section by functional area.

Staff representatives do more than collect and store information; they process it into knowledge and apply judgment to get that knowledge to those requiring it. The staff section's running estimate is one product of this effort. By synthesizing relevant information from a variety of sources into a prescribed format, the staff and commander can quickly identify and summarize how a given functional area impacts other functional areas and the operation as a whole.

1. Requirements

The first step in the log estimate process is to determine the logistical requirements for the mission. To determine the requirements, you may use a number or combination of methods. Automated systems such as OPLOG PLANNER are good tools to use to estimate requirements. If you prefer to do the number crunching yourself, planning factors from the CGSC ST 101-6 (G1/G4 BATTLEBOOK) or the FM 101-10-1/2 may be used to determine estimates; historical data from previous missions should also be used.

2. Capabilities

To correctly determine the logistics capability of your unit, you must consider the capabilities of all the available Sustainment assets at your disposal. This includes all available Sustainment units assigned, attached, or OPCON, and the sustainment capability organic to the maneuver units themselves. Be sure to consider the unit's current status in terms of personnel and equipment, as well as the projected status at mission execution.

3. Comparison/Shortfall

Once you have determined your estimated requirements and the unit's sustainment capability, compare them to determine any logistical shortfalls. If there are no shortfalls, go to the analysis step of this methodology. Shortfalls may occur in terms of storage, distribution, and transportation capability or may be caused by personnel or equipment shortfalls based on current on hand shortages or maintenance status. A shortfall may also occur if required facilities or terrain are not available or the plan does not provide enough time to prepare. If there is a shortfall, determine what the shortfall is in terms of short tons, gallons, or other units of measurement and when or where during the operation the shortfall occurs.

4. Analysis

Whether or not there is a shortfall, the analysis process must occur for all support operations. The sustainment planner must determine: when the support operation must begin, how much time there is to prepare for the mission, the purpose of the support mission, the duration of the mission, and whether the mission can be supported from a fixed location or whether to echelon support forward in some way. If there is a shortfall identified in the comparison of requirements and capabilities, you must also determine its cause, its significance and its potential impact on the tactical operation.

5. Solutions

Determine the most workable solutions based on your analysis. Ensure that all solutions are integrated into the MDMP to enhance continuity between the tactical decision making and logistical planning.

A comprehensive running estimate will address all aspects of operations and contain both facts and assumptions based on the staff's experience within a specific area of expertise. A base document for a running estimate is provided below. Each staff section modifies it to account for their specific functional area. All running estimates should cover essential facts and assumptions, a summary of the current situation by the mission variables, conclusions, and recommendations. The mission variables of METT-TC are mission, enemy, terrain and weather, troops and support available, time available, civil considerations.

1. Mission.
Show the restated mission resulting from mission analysis

2. Situation and Considerations

a. Characteristics of the Operational Environment. Identify area structures, capabilities, organizations, people, and events that impact or affect functional area considerations.

b. Characteristics of the Area of Operations.

(1) Weather. State how weather affects staff functional area's capabilities

(2) Terrain. State how terrain affects staff functional area's capabilities

(3) Enemy Forces. Describe enemy disposition, composition, strength, capabilities, systems, and possible courses of action (COAs) with respect to their effect on functional area.

(4) Friendly Forces. List current functional area resources in terms of equipment, personnel, and systems. Identify additional resources available for functional area located at higher, adjacent, or other units. Compare requirements to current capabilities and suggest solutions for satisfying discrepancies.

(5) Civilian Considerations. Describe additional personnel, groups, or associations that cannot be categorized as friendly or enemy. Discuss possible impact these entities may have on functional area.

(6) Assumptions. List all assumptions that affect the functional area. Best practice is to prioritize assumptions based on operational requirement or to divide assumptions (if many) into categories such as —friendly assumptions and —enemy assumptions.

3. Courses of Action (COAs)

a. List friendly COAs that were war-gamed.

b. List enemy actions or COAs that were templated that impact functional area.

c. List the evaluation criteria identified during COA analysis. All staff use the same criteria.

4. Analysis.
Analyze each COA using that evaluation criteria from COA analysis. Review enemy actions that impact functional area as they relate to COAs. Identify issues, risks, and deficiencies these enemy actions may create with respect to functional area.

5. Comparison.
Compare COAs. Rank order COAs for each key consideration. Use a decision matrix to aid the comparison process.

6. Recommendations

a. Recommend the most supportable COAs from the perspective of the functional area.

b. Prioritize and list issues, deficiencies, and risks and make recommendations on how to mitigate them.

II. The Personnel Estimate

Ref: Adapted from FM 1-0, Human Resources Support (Feb '07), pp. I-3 to I-5. [Not provided in Apr '14 version of FM 1-0.]

G-1s and S-1s at all levels, HRSC divisions, Expeditionary Sustainment Command (ESC) and Sustainment Brigade SPO HR Operations Cells, HR Company headquarters, TG R5 Teams, MMT Teams, R5 Plans and Operations Teams, and Postal Plans and Operations Teams are responsible for executing the core competency of HR Planning and Operations to ensure that effective and efficient HR support is provided to both units and individuals.

Mission analysis is a critical early step in the MDMP. The HR planner/G-1/S-1 has specific responsibilities that establish favorable conditions for the remainder of the MDMP from the HR perspective. The following are G-1/S-1 mission analysis responsibilities:

- Analyze personnel strength data to determine current capabilities and project future requirements
- Analyze unit strength maintenance, including monitoring, collecting, and analyzing data affecting Soldier readiness
- Prepare estimates for personnel replacement requirements, based on estimated casualties, non-battle losses, and foreseeable administrative losses to include critical MOS requirements
- Determine personnel services available to the force (current and projected)
- Determine personnel support available to the force (current and projected)
- Determine HR specified, implied, and essential tasks (if any)
- Determine any HR constraints which may impact successful completion of the mission
- Identify critical HR facts and assumptions which will impact the mission
- Determine any HR Commander's Critical Information Requirements (CCIR) and Essential Elements of Friendly Information (EEFI) which may impact the mission

One of the primary tools used by HR planners is the Personnel Estimate, and like any staff estimate it: "Is an assessment of the situation and an analysis of those COAs a Commander is considering that best accomplishes the mission. It includes an evaluation of how factors in the HR staff section's functional area influence each course of action and includes conclusions and a recommended COA to the Commander" from the perspective of the HR supporter. The personnel estimate is a continuous process that evaluates current and future operations from the functional perspective of the HR provider. Once the commander has selected a COA, the Personnel Estimate becomes an assessment tool to evaluate the relative success of the COA and a means for HR operations cells to track the progress of the operation from an HR support perspective. Format is not the critical element of the personnel estimate. It is a tool that facilitates the collection and processing of key HR information, and may be adapted by HR planners for their own organization or purposes. The following are general characteristics of the personnel estimate:

- Prepared as thoroughly as time allows in either a simple form or a preformatted digital product
- Revised continuously as planning factors, manning levels, facts and assumptions change
- Prepared at all levels of command from Battalion to Army/ASCC by S-1s/G-1s and HR Operations Cell planners
- Not necessarily prepared in a fixed sequence
- Provides a thorough, clear, unemotional analysis of all pertinent data

1. MISSION: Commander's restated mission resulting from mission analysis

2. SITUATION AND CHARACTERISTICS:

- **a. Characteristics of the AO:** Discuss how the weather, terrain, civil considerations, and other AO-specific conditions affect the HR support to the planned operation
- **b. Enemy Forces:** Discuss any affect enemy dispositions, composition, strength, capabilities and COAs may have on the HR support to the planned operation
- **c. Friendly Forces:** Discuss current status of resources available from the HR perspective and other resources which impact HR support and compare requirements and capabilities to develop recommendations to solve discrepancies.
  - 1) Current subordinate unit manning levels/critical MOS shortages (consider all manning influences – Task Organization, R&R flow, Boots On Ground data, etc.)
  - 2) New personnel requirements resulting from the operation (language skills, ASI, etc.)
  - 3) Casualty estimates (as developed)
  - 4) Forecasted replacement availability
  - 5) Evacuation policy for the operation
  - 6) Supporting HR organizations (location, command/support relationship, controlling element)
  - 7) Supporting medical elements
  - 8) Replacement fill priority (Coordinate with S-3/G-3)
  - 9) Crew/key leader replacement
  - 10) Projected postal flow/limitations
  - 11) Change to established PASR reporting flow/times
  - 12) Change to casualty reporting flow (CLT change, reporting changes)
  - 13) RSOI Reception HR impacts
  - 14) R&R schedule/flow operational impact (during operations conducted during sustained operations)
  - 15) Civilian and JIM manning impacts on the operation focusing on strengths and personnel service requirements
  - 16) Specific Army G-1 Personnel Planning Guidance (PPG) impacts on the operation
  - 17) Status of other personnel service support factors (postal, MWR, legal, MILPAY support)
  - 18) Connectivity and NIPR/SIPR and HR system access requirements and availability
- **d. Sustainment Situation:** Discuss the sustainment supporting organizations (TSC, ESC, Sustainment Brigades) and possible impacts to include: supported/supporting relationship, reporting requirements, transportation planned, MSR status, etc.
- **e. Assumptions:** List any assumptions that affect the delivery of HR support.
- **f. Other Considerations:** List everything not covered elsewhere in the estimate.

3. COURSES OF ACTION:

- a. COAs: List friendly COAs to be wargamed
- b. Evaluation Criteria: List HR evaluation criteria determined during COA analysis

4. ANALYSIS: Analyze each COA using the HR evaluation criteria established during the COA analysis determining advantages and disadvantages of each COA

5. COMPARISON: Compare COAs and rank order from the perspective of HR supportability based on Step 4 (Analysis).

6. RECOMMENDATION and CONCLUSIONS:

- a. Recommendation: Recommend the most supportable COA from HR support perspective.
- b. Issues/Risks/Mitigations: List the issues, risks, or deficiencies associated with the recommended COA and proposed mitigation to reduce the impact.

APPENDIXES. Use appendixes when information/graphs is of such detail and volume that inclusion in the body makes the estimate too cumbersome.

III. Casualty Estimation

FM 1-0, Human Resources Support (Apr '14), pp. 4-30 to 4-31.

The Army G-1 is the functional proponent for Army total mission casualty estimation (KIA, captured, MIA, and WIA). The Army Surgeon General is the functional proponent for Army disease and non-battle injury (DNBI) casualty estimation. The Army G-1 is the functional proponent for overall casualty estimation, covering both total mission casualty and DNBI, in support of projected manning requirements. The Director, Military Personnel Management oversees policy and oversight of Army casualty estimation (to include methods and procedures) for the Army G-1.

Casualty estimation is conducted at division-level and above as part of the planning process for contingency operations. Casualty estimates support operations planning, future force planning, and staff training. Supported functions include:

- Commander's evaluation of a COA by assessment of force strength for missions within the concept of operations and scheme of maneuver.
- Personnel replacements flow planning, and allocation among forces (if casualty shelves are used).
- Medical support planning for both force structure and sustainment support.
- Transportation planning, including both inter and intra-theater requirements, to deliver medical force structure and to evacuate and replace personnel.
- Evacuation policy options to sustain the force by balancing minimal support force footprint, maximum in-theater RTDs, and stable personnel rotation.

The G-1/AG, as the principal staff officer for manning the force (personnel readiness requirements, projections, and recommendations), prepares the casualty estimate as part of the operations order. The G-1/AG estimates mission casualties and administrative losses and combines this with the medical staff's DNBI estimate.

Achieving Reasonable Casualty Estimates

Ensuring reasonable casualty estimates requires more than a numeric estimating procedure or set of rates; a rates frame of reference is critical to show which rates relate to which variables, and how. The currently approved and mandated methodology for deriving casualty rates for operational planning is the Benchmark Rate Structure (BRS) which orients the planner by showing how rates vary as forces (size, type), time (duration of rate application), and operational settings vary. The BRS describes rate ranges and patterns seen in actual operations for both maneuver forces (battalions, brigades, divisions) and support forces across the spectrum of conventional operations, including major combat and stability operations. The operating environments described range from peer or near-peer confrontations, to overwhelming dominance by one side, to isolated asymmetric events. There is currently no approved estimation tool at this time so planners must temper their estimates with these limitations in mind.

The BRS permits bracketing optimistic-to-pessimistic rate possibilities and settling on a risk-based recommendation. An estimate may be built directly from the BRS, or the planner may use it to assess the reasonableness of rates from any source or method. Functional area planning rests, ultimately, on projected numbers of casualties. However, the reasonableness of the numbers must first be established. This requires use of casualty rates, which reference (normalize) the numbers to the population-at-risk and the time (number of days) during which the casualties are generated. Rates are expressed in numbers of casualties per 1,000 personnel (population-at-risk) per day. Standard notation is [number]/1,000/day.

IV. The Concept of Support (para. 4a)

Ref: Adapted from FM 4-90.7, Stryker Brigade Combat Team Logistics (Sept '07) and related sources (ST 101-6, chap. 3 and app. C-G). See also p. 4-17.

After the commander selects a specific COA, the staff communicates this decision by publishing the operation plan/operation order (OPLAN/OPORD). The G4, with input from the other logistic staff elements (G1, G5, surgeon, finance and personnel officers, and the support command), will prepare paragraph 4 of the plan.

Paragraph 4a is the support concept. This concise, but comprehensive, paragraph tells the maneuver commander and his primary staff those critical or unusual logistic actions that will occur by phase or before, during, and after the battle to support the concept of the operation.

Additional subparagraphs can be used to provide more detailed sustainment information by functional area. Usually, these subparagraphs are omitted, and this detailed information is published as part of the service support annex to the plan. The G4 prepares this order with input from the other logistic staff elements. The G4 can also prepare a Sustainment overlay to show supported units' supply route locations and supporting logistic organizations. Finally, routine, doctrinal, or constant information is incorporated into the unit tactical standing operating procedures (TSOP) to avoid repetition.

I. Developing the Sustainment Concept

The logistician actively participating in the decisionmaking process facilitates the support concept's development. Specifically, during mission analysis, the Sustainment planner determines the units' current materiel and personnel posture before the operation begins. This, with the commander's priorities, determines which units and items of equipment should receive priority before the operation.

The wargaming and quantitative analysis portions of COA analysis highlight critical and/or unusual logistic requirements and determine support priorities during each phase of the operation. By its very nature, wargaming facilitates logistic synchronization with the concept of the operation.

There are numerous other information sources for the support concept:

- Commander's guidance and intent
- Concept of the operation
- Higher HQ support concept, service support order or plan (if applicable), and Sustainment overlay
- Maneuver control system screens and/or other locally generated status charts
- Lessons learned data and historical perspectives to view how others successfully, or unsuccessfully, supported other similar operations
- The unit's battle book

II. The Sustainment Overlay

The sustainment overlay is a graphic representation of the tactical array of support areas and units. Ideally, it accompanies copies of the OPLAN and/or OPORD distributed to subordinate HQ and is used as a graphic backdrop to the support concept briefing.

The sustainment overlay should include (as a minimum):

- Locations of current and proposed support areas
- Boundaries for sustainment responsibilities
- MSRs
- Locations of major HQ
- Locations of sustainment installations and units
- Locations of critical resources [potable water, maintenance collection points, ATPs, mortuary affairs (MA) collection points, ambulance exchange points (AXPs), etc.]

The sustainment overlay will not only depict the tactical array of sustainment units/nodes, but it is also an integral part of the overall OPLAN/OPORD graphics and must be synchronized with the operations overlays.

The **BRIGADE sustainment overlay** would include (as a minimum)—

- The brigade support area (BSA) location and, using type unit symbols, the sustainment units and HQ located therein
- Locations of alternate/proposed BSAs
- Locations of forward logistics elements (FLEs)
- The supply routes from the BSA to the logistic release points and/or maintenance collection points
- The MSR from the division support area (DSA) to the BSA

The **DIVISION sustainment overlay** would include (as a minimum):

- The division support area (DSA) location and using type unit symbols, the sustainment units and HQ contained there-in, whether they are divisional or nondivisional
- Locations of alternate and/or proposed DSAs
- The MSRs from the corps rear area to the DSA and from the DSA to each BSA

The **CORPS sustainment overlay** may have to encompass the entire corps area of operation (AO) as well as a part of the communication zone (COMMZ) and, as a minimum, would depict:

- The logistic support areas (LSAs) and, using type unit symbols, the Sustainment units and HQ located therein, and the locations of any other critical sustainment nodes not located in an LSA
- The MSRs leading into the corps rear area from the COMMZ and the MSRs leading from the corps rear area to each DSA (or, as a minimum, to the division rear boundary) and to other critical logistic nodes
- Locations of alternate and/or proposed LSAs
- Locations of corps sustainment units operating forward of the divisional rear boundaries

III. The Sustainment Matrix

The oral support concept briefing will allow the commander and his subordinates to visualize how the operation will be logistically sustained. The sustainment planners' oral briefing, using the sustainment overlay, is useful in communicating the support concept to the commander. In addition, a support concept matrix can be used to make complex logistic concepts more easily understood.

The sustainment matrix's design is aligned with the support concept format. The logistic functions are in the "by phase" context. The matrix can also be modified to reflect before, during, and after phases. The matrix will highlight those critical aspects of each sustainment function. It can also depict other critical information such as priorities, shifts in priorities, problem areas, critical events, and other critical action. The matrix is not intended to stand alone or to replace the support concept briefing. It should complement and supplement the support concept briefing.

Concept of Support (Format & Briefing)

Ref: Adapted from ST 101-6, chap. 3, pp. 3-5 to 3-7. See also p. 4-17.

The logistician's role in the overall OPLAN/OPORD briefing is to brief the support concept, but he must first understand the general concept of the operation and the commander's intent. The support concept briefing should address the critical, non-SOP, or unusual aspects of logistic support by phase of an operation by critical Sustainment functions. Doctrinal, usual, or SOP matters should not be addressed unless there is a deviation in support relationships or normal methods.

4. Concept of Support (Paragraph 4a)

a. Support Concept. Paragraph 4a will provide an overall view of the support concept. Its intent is to provide the non-Sustainment commanders and their primary staffs an image of how the operation will be logistically supported. If the information pertains to the entire operation, or if it pertains to more than one unit, include it in the introductory portion of paragraph 4a. Change it in the ensuing subparagraphs when needed. This could include: a brief synopsis of the support command mission; support command headquarters and/or support area locations, including locations of next higher logistics bases if not clearly conveyed in the Sustainment overlay; the next higher level's support priorities and where the unit fits into those priorities; priorities that remain unchanged throughout the operation; units in the next higher Sustainment organization supporting the unit; significant and/or unusual Sustainment issues that might impact the overall operation; the use of host nation support; and any significant sustainment risks

(1) PHASE I (repeat for each phase)

- Logistics focus.
- Priorities: By unit; for personnel replacements; maintenance and/or recovery and evacuation priorities (by unit and equipment type); mvmt; class of supply.
- Critical events. Use the Sustainment functions for information.
- Sustainment risks.

(4) Paragraphs 4b through 4e are normally more detailed and are included in the service support annex. They are not part of the support concept.

(5) Concept of support written before, during, and after format. Follow the same guidance as by phase.

Concept of Support Briefing Format

1. Introduction (overview of the support concept and orientation to the map, if required). Orientation to the map is not required if another briefer has done so previously. Do not assume the commander totally knows the terrain. Focus on locating critical Sustainment nodes, MSRs, etc.

2. Brief the support concept starting with critical actions that must be accomplished in the first phase of the operation and concluding with critical actions to be accomplished in the last phase. This will prepare for future operations using the Sustainment functions as a guide.

3. Identify which units have priorities for each critical sustainment function (this should correlate with the commander's priorities; e.g., main effort).

4. Identify the next higher echelon unit providing support and/or backup support.

5. Identify any critical shortages/problem areas for each sustainment function and solution. For example, this can be supported, but _____ , or it can be done, but not without risk in_____ .

6. Identify any other sustainment problem areas, arrangements, special requirements, or any other critical aspects addressed elsewhere in the briefing.

BCT Sustainment Planning (The BCT S-4)

Ref: Adapted from FM 4-90.7, Stryker Brigade Combat Team Logistics (Sept '07) and related sources.

The lead planner for sustainment in the BCT is the BCT S-4, assisted by the BCT S-1, the BCT surgeon, and the BSB support operations officer. Representatives from these and other sections form a sustainment planning cell at the BCT main CP to ensure sustainment plans are fully integrated into all operations planning. The unit SOP should be the basis for sustainment operations, with planning conducted to determine specific requirements and to prepare for contingencies. BCT and subordinate unit orders should address only specific support requirements for the operation and any deviations from SOP.

Although the sustainment planners at the BCT main CP control and coordinate sustainment for specific BCT operations, routine operations usually are planned in the BSA. The BCT S-1 may have representatives at or near the BSB to handle various personnel functions (e.g., replacement or monitor casualty operations). The BCT S-4 might choose to locate the property book or movement sections with the BSB support operations section.

To provide effective support, sustainment planners and operators must understand the mission statement, intent, and concept of the operation which in turn will lead to developing a concept of support that will be provided in the BCT operation order (OPORD). The BCT S-4 is responsible for producing the sustainment paragraph and annexes of the OPORD, which should include the following:

- Commander's priorities
- Class (CL) III/ V resupply during the mission, if necessary
- Movement criteria
- Type and quantities of support required
- Priority of support, by type and unit
- Sustainment overlay
- Supply routes
- Logistic release points (LRPs)
- Casualty evacuation (CASEVAC) points
- Maintenance collection points

The remainder of this section provides a description of the concepts that apply to planning sustainment support within the BCT. It also describes how the BCT organizes the sustainment staff and organizations during full spectrum operations.

BCT Concept of Support

The concept of support (Paragraph 4 of the Operations Order) establishes priorities of support (by phase or before, during, and after) for the operation. These priorities are set by the BCT commander at each level in his intent statement and the concept of the operation. Priorities include such things as personnel replacements; maintenance and evacuation, by unit and by system (aviation and surface systems would be given separate priorities); fuel and/or ammunition; road network use by unit and/or commodity; and any resource subject to competing demands or constraints. To establish the concept of support, the BCT sustainment planners must know:

- Subordinate units' missions
- Times missions are to occur
- End states
- BCT scheme of movement and maneuver
- Timing of critical events

Chap 4

V. Army Health Service (AHS) Planning

Ref: FM 4-02, Army Health System (Nov '20), chap. 4. See also pp. 3-47 to 3-50.

I. Support to Decisive Action

Decisive action is the simultaneous combination of offense, defense, and stability or defense support of civil authority tasks. These tasks require versatile, adaptive medical support, and flexible leadership.

The traditional and primary Army Medicine mission is to conserve the fighting strength of the tactical commander. The Army Medicine rhythm of military operations is that of the operational commander. Casualties begin to occur immediately upon engagement with the enemy. Due to the necessity to perform lifesaving interventions for Soldiers suffering combat trauma within minutes of wounding or injury, AHS resources must be arrayed in close proximity to the forces supported. This also permits the AHS assets to rapidly clear the battlefield of casualties and enhances the CCDRs ability to quickly take advantage of opportunities which present themselves during the operation.

Army Health System planners must be included early-on in the planning cycle for tactical operations and must fully participate in rehearsals conducted by the operational Army being supported. Within noncontiguous operations, the linear array of AHS units will not always occur and AHS units must fully understand the various support relationships described in the OPORDs to ensure that a seamless continuum of health care is established and can be maintained.

The MEDEVAC plan for the tactical operation includes both rotary-wing air ambulances and ground ambulances. The preferred means of evacuation is the air ambulance; however its availability can be affected by air superiority issues and environmental factors such as visibility, winds, and dust. The evacuation plan must address the use of ground ambulances when feasible and/or the simultaneous use of both platforms.

Sustainment Planning

II. Operational Variables (PMESII-PT)

As the operational environment (OE) is comprised of all of the factors, both military and civilian, that affect the conduct of military operations in an operational area, the medical commander must define how the different elements will impact on the concept of operations. The operational variables are a means for exploring and describing an OE that focuses on the human aspects of the environment. Commanders and planners can use political, military, economic, social, information, infrastructure, physical environment, and time (operational variables) to ensure all elements are considered. The operational variables are used by strategic planners in the development of plans and information may be broader than required for mission analysis at the tactical level, however, as medical issues often have a regional focus and may be the result of environmental,socioeconomic, political, and religious practices, it is essential for the AHS planner to consider the medical aspects of an operation on a much broader scale than the immediate AO. As the theater medical command, the MEDCOM (DS) provides this regional focus in support of the CCDRs theater engagement strategy.

See following pages (4-28 to 4-30) for listing and discussion of medical factors for consideration in relation to the operational variables.

In addition to FM 4-02 (Nov '20), Medical planners should refer to ATP 4-02.55, Army Health System Support Planning (Mar' 20). Refer also to the Joint Medical Planning Tool, the Medical Planners' Toolkit, and the Medical Contingency Requirements Workflow for calculation of medical requirements.

Medical Aspects of the Operational Variables (PMESII-PT)

Ref: FM 4-02, Army Health System (Nov '20), table 4-1, pp. 4-2 to 4-5.

Table 4-1 provides medical aspects for consideration in relation to the operational variables and subvariables. This table is not an all-inclusive listing but does provide the AHS planner with some initial considerations.

Variable	Subvariables	Medical aspects
Political	Attitude toward the United States. Centers of political power. Type of government. Government effectiveness and legitimacy. Influential political groups. International relationships.	Health status of population. Public health issues. Accessibility to health care. Nutritional status of the population and/or subgroups of the population.
Military	Military forces. Government paramilitary forces. Nonstate paramilitary forces. Unarmed combatants. Nonmilitary armed combatants. Military functions. • Command and control. • Maneuver. • Information operations. • Reconnaissance, security, and surveillance capabilities acquisition. • Fire capabilities. • Protection. • Sustainment. • Cyberspace operations and electronic warfare capabilities. • Special operations capabilities.	Development of military medical infrastructure. Level of education and training of military medical personnel. Trauma care capabilities. Medical evacuation (ground and air). Forward surgical/damage control surgical capabilities. Hospitalization capabilities. Disease and nonbattle injury rates. Identification and treatment of mild traumatic brain injuries and traumatic brain injuries. Dental care services. Blood supply and blood-banking capabilities. Organic medical assets. Area medical support capabilities. Availability of medical supplies and equipment. Medical equipment maintenance and repair. Medical logistics system to include medical gases and optical fabrication and repair. Behavioral health and treatment of combat and operational stress reaction capabilities. Rehabilitative and convalescent care capabilities to include prosthetics. Food inspection and laboratory analysis. Veterinary care for military working dogs and other government-owned animals and veterinary public health capabilities including zoonotic diseases and food protection infrastructure/programs.
Economic	Economic diversity. Employment status. Economic activity. Illegal economic activity. Banking and finance.	The economic base can affect health care for both the human and the animal populations in the nation. The types of injuries and health issues may vary significantly based upon whether it is an agricultural society or an industrialized nation and/or region. This affects the types of health care available including restorative and rehabilitative services and programs and the availability of health care to the populace. The gross national product and the per capita income of the population affect the availability of resources for the government to expend on public health concerns and health care in general. When the Army Health System planner examines the economic factors of a nation or region, it is important to determine what influence it has on how much money is expended in the health sector (both private and public) as this will affect health care, medical equipment, and pharmaceuticals availability.

Variable	Subvariables	Medical aspects
Social	Demographic mix. Social volatility. Education level. Ethnic diversity. Religious diversity. Population movement. Common languages. Criminal activity. Human rights. Centers of social power. Basic cultural norms and values.	Age, gender, and genetics affect how individuals are affected by disease and existing environmental factors. Religion affects how people view medical intervention; it can affect how a person will comply with medical treatment regimens and whether they will accept recommended treatments (such as the use of blood transfusions). Persons who are uprooted may be more susceptible to disease because of lowered immunity status due to fatigue, restricted food intake, poor living conditions, inadequate shelters, and poor sanitation. If public health and disease prevention programs are not instituted, the general health of the population or the affected subpopulation will decrease. Populations where education and literacy are not widespread will often have a lower standard of living, less appreciation for public health and disease prevention practices, less skilled workers, and be more difficult to reach with public health alerts and programs. Cultural, ethnic, and religious beliefs often influence who will seek medical care and who will not. Privacy issues may require that consideration of the provider's gender is relevant in addressing women's health issues. Providers must be cautious in using graphic aids to communicate with their patients, as the explicit graphics may be considered offensive. Medical personnel should develop a guide for asking medical questions in the local language dialect.
Information	Public communications media. • Telephone/cell phones. • Print Media. • Broadcast media (TV, radio). • Public Webpages.	Availability of mass communications enablers for public health warnings, alerts, and information. • Telephones. • Televisions. • Radios. • Newspapers/periodicals. • Computers/e -mail.
Infrastructure[1]	Construction pattern. Urban zones. Urbanized building density. Utilities present. Utility level. Transportation architecture.	Availability of electricity and potable water. Number of medical providers (by category). Numbers of primary, secondary, and tertiary medical treatment facilities. Status of waste disposal. Sanitation practices and standards (availability of toilets, showers, and bathing facilities). Urbanization can increase the spread of infectious diseases due to inadequate living space, improper ventilation, poor sanitation practices, and lowered immunity. Accessibility issues (roads [paved and unpaved], commercial transportation systems [buses, taxis, rail, and air], vehicles and/or pack animals, and natural barriers [mountains, streams, jungles, and deserts]). Availability of transportation assets for medical evacuation or other medical purposes in the event of natural or man-made disaster or other mass casualty situation.

Continued on next page

Continued from previous page

Medical Aspects of the Operational Variables (Cont.)

Variable	Subvariables	Medical aspects
Physical environment	Terrain • *Observation and fields of fire. • *Avenues of approach. • *Key terrain. • *Obstacles. • *Cover and concealment. • Landforms. • Vegetation. • Terrain complexity. • Mobility classification. Natural hazards. Climate. Weather. • *Precipitation. • *High temperature—heat index. • *Low temperature—chill index. • *Wind. • *Visibility. • *Cloud cover. • *Relative humidity.	Are brick and mortar structures available for use as medical treatment facilities? Climate and weather effects on— • Disease vectors. • Categories and types of injuries. • Acclimatization issues pertaining to heat, cold, or altitude. • Medical evacuation operations. Topography and hydrology considerations include: • Character and types of injuries to be encountered. • Natural barriers to medical evacuation. • Lines of patient drift. • Suitable for farming and for grazing animals. Natural resources to include the availability of medicinal herbs. Presence of toxic plants and animals and whether they pose a health hazard to deployed troops.
Time	Cultural perception of time. Information offset. Tactical exploitation of time. Key dates, time periods, or events.	Time affects not only the provision of medical care, but also may affect the types of diseases and injuries which may occur. Short duration operations require emphasis on rapidly treating Soldiers with traumatic injuries, while longer duration operations require emphasis on disease prevention and the management of chronic medical conditions.
NOTE: Subvariables marked with an asterisk (*) are also the military aspects of terrain and weather used in analyzing mission, enemy, terrain and weather, troops and support available, time available, and civil considerations.		

Sustainment Planning

III. Mission Variables (METT-TC)

Mission variables (mission, enemy, terrain and weather, troops and support available, and time available) are used by AHS planners to determine the impact they will have on medical operations. Mission variables describe characteristics of the operational area, focusing on how they might affect a mission. The mission variables are discussed below. In Table 4-1 above, the subvariables which are the same as mission variable considerations are marked with an asterisk (*).

Continued from previous page

Refer to BSS6: The Battle Staff SMARTbook, 6th Ed. (Plan, Prepare, Execute, & Assess Military Operations) for discussion of ***operational and mission variables*** *from ADP 5-0. Additional related topics include the operations process, the three Army planning methodologies, integrating processes and continuing activities, plans and orders, mission command, rehearsals and after action reviews, and operational terms and military symbols.*

Chap 5

Brigade Support

Ref: FM 4-0, Sustainment (Jul '19), pp. 2-54 to 2-56 and ATP 4-90, Brigade Support Battalion (Jun '20), chap. 1.

Brigades include both Brigade Combat Teams (BCTs), multifunctional, and functional brigades. BCTs, as notionally depicted in figure 2-32, are the Army's primary combined arms, close-operations force, and principal ground maneuver units of the division. BCTs have organic capabilities, including battalion-sized maneuver, field artillery, reconnaissance, and sustainment units. Each BCT has organic medical support for Roles 1 and 2 medical operations. The three types of BCTs are armored, infantry, and Stryker. BCTs normally operate as part of a division.

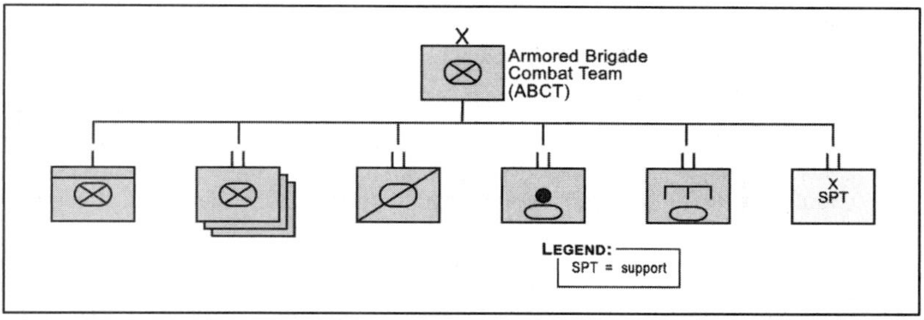

Ref: FM 4-0, fig. 2-32. Notional brigade combat team (with associated Brigade Support Battalion).

Brigade combat teams provide the division commander close combat capabilities to execute missions to achieve military objectives during engagements, battles, and campaigns. There are three standard types of BCTs – the armored brigade combat teams (ABCT), the infantry brigade combat team (IBCT), and the Stryker brigade combat team (SBCT). All six warfighting functions – command and control, movement and maneuver, intelligence, fires, sustainment, and protection – are organic to a BCT. Multifunctional support brigades reinforce brigade combat teams by providing capabilities that enable them to fight as formations: sustainment, field artillery, maneuver support, protection, and aviation.

I. Brigade Support Battalion (BSB)

The **brigade support battalion (BSB)** provides logistics and medical support to a brigade combat team (BCT) and multifunctional support brigades. Brigade combat teams are the primary combined arms force that provide combat power to execute close combat during large-scale combat operations. The BSB is capable of employment across the range of military operations and in any environment. This chapter provides an overview of the BSB's role and describes how the BSB provides support to brigade combat teams and multifunctional support brigades. It is imperative for BSB commanders and staff to understand the BSB role and responsibilities within the overall combined arms approach to operations.

II. BSB Role, Core Competencies, & Functions

The BSB is the most important sustainment organization in the Army. It supports the brigade combat team and the other brigade formations that constitute the majority of close combat capability in the Army. The BCT area of operations is expansive and its missions diverse. The BSB and its subordinate companies normally operate within the close area of the operational construct closer to the forward line of troops than any other battalion-sized sustainment organization. This places the BSB in operational environments that are highly lethal, rapidly changing, and extremely demanding. This operational environment (OE) requires frequent movement, strong protection, and perseverance. In all, success relies on disciplined leadership founded on mission command, well-trained troops that display strong resilience, and closely synchronized staff operations.

The role of a BSB is to provide sustainment support (logistics and medical support) to a BCT. The BSB core competencies are planning, synchronization, and execution of sustainment to support BCT operations. The BSB performs the following functions: distribution management and operations, transportation, supply support, field maintenance, and Role 2 medical care.

III. BSB Characteristics and Capabilities

The BSB is a multifunctional logistics battalion capable of operating at the tactical level to support a BCT. The BSB organic design and core competencies allow it to provide multi-class supply, field maintenance, and medical support. The BSB has a very broad span of control with ten subordinate companies, six of which operate in the other battalions' areas. The BSB has organic medical support capability, medical logistics, and medical operations personnel in the support operations (SPO) staff. It is dependent on the division sustainment support battalion (DSSB) for non-mobile class III(B) storage, light infantry troop transportation, and water treatment. These capabilities, when required, must be coordinated with the division sustainment brigade (DSB).

The BSB is organic to and is employed by the BCT. It operates in conjunction with other BCT battalions. The BSB commander executes command and control (C2) of BSB units based on mission orders issued by the BCT commander. BSB operations are based on and nested with the BCT concept of operations. Although the BSB collaborates closely with the DSB it does not receive mission orders from the DSB. The DSB may influence BSB activities in support of division priorities through the operations process but in all cases the BSB receives its orders from the BCT or other brigade it supports. The BSB commander maintains continuous dialog through command and staff channels with higher sustainment echelons (DSB, DSSB) in order to provide situational awareness and facilitate anticipation of future requirements.

The BSB is an expeditionary formation and deploys with the BCT unless otherwise directed by higher headquarters. The BSB in IBCTs is designed to deploy forces on short notice to austere locations and perform sustainment operations immediately upon arrival. Airborne BSBs participate in parachute assault operations. The BSB integrates joint, inter-organizational, and multinational capabilities as needed. The battalion is capable of sustaining the BCT across the range of military operations.

The BSB operates from a base or base cluster in the brigade support area (BSA). From this location the BSB executes C2 over its organic companies conducting sustainment support. The BSB distribution company, field maintenance company, and medical company operate in the BSA. These companies may be collocated with the BSB command post (CP), but during large-scale combat operations are tactically dispersed within the BSA in a way that facilitates sustainment operations, mutual support, and protection. The forward support companies (FSC) collocate with the combat trains command post and operate in close proximity with the supported battalions in the BCT close area. The FSC position capabilities in the BSA as part of field trains to expedite distribution support to the maneuver battalions.

The BSB is responsible for protecting the BSA from level I and II threats using organic equipment.

See pp. 5-18 to 5-19 for further discussion of BSB protection and area security.

The BSB provides area support to units, not organic to the BCT, when tasked by the brigade commander. Area support is a task assigned to a sustainment unit directing it to support units in or passing through a specified location.

IV. BSB Organization in Brigade Combat Teams

The BSB is organic to and provides sustainment support to a BCT. BSBs in IBCT, SBCT, and ABCT differ in terms of specific equipment and manning, reflecting the characteristics of the BCTs they support.

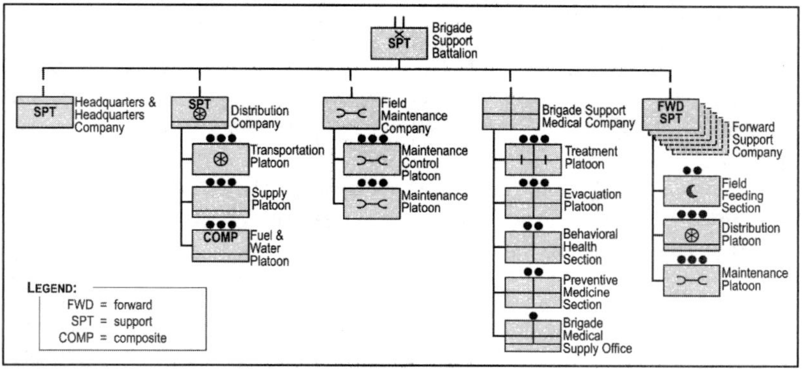

Ref: ATP 4-90 (Jun '20), fig. 1-1. Brigade support battalion task organization.

BSBs have the capability to operate a supply support activity (SSA), operate a modular ammunition transfer point (MATP), perform field-level maintenance support and distribution operations, as well as providing medical support. BSBs that contain a medical company provide Role 2 medical care to all units in the brigade. FSCs provide distribution, field feeding, and field-level maintenance support to their supported battalions. Other capabilities that BSBs provide include:

- Field-level maintenance support to the BSB and the brigade headquarters.
- Field feeding support to the BSB headquarters company, distribution company, field maintenance company, and medical company.

The DSB assigned to a division can provide logistics capability not organic to the BSB or provide additional capacity to support the BSB. The DSB has an organic DSSB comprised of organic companies: a headquarters company, a composite supply company, a composite truck company, and a support maintenance company. The DSSB provides water treatment and storage, non-mobile petroleum storage, and troop transportation support to the BCT.

Support Battalions of Multi-Functional Brigades

A mix of multifunctional brigades support theater army, corps, and divisions. These supporting brigades include the field artillery brigade, maneuver enhancement brigade (MEB), combat aviation brigade (CAB), and sustainment brigade. Additionally, functional support brigades can include engineer, medical, military intelligence, and military police brigades. A mix of these functional and multifunctional brigades are task organized under corps and division control, enabling them to conduct LSCO.

Most support brigades are not fixed organizations with a set list of organic subordinate units. Each support brigade, except the CAB, is designed around a small base of organic elements to which a mix of units and additional capabilities are added based on mission variables.

Brigade Support

A. Distribution Company

Ref: ATP 4-90, Brigade Support Battalion (Jun '20), p. 1-12.

The role of the BSB distribution company is to plan, direct, and supervise supply support operations and supply distribution to the brigade combat team or multifunctional brigade units.

BSB Distribution Company

- **Role**: The BSB distribution company plans,directs, and supervises supply support and distribution in support of a brigade combat team or multifunctional brigade.
- **Capability**: The distribution company manages the stockage of supplies for the brigade and provides distribution capability for all classes of supply.
- **Parent**: Brigade support battalion.
- **Command relationship**: Organic to a brigade support battalion.
- **Support relationship**: General support to BSB and brigade units.
- **Span of operations**: Brigade combat team area of operations.

The BSB distribution company is a multi-functional distribution and supply company operating as directed by the BSB commander. The BSB employs the distribution company in the BSA and its subordinate units operate throughout the supported brigade area of operations. The company executes a supply support activity that includes managing the daily receipt, storage, and issue of supply class I, II, III, IV, V and IX. The company also executes distribution, distribution integration, and transportation operations to ensure timely supply support to the BCT.

The distribution company commander, supported by subordinate leaders, uses troop-leading procedures to plan, prepare, and execute the BSB commander's intent. The company commander also uses the procedures to assess the effectiveness of the company plan and adjusts the plan as required.

A BSB distribution company has three platoons: a transportation platoon, a supply platoon, and a fuel and water platoon. The distribution company leadership (commander, executive officer, and first sergeant) is responsible for company support operations. These personnel ensure the distribution management tasks are executed effectively. The company commander may designate responsibility for the distribution management tasks between the executive officer and the first sergeant as necessary.

The platoon leaders and platoon sergeants of the supply and fuel and water platoons execute a portion of the 14 materiel management functions as required to ensure supply stocks are available and ready for distribution.

Materiel Management

Ref: ATP 4-90, Brigade Support Battalion (Jun '20), pp. 3-2 to 3-3.

Materiel management is the continuous situational understanding, planning, and execution of supply and maintenance capabilities to anticipate, synchronize, and direct all classes of supply to maximize combat power and enable freedom of action in accordance with the supported commander's priorities. Materiel management determines the materiel requirements of the BCT by class of supply, determines availability of the materiel from either on-hand or higher source-of-supply stocks, obtains the materiel, and coordinates the movement of the distribution of the materiel with distribution integration personnel. It identifies the commodity to be distributed, the quantity of that commodity to be distributed, and priority of distribution by unit. Material management addresses all internal and external logistical processes, information, and functions necessary to satisfy an operational supply requirement. Effective materiel management enables the commander greater situational awareness informing decision making and enhancing control and flexibility. The BSB SPO, distribution company supply platoon, medical company (class VIII only) and the FSC distribution platoons execute materiel management functions. They perform materiel management simultaneously and continuously, with or without automation, during all decisive action tasks. The BSB can also integrate supply capability provided by joint services, allied partners, and host nations to the fullest extent possible within mission and operational variables.

Units execute materiel management through the following functions. These functions may be executed all or in part based on operational and mission variables.

- Supply planning (SPO).
- Requirements determination (SPO, medical company, FSC).
- Requirements validation (SPO, medical company, and FSC).
- Resupply (Distribution company, medical company, and FSC).
- Funds management (SPO).
- Storage (Distribution company, medical company, and FSC).
- Stock control (Distribution company, medical company, and FSC).
- Supply (Distribution company, medical company, and FSC).
- Maintenance.
- Asset visibility (SPO, distribution company, medical company, and FSC).
- Asset reporting (SPO, distribution company, medical company, and FSC).
- Retrograde of materiel (Distribution company, medical company, and FSC).
- Disposal (Distribution company, medical company, and FSC).

Materiel management functions are described in ATP 4-90, chapters 3 and 6.

Transportation Operations

Ref: ATP 4-90, Brigade Support Battalion (Jun '20), pp. 3-4 to 3-5.

Transportation is a logistics function that includes movement control and associated activities to incorporate military, commercial, and multinational motor, rail, air and water mode assets in the movement of units, personnel, equipment, and supplies in support of the concept of operations. Transportation operations are executed by the BSB SPO transportation personnel, distribution company distribution platoon, the medical company (class VIII only), and the FSC distribution platoon as shown in paragraph 3-28. Transportation allocates specific modes for specific commodities and coordinates distribution and routing to meet command priorities.

The transportation platoon leader(s) and platoon sergeant(s) execute both distribution integration and transportation operations functions as required to ensure supplies are assigned a transportation mode, route, and movement time. These functions are also discussed in ATP 4-90, chapters 3 and 6.

Distribution Management

Ref: ATP 4-90, Brigade Support Battalion (Jun '20), pp. 3-4 to 3-5.

The BSB commanders and staff integrate the distribution management process into the operations process to develop a distribution plan that is synchronized with operations and all warfighting functions. Failure to execute this integration leads to operational conflicts that hinder or prevent timely supply distribution. The commander and staff realize that support operations offices and the plans developed therein are fundamentally an integral part of BSB operations. Although the S3 and SPO are separate offices, they are inextricably linked during the operations process and MDMP. The BSB S3 and SPO collaborate continuously through the phases of planning, preparation, execution and assessment.

ATP 4-90, chapter 6 further describes distribution operations and other units involved in the distribution process to support the BCT.

B. Field Maintenance Company (FMC)

Ref: ATP 4-90, Brigade Support Battalion (Jun '20), p. 1-13.

The role of a field maintenance company is to provide field-level maintenance support to the BSB and BCT. The field maintenance company provides repair capability for automotive, ground support, communications and electronics, land combat missile systems, and armament systems. The company provides services including welding and machine shop support and recovery support. The field maintenance company also provides limited field-level maintenance reinforcement support to the FSCs for low-density commodities such as—communications, electronics, and armament equipment. The BSB typically employs the field maintenance company in the BSA.

BSB Field Maintenance Company

- **Role**: The BSB field maintenance company provides field-level maintenance support to the BSB and brigade combat team.
- **Capability**: The field maintenance company performs field-level maintenance on or near the unserviceable piece of equipment or weapon system utilizing line replaceable units or modules and component replacement or repair.
- **Parent**: Brigade support battalion.
- **Command relationship**: Organic to a brigade support battalion.
- **Support relationship**: General support to the BSB companies and brigade headquarters.
- **Span of operations**: Brigade combat team area of operations, typically located in the BSA.

The field maintenance company performs field-level maintenance on or near equipment or weapon systems utilizing line replaceable units or modules and component replacement or repair. Field-level maintenance is not limited to the removal and replacement of components but also allows for the repair of components or end items on or near the system. Field-level maintenance also includes adjustment, alignment, services, applying approved field-level modification work orders as directed, fault/failure diagnoses, battle damage assessment, repair, and recovery. Field-level maintenance is always repaired and returned to the user and includes maintenance actions performed by operators.

Large-scale combat operations demands a maintenance support focused on returning systems to operational status quickly and as close as possible to the point of failure or damage. This requires that maintenance be conducted forward in the brigade and battalion areas. Maintenance assets move as far forward as the tactical situation permits to return inoperable and damaged equipment to the battle as quickly as possible. Offensive operations may require that inoperable equipment be collected in central points for later repair, since damaged or inoperable equipment would encumber a supported unit if hauled forward.

The field maintenance company commander, supported by subordinate leaders, uses troop-leading procedures to plan, prepare, and execute the BSB commander's intent. The company commander also uses the procedures to assess the effectiveness of the company plan and adjusts the plan as required.

Maintenance Operations

Ref: ATP 4-90, Brigade Support Battalion (Jun '20), pp. 7-1 to 7-2.

The modern battlefield demands a maintenance system that is flexible, responsive, and focused on returning systems to operational status quickly and as near as possible to the point of failure or damage. This requires a forward presence of maintenance in brigade areas. During large-scale combat operations this requires the echeloning of maintenance capabilities throughout the depth of the BCT AO. This chapter provides an overview of maintenance fundamentals at the BCT level and describes the role, organization, and operations of the BSB's field maintenance company and FSCs to support the BCT.

The primary purpose of maintenance is to ensure equipment operational readiness. The goal for every unit is to have all of its equipment fully operational for its intended purpose. The second purpose of maintenance is to generate combat power by repairing damaged equipment as quickly and as close to the point of failure as possible. Repairs should return the damaged equipment to fully mission capable status or to a state that allows mission accomplishment.

Once units enter combat operations, maintenance is critical to maintain combat power and momentum. Replacement systems are seldom immediately available. This is especially true during the early stages of an operation and when units suffer heavy combat losses. Units maintain existing systems to ensure it is mission capable for the duration of the operation or until the system is clearly damaged beyond field-level maintenance repair capability.

The Army's maintenance structure in the BCT is highly adaptable and flexible. BCTs have the capability to repair all brigade systems with BSB organic maintenance units, crews and operators. As such, there is no need to evacuate BCT equipment that requires field-level maintenance to another organization or echelon of support. Furthermore, in the absence of a maintenance surge team, there is no repair capability outside of the BCT for the main battle tank, infantry fighting vehicles, or Stryker systems. A maintenance surge team provides maintenance capability for these primary weapon systems when maintenance requirements exceed the capability of the BSB to support. The BCT will evacuate equipment that requires sustainment-level maintenance.

Maintenance and recovery planning is integrated into all aspects of the MDMP to ensure synchronization and unity of effort. Planning includes identifying requirements, reviewing available assets, preparing a maintenance estimate, comparing requirements to capabilities, and making adjustments to maintenance priorities to meet the mission requirement. Maintenance planning is included in the overall sustainment concept of support. Maintenance planners understand the overall mission and concept of operations for maneuver forces in order to prioritize and weight maintenance support to the main effort. Maintenance planners recommend to the BSB commander, BCT XO, and BCT commander how to task organize for optimal maintenance capability. They recommend the cross leveling of system maintainers to ensure adequate maintenance capability is available to support the main effort. This is a dynamic process that requires daily adaptation to existing circumstances.

The BCT S-4 and SPO collaborate to determine how many key systems identified are mission ready and then work with the FSCs to prioritize the maintenance effort. The staffs war game the operation during the MDMP and prepare an estimate from projected system losses and gains during each stage of the operation. It is imperative that maintenance planners understand that in the absence of a maintenance surge team there is no repair capability outside of the BCT for the main battle tank, infantry fighting vehicles, or Stryker systems.

See pp. 3-5 to 3-10 for additional discussion. *ATP 4-90, chapter 7 further describes maintenance operations in decisive action, the operations of the BSB's field maintenance company, and other units involved in the maintenance operations to support the BCT.*

Brigade Support

C. Brigade Support Medical Company (BSMC)

Ref: ATP 4-90, Brigade Support Battalion (Jun '20), pp. 1-14 to 1-15.

The role of a brigade support medical company (BSMC) is to provide medical support to a BCT. The BSMC provides medical area support to all BCT units that do not have organic medical assets.

Brigade Support Medical Company

- **Role**: The brigade support medical company provides medical support to a brigade combat team.
- **Capability**: The BSMC provides Role 1 and Role 2 in support of a brigade combat team and medical support on an area basis to all BCT units that do not have organic medical assets.
- **Parent**: Brigade support battalion.
- **Command relationship**: Organic to a brigade support battalion.
- **Support relationship**: General support to the BSB and BCT; general support to others on a limited basis by exception.
- **Span of operations**: Brigade combat team AO, typically located in the BSA.

The BSMC consists of a company headquarters, preventive medicine section, mental health section, medical treatment platoon, medical evacuation platoon, and a brigade medical supply office. The BSMC commander leads and supervises its organic and attached medical augmentation elements. Planners may augment the BSMC with a forward surgical capability when required based upon mission requirements or mission variables.

The BSMC locates and establishes its company headquarters and a brigade Role 2 medical treatment facility in the BSA. It typically collocates with the BSB headquarters on the BSB base since the BSMC has limited protection capability.

Refer to ATP 4-02.3, Army Health System Support to Maneuver Forces for more information.

The company headquarters provides C2 for the company and attached units. The brigade support medical company commander, supported by subordinate leaders, uses troop-leading procedures to plan, prepare, and execute the BSB commander's intent. The company commander also uses the procedures to assess the effectiveness of the company plan and adjusts the plan as required. It provides unit-level administration, general supply, and chemical, biological, radiological, and nuclear (CBRN) defense support. The company headquarters has a command element, a supply element, and CBRN operations element.

The medical treatment platoon receives, triages, treats, and determines the disposition of patients in the brigade AO. The platoon provides for advanced trauma management, tactical combat casualty care, general medicine, general dentistry, and physical therapy. In addition, the medical treatment platoon has limited radiology, medical laboratory, and patient holding capabilities. The medical treatment platoon is organized with a headquarters, a medical treatment squad, a medical treatment squad (area), and patient holding squad.

The evacuation platoon performs ground evacuation and enroute patient care for supported units. The platoon employs ten evacuation teams, utilizing wheeled or tracked ambulances. The evacuation platoon provides ground medical evacuation support from the Role 1 battalion aid stations of the maneuver battalions of the BCTs to the Role II care of the BSMC. In addition, it provides ground medical evacuation support to units receiving area medical support from the BSMC. The BCT surgeon cell coordinates medical evacuation to higher roles of care (Role 3), including ground and air ambulance.

The preventive medicine section provides advice and consultation in the area of health threat assessment, force health protection, environmental sanitation, epidemiology, sanitary engineering, and pest management. Through routine surveillance, inspection of potable water supplies, inspection of field feeding facilities, and the application of pest management practices, they identify actual and potential health hazards, recommend corrective measures, and help in training BCT Soldiers in disease and non-battle injury prevention programs. The preventative medicine section provides unit field sanitation team training.

The behavioral health section supports commanders in the prevention and control of combat and operational stress reaction through the brigade's behavioral health activities by the provision of advice and assistance in the areas of behavioral health and combat and operational stress control. The section collects and records social and psychological data and counsels personnel with personal, behavioral, or psychological problems.

The BSMC's medical supply element provides class VIII support to the BSMC and battalion aid stations.

Medical Support

Ref: ATP 4-90, Brigade Support Battalion (Jun '20), p. 8-1.

Medical support to maneuver forces is provided by both organic medical assets and echelons above brigade assets providing direct or general support to the BCT. The focus of the medical effort at the BCT-level and below is to rapidly locate, acquire, treat, stabilize, and evacuate patients. The brigade support medical company provides Role 1 and Role 2 medical support to all BCT units operating in the brigade AO. The company also provides Roles 1 and 2 medical support on a coordinated and synchronized area basis to units operating in or traversing through its AO that do not have organic medical assets. This chapter provides an overview of medical support at the BCT level and describes the role, organization, employment, and operations of the BSMC as well as other medical units that a BCT may interact with on the battlefield.

Medical support includes all services performed, provided, or arranged to promote, improve, conserve, or restore the mental or physical well-being of personnel, which include preventive and curative health measures, evacuation of wounded, injured, or sick; selection of the medically fit and disposition of the medically unfit; blood management; medical supply, equipment, and maintenance thereof; combat and operational stress control; medical, dental, veterinary, laboratory, optometry, nutrition therapy, and medical intelligence services.

In order for medical support to be successful for the BCT, several medical planners across the brigade must synchronize and oversee the organization, employment, and operations of medical units across the BCT's AO. These include the brigade surgeon section, BSB SPO medical planners, BSMC commander, and task force surgeons and medical platoon leaders.

The focus of the medical effort at the BCT level and below is to rapidly locate, acquire, treat, stabilize, and evacuate patients. Medical personnel treat and stabilize patients for medical evacuation to a medical treatment facility staffed and equipped to provide the required medical treatment or returned to duty within 72 hours. Roles 1 and 2 medical assets also support force health protection measures to avoid or mitigate the adverse effects of disease and non-battle injuries and promote the health of deployed forces.

ATP 4-90, chapter 8 further describes the operations of the brigade support medical company, medical operations in decisive action, the operations of the BSMC, and other medical units involved in the medical operations to support the BCT.

D. Forward Support Companies (FSCs)

Ref: ATP 4-90, Brigade Support Battalion (Jun '20), pp. 1-15 to 1-16.

The role of the FSC is to provide logistics in direct support to maneuver, fires, and engineer battalions. The FSC provides the supported commander with dedicated logistics assets organized to meet the battalion's requirements. An FSC provides field feeding, bulk fuel, general supply, ammunition, and field-level maintenance to the maneuver battalion.

The FSCs provide the brigade, battalion, and BSB commanders flexibility for providing logistics support. FSCs provide the BSB commander the ability to task organize and prioritize the logistics effort to support large-scale combat operations.

Forward Support Company

- **Role**: The forward support companies (FSC) provides logistics in direct support to its specific supported battalion.
- **Capability**: The FSC provides field feeding, bulk fuel, general supply, ammunition, and field-level maintenance in direct support of a supported battalion.
- **Parent**: Brigade support battalion.
- **Command relationship**: Organic to a brigade support battalion; may be attached or OPCON to its supported battalion for a limited duration.
- **Support relationship**: Direct support to its specific maneuver battalion; general support to other units in the BCT; general support to others on a limited basis by exception.
- **Span of operations**: Brigade combat team area of operations from the BSA to the forward line of troops.

The FSCs are organic to the BCT's BSB and may be attached to or under operational control (OPCON) to the supported battalion for a limited duration; a mission or phase of an operation. The decision to establish these types of command relationships is made by the brigade commander upon the advice of the BSB commander after thorough mission analysis. FSCs are also located in aviation support battalions and special forces group support battalions.

The FSC commander, supported by subordinate leaders, uses troop-leading procedures to plan, prepare, and execute the BSB commander's intent. The company commander also uses the procedures to assess the effectiveness of the company plan and adjusts the plan as required. The FSC command relationship determines the organization that issues orders to the FSC. If the FSC is attached to the maneuver battalion, that battalion will issue the order. If the FSC is not attached, the BSB issues the order.

The FSCs in the various BCTs are structured similarly with the most significant differences being maintenance capabilities and bulk class III distribution capability. The FSC depends upon the brigade support battalion or its supported battalion if attached or OPCON for the support areas. In addition to those already highlighted the field maintenance and distribution companies provide training and senior warrant officer/noncommissioned officer (NCO) mentorship for low-density military occupational specialties in the FSC. As the FSC's command relationship or support relationship shifts from the BSB to the supported battalion, planners address, agree upon, and formalize these dependencies at the brigade level.

- Human resources support.
- Religious support.
- Common operational picture, also known as COP, input.
- Combat information and intelligence.

- Roles 1 and 2 medical support.
- Water distribution.
- Reporting requirements.

The FSCs have a headquarters section, a distribution platoon, and a maintenance platoon. The headquarters' food service section provides class I support, food service, and food preparation for the company and its supported battalion. The food service section prepares, serves and distributes the full range of operational rations.

The distribution platoon consists of a platoon headquarters and four squads that can be task organized to distribute classes II, III, IV, V, and VII. The distribution platoon leader and platoon sergeant execute all or a portion of the 14 materiel management functions as required to ensure supply stocks are available and ready for distribution. These personnel also execute both distribution integration and transportation operations functions, as required, to ensure supplies are assigned a transportation mode, route, and movement time.

FSC maintenance platoons vary based upon the equipment and major weapon systems of the supported battalion. Generally, the maintenance platoon of an FSC consists of a platoon headquarters, maintenance control section, field maintenance section, service and recovery section, and field maintenance teams.

Methods of Distribution

Ref: ATP 4-90, Brigade Support Battalion (Jun '20), p. 6-2.

In the BCT, distribution builds and maintains combat power through the delivery of supplies, personnel, and equipment to extend the operational reach of maneuver forces, maintain freedom of action, and prolong endurance. Sustainment units select a distribution method appropriate to the mission, tactical situation, the supported unit's priority of support, time/distance, and other factors of mission and operational variables. There are two methods of distribution:

- **Unit Distribution.** Unit distribution is the routine distribution method the BSB uses to support the BCT. Unit distribution is a method of distributing supplies by which the receiving unit is issued supplies in its own area, with transportation furnished by the issuing agency (FM 4-40). In unit distribution, logisticians organize supplies in configured loads and deliver supplies to one or more central locations. Supply personnel can create unit load configurations to resupply specific battalion, company, or platoon-sized elements depending on the level of distribution needed and mission variables. Unit distribution maximizes the use of the BCT lift capacity of its transportation assets and minimizes the delivery and turnaround time.
- **Supply Point Distribution.** Supply point distribution is a method of distributing supplies to the receiving unit at a supply point. The receiving unit then moves the supplies to its own area using its own transportation (FM 4-40). Supply point distribution requires unit representatives to move to a supply point to pick up their supplies. Units most commonly execute supply point distribution by means of a LRP. In each method of distribution, there are multiple techniques for the distribution of supplies, personnel, and equipment. Examples of techniques for supply point distribution used by units are refuel on the move (ROM), pre-positioned supplies.

Throughput Distribution. Throughput distribution is a method of distribution that bypasses one or more intermediate supply echelons in the supply system to avoid multiple handling (ATP 4-11). The BSB or a DSSB may execute throughput distribution in the BCT's area of operations when needed. An example of throughput distribution would be the BSB's distribution company bypasses the FSC to distribute supplies directly to maneuver units.

ATP 4-90, chapters 6 and 7 describe the distribution and maintenance operations of forward support companies as well as other units conducting these operations to support the BCT.

Brigade Support

V. Aviation Support Battalion (ASB)

Ref: ATP 4-90, Brigade Support Battalion (Jun '20), pp. 1-20 to 1-22 and FM 4-0, Sustainment Operations (Jul '19), p. 2-56.

The aviation support battalion (ASB) is organic to the Combat Aviation Brigade (CAB), the expeditionary CAB, and the theater aviation brigade. It consists of a HSC, a distribution company, an aviation support company, and a brigade signal company. The ASB provides aviation and ground field maintenance, ground and aviation recovery, network communications, resupply, and Role 1 medical support for the CAB. The ASB provides maintenance augmentation to aviation battalions when required The ASB supports the CAB's forward support companies/troops, aviation maintenance companies/troops, the brigade's HHC, and the unmanned aircraft systems, if applicable. The ASB provides distribution of classes I, II, III (B), IV, V, and IX. It provides water storage and operates both MATP and supply support activity.

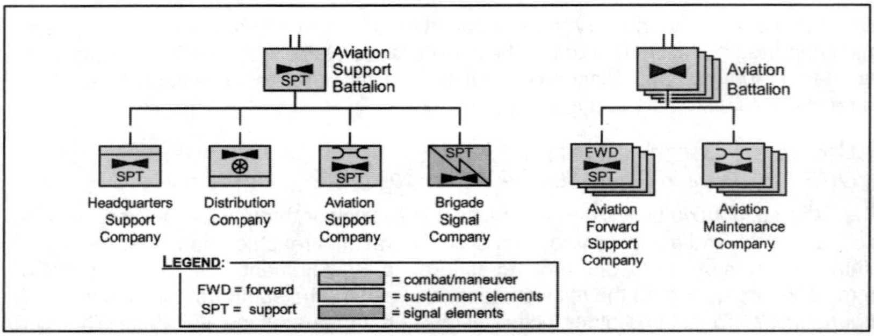

Ref: FM 4-0, fig. 2-34. Aviation support battalion.

The ASB consists of four companies: headquarters and support company, distribution company, aviation support company (ASC), and a brigade signal company. The ASB is configured differently from other BSBs; the ASB does not have an assigned medical company but has a medical platoon that provides Role 1 medical care to the ASB and its units. Additionally, the forward support companies and aviation maintenance companies are organic to aviation brigade's battalions and not the ASB.

There are three primary differences between a BSB and an ASB:

- The brigade medical support company becomes a medical company, air ambulance in the GS aviation battalion in the same brigade. The company has significantly less medical capability than a typical brigade medical support company.
- The forward support companies are distributed to helicopter battalions in the brigade.
- A signal company is added, similar to a MEB.

The ASB performs the following tasks:

- Ground vehicle and aviation maintenance and recovery operations.
- Signal and network security to the CAB for command and control.
- Aviation and ground sustainment operations for the aviation brigade.
- Distribution management operations within the aviation brigade.
- Role 1 Army health service support for the aviation brigade.

Additional sustainment to aviation battalions and squadrons is provided by an organic aviation maintenance company and FSC.

Refer to FM 3-04 more information on aviation sustainment.

Brigade Support

Chap 5

II. Brigade Support Area (BSA)

Ref: ATP 4-90, Brigade Support Battalion (Jun '20), chap. 4.

I. BSA Operations

The BSA is a designated area in which sustainment elements locate to provide support to a brigade. The BSA typically encompasses a unit base or base cluster, landing/pickup zones, and field trains elements.

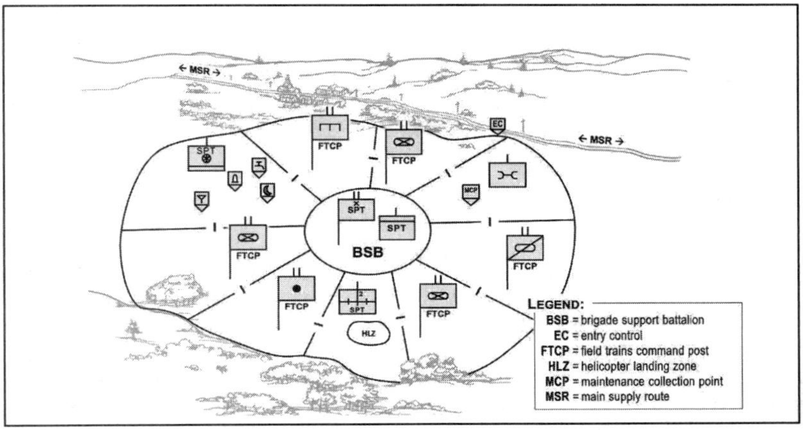

Ref: ATP 4-90, fig. 4-1. Example of a notional brigade support area (BSA) layout.

The BCT subordinate units that normally occupy a BSA are the BSB and the brigade engineer battalion. The brigade commander must determine which of these subordinate units is responsible for controlling the BSA. BSA control requires performing area security and stability tasks, employing and clearing fires, and controlling airspace. The unit designated to control the BSA will require BCT staff augmentation to control fires and airspace.

The BCT commanders designate close, deep, support, and consolidation areas to describe the physical arrangement of forces in time and space. The commander must designate a close area and a support area for every decisive action operation. They designate a deep area and consolidation area as required.

Support Area

A support area is the portion of the BCT commander's AO designated to facilitate the positioning, employment, and protection of sustainment assets required to sustain, enable, and control operations. BCTs assign a BSA to the BSB.

Consolidation Area

The consolidation area is the portion of the commander's area of operations designated to facilitate the security and stability tasks necessary for freedom of action in the close area and to support the continuous consolidation of gains. If the BCT AO is designated a division consolidation area the BSB still operates from a BSA. The focus of sustainment support over time is likely to shift from offensive and defensive operations to stability operations as enemy forces are defeated and the security situation improves.

II. BSA Establishment and Occupation

Ref: ATP 4-90, Brigade Support Battalion (Jun '20), pp. 4-2 to 4-3.

Site Selection

Many factors govern BSA site selection and all should be considered when establishing the BSA. The BSA is normally near or in close proximity to an MSR. It is large enough to allow adequate space for unit occupation and to execute sustainment operations but not too large to hinder effective security and control. If line-of-sight communications are required, the site terrain must be conducive to it.

Commanders evaluate the worthiness of a site with respect to mission accomplishment and then consider camouflage, concealment, and survivability. Ideally, the BSA is out of the range of the enemy's medium artillery. BSB planners consider trafficability and soil composition when selecting a BSA location.

Dispersion requirements often dictate the size of a site. A site has limited usefulness if it will not permit enough dispersion for survivability and effective operations. Support assets from a DSSB should be able to maneuver through the traffic pattern without causing unnecessary massing of vehicles.

The establishment and occupation of a BSA is deliberately planned and executed. During initial planning, the BCT staff and the BSB staff perform a map reconnaissance of the proposed BCT AO. During this reconnaissance, the staffs identify the area for the initial BSA, planned base location(s) in the BSA, and unit occupation of the base(s). All units that will occupy the BSA, to plan initial establishment and occupation of the BSA, use this information. Similar actions occur for BCT movement and subsequent BSA establishment and occupation.

Upon arrival at the BCT AO, the BSB uses quartering party operations for initial occupation. The quartering party is key to the initial establishment of the BSA. The purpose of the quartering party is to verify the site selection of the BSA and make limited preparations for receiving units that occupy the BSA. It consists of representatives from the BSB's S-3, S-2, and SPO sections. The quartering party is typically a small portion of each unit empowered by its commander to establish locations for personnel and equipment. Depending on units to be located in the BSA, the quartering party may also contain elements of the BCT CP, each maneuver battalion's field trains, and attached units located in the BSA. If a single base is used to contain all units in the BSA, the quartering party locates that base position. If a base cluster is used, the quartering party locates each base position. The arrival of the quartering party is the first opportunity to see the terrain and make adjustments to the BSA layout and defenses as necessary.

On arrival at the BSA site, the quartering party begins its priorities of work. Priority of work is a set method of determining the precedence of tasks when establishing a new location and conducting a defense of a location such as—the BSA. A unit's SOP will dictate the exact steps in that specific BSB's priorities of work. The commander may change priorities based on the situation and mission variables. Although listed in sequence, the BSB may perform several tasks in their priorities of work at the same time. An example priority of work sequence is—

- Establish local security.
- Check for CBRN contamination and unexploded ordnance.
- Position vehicles, crew served weapons, and Soldiers; assign sectors of fire.
- Establish communications.
- Position other assets, for example CPs.
- Designate final protective fires.

- Clear fields of fire and prepare range cards and sector sketches.
- Prepare fighting positions.
- Emplace obstacles.
- Identify artillery target and direct fire control measures (day/night).
- Improve primary fighting positions with overhead cover.
- Prepare alternate and supplementary positions.
- Establish observation posts and listening posts.
- Adjust positions and control measures as required.
- Assess ammunition, food, and water stockage levels.
- Reconnoiter surrounding area.
- Prepare commodity supply points as necessary.
- Establish a sleep and rest plan.
- Continue to improve positions.

The quartering party establishes initial communications to begin the transfer of C2 from a tactical assembly area, intermediate staging base, or previous BSA to the new BSA location. The quartering party establishes a tactical CP. A tactical command post is a facility containing a tailored portion of a unit headquarters designed to control portions of an operation for a limited time (FM 6-0). Commanders employ the tactical CP as an extension of the main CP. The functions of a tactical CP typically include the following—

- Controlling the overall unit's operations for a limited time when the main CP is displacing or otherwise not available.
- Controlling a specific task within larger operations such as—the establishment of a BSA, a gap crossing, a passage of lines, a relief in place, or air assault operations.
- Providing a forward location for issuing orders and performing rehearsals.
- Controlling decisive operations or specific shaping operations.
- Performing short-range planning.
- Contributes to future operations planning.

The tactical CP also continues to plan the positioning of units in the BSA and begins to outline the base perimeter in more detail than original plans. Representatives of maneuver battalion field trains and other units begin preparations for occupation, which include selecting sites for crew-served weapons. The tactical CP reports to the BSB CP when it is prepared to begin operations. It also relays any information the commander will need to change movement plans.

After the quartering party, the commander divides the other elements of the BSA into serials to perform a tactical road march to the new site. The battalion S-3 divides these vehicle movements into the advance party, a number of main body movements, and a trail party. The main body begins the move in accordance with the operations order issued by the BSB main CP and the unit's tactical SOP. The BSB S-3 plans the movement of the BSA's main body echelons by serial. The BSB planners should not include an entire BSB subordinate company's sustainment capability in a single serial since loss of a serial would eliminate all of the BSB's capability in a functional area. In addition, if an entire company is moving at the same time, it will have difficulty maintaining continuity of support. Instead, each BSB company moves by echelon.

III. BSA Layout

Ref: ATP 4-90, Brigade Support Battalion (Jun '20), pp. 4-3 to 4-5.

The BSB commander, brigade engineer battalion commander, and the BCT S-3 collaborate to determine if all elements in the BSA occupy a single large base or a cluster of smaller bases. The commanders must consider the advantages and disadvantages of each option. Terrain features, terrain patterns, natural concealment, and soil composition must be considered from practical and security points of view. Bases must be located in areas suitable to execute sustainment operations as well as being defensible. A key factor to consider is whether operations will disturb the terrain pattern enough to make it discernable to enemy surveillance. The goal is not to disturb the terrain pattern at all. The BSB and brigade engineer battalion commander will develop an informed recommendation and the commander designated to control the BSA will present the recommendation to the brigade commander who ultimately decides on the best option to support operations.

Single Base Operations

If a single large base is used, the unit designated to control the BSA will control the base. That unit commander will determine unit placement within the base. The BSB HHC and BSB A, B, and C companies occupy the base with the brigade engineer battalion and its subordinate elements. In addition, elements from the BCT staff, the brigade alternate CP, signal assets, field trains, and sustainment units from higher headquarters may occupy the base. Depending on mission variables, there could also be elements from Army Materiel Command present. Units occupying the BSA will change in accordance with the BCT mission and task organization. The brigade commander may place attached elements in the BSA during certain phases of an operation or for its duration.

See fig. 4-1 on p. 5-13 for an example notional brigade support area (BSA) layout.

A single base is advantageous in terms of simplified C2 and perimeter security. However, a single base containing the number of units that reside in the BSA will be very large and easily detected by enemy reconnaissance. A single base simplifies targeting and attack by enemy artillery, attack aircraft, and ground forces. It also puts a significant portion of the BCT sustainment and support structure at risk from a single attack.

Locations of elements in the base will vary depending on mission variables. The BSB commander and S-3 use their best judgment in positioning BSB units and assigning sectors for security of the BSB portion of the base. Troop safety guidelines also influence unit placement. The MATP is often outside the base due to net explosive weight distance factors.

In addition to staff elements and units, there are multiple supply points associated with the base. The distribution company has an SSA, a fuel point, class I break bulk point, and an MATP in addition to a convoy staging area. The field maintenance company may establish a MCP, and the BSMC establishes its Role 2 medical treatment facilities. If the sustainment brigade provides transportation or water purification support, those units will also need space for a water point and truck parking in or adjacent to the BSA.

Some considerations to locating units and commodities within the base—

- Make supply points accessible to both customers and resupply vehicles and helicopters. Keep class III points away from supplies and at least 100 feet from water.
- Locate the medical treatment areas away from likely target areas (such as the MATP, class III point, and road junctions).
- Ensure evacuation routes and an open area for landing air ambulances are readily accessible by medical personnel.
- Position the MATP near, but off the MSR, so that resupply vehicles bringing ammunition into the area do not block the MSR. The MATP requires sufficient area to perform transload operations without interfering with traffic.
- Establish separated entry and exit entry control points if possible to control the flow of traffic for the base.

- Position CPs near the center of the base for C2, data and voice connectivity, and security reasons.
- Position units with greatest firepower (such as the maintenance company) along the most threatening avenues of approach.

BSB commanders and S-3 sections shift security responsibilities as tenant units enter and leave the base. The BSB command post standard operating procedures cover the overall base layout with and without the presence of FSC personnel and the BCT's battalion field trains command posts.

Base Cluster Operations

If a base cluster is used, the commander designated with the control of the BSA executes terrain management to locate a base for each element in the BSA and designates a unit responsible for controlling each base. The commander controlling the BSA does so from the battalion main CP. Security, communications capability, proximity to road network, and other factors influence base placement.

A base cluster dispersion complicates enemy detection and targeting. A base cluster spreads sustainment and support assets over a larger geographic area that minimizes the effects of artillery, air, or ground attack. However, a base cluster complicates C2 and perimeter security. Each base requires perimeter defense to allow adequate and mutual protection of other bases. This is accomplished while minimizing the likelihood of striking an adjacent base with small arms fire. Base cluster considerations are—

- Coordinating with tenant and transient units/organizations, subordinate base clusters, adjacent base camps, and higher headquarters.
- Reasonable span of control based on the number and echelon of tenant and transient units or subordinate base clusters.
- Transporting personnel, equipment, supplies, and waste in the bases and between base camps as part of a base cluster as part of base camp functions, services, and support.

A base cluster often lacks a well-defined perimeter or established access points. Although individual bases in the cluster maintain perimeter security, entry, and access control, security requires more personnel than with a single base.

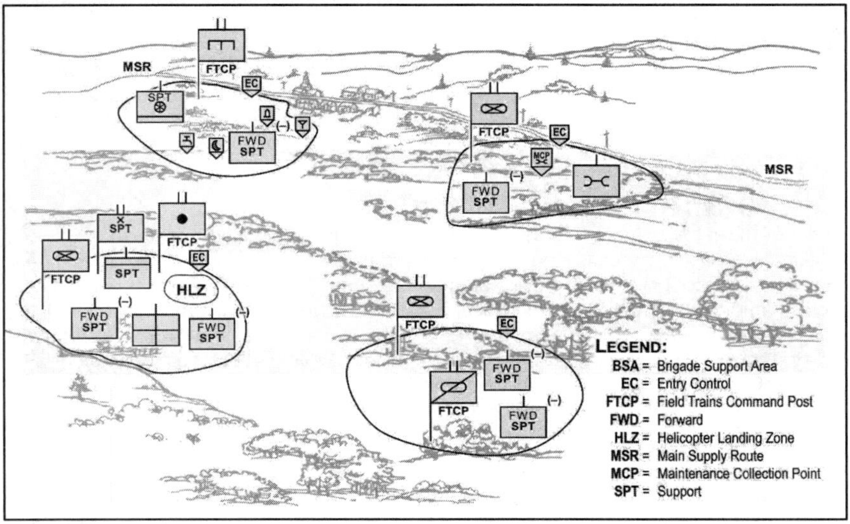

Ref: Figure 4-2. Example of a notional brigade support area (BSA) layout using a base cluster.

IV. BSA Security

Ref: ATP 4-90, Brigade Support Battalion (Jun '20), pp. 4-5 to 4-12.

Security operations undertaken by a commander provide early and accurate warning of enemy operations. The unit uses this to react to the enemy and to develop the situation to prevent surprise. In the BSA, units execute local and area security.

Area security neutralizes or defeats enemy operations in a specified area. It includes counter-reconnaissance and security of personnel, airfields/landing zones, facilities, MSRs, lines of communication, equipment, and critical points.

The BSB is responsible for its own local security against level I and II threats. The BCT may provide a tactical combat force to augment the BSB for defense against level III threats. When the BEB is collocated with the BSB in the BSA both units coordinate defense responsibilities under the lead of the overall designated commander of the BSA

Local security actions are those that are an inherent part of self-protection and mission assurance measures. Local security consists of base perimeter security, manning observation and listening posts, local security patrols, access control, barriers, a reserve of personnel to augment perimeter security, and other measures to provide security for a base. Every unit assigned a base or collocated on a base with another unit has local security responsibility.

The BSB commander is responsible for securing the base (or bases if a base cluster is used) to which it and its subordinate units are assigned. Area security measures are necessary for convoys and other logistics sites and actions that do not occur on a base including MATP or FARP operations. Continuous mission accomplishment is critical despite security operations.

As the threat increases, the BSB commander may stop sustainment support to adequately protect personnel and equipment. Failure to provide adequate protection may cause personnel and equipment losses to be so significant that they prevent sustainment support to the BCT. The BCT commander and the BSB commander must have previously discussed what risks are reasonable to accept and what risk mitigation measures to implement based on requirements and priorities.

See related discussion of protection considerations on pp. 6-31 to 6-38.

Level I Threat

The BSB commander ensures logistics missions and associated activities continue without restriction and that all logistics units can perform protection operations against a level I threat. A level I threat is a small enemy force that can be defeated by those units normally operating in the echelon support area or by the perimeter defenses established by friendly bases and base clusters.

Level II Threat

A level II threat is an enemy force or activities that can be defeated by a base or base cluster's defensive capabilities when augmented by a response force. A typical response force is a military police platoon with appropriate supporting fires or a larger combined arms maneuver element. Level II threats consist of enemy special operations teams, long-range reconnaissance units, mounted or dismounted units, and bypassed combat units. Typical objectives for a level II threat include the destruction and disruption of friendly C2 nodes and logistics facilities. Level II threats; interdict friendly lines of communications. A level II threat may have a measurable effect on sustainment accomplishment.

Level III Threat

A level III threat is an enemy force or activities beyond the defensive capability of both the base and base cluster and any local reserve or response force. Possible objectives for a level III threat include seizing key terrain, interfering with the movement and commitment of reserves and artillery, and destroying friendly combat forces. Its objectives include destroying friendly sustainment facilities, supply points, command post facilities, airfields, aviation assembly areas, arming and refueling points, and interdicting lines of communications and major supply routes. A level III threat is capable of causing sustainment mission failure.

Although the threat is described by levels (level I, II, and III threats) as a planning guide, these threat levels do not restrict responses.

Components of BSB Security

The BSB commander and S3 analyze the terrain in detail from all perspectives and then verify on the ground to select engagement areas and positions that allow for the massing of fires and the concentration of forces on likely enemy avenues of approach. Emphasis is on preparing and concealing positions, routes, obstacles, logistical support, and C2 facilities and networks. The BSB commander plans, coordinates, and uses rehearsals to ensure subordinates understand the base defense concept of operations. To be effective, a BSA defensive plan must include four components:

1. Protect the Base

The BSA requires continuous protection from occupation until displacement. Properly designed perimeter security is the base's first line of defense. Perimeter security is designed to incorporate layered defense in depth and integrate security elements including: cleared fields of fire, interlocking fires, a final protective line, barriers, surveillance, and access control. Units organize a perimeter defense to accomplish a specific mission (such as protecting a base or providing immediate self-protection). Except for very small, dense BSAs no true perimeter defense is possible. The BSB depends upon early warning, key defensive positions, and a quick reaction force.

2. Detect the Enemy

Detection includes discovery of enemy forces in the immediate vicinity and knowing as much about their positions and intentions as possible. Detection is critical in achieving timely response to an enemy threat and disrupting the threat quickly before it affects sustainment operations. Units on the BSB base employ early warning systems, trip flares, and listening posts/observation posts in sufficient quantities to provide adequate coverage of the base perimeter. The BSB assigns areas around the base for random reconnaissance patrols. Active patrolling, unit standard operating procedures, and continuous reconnaissance are active measures that help provide detection. The BSB also employs its chemical detection equipment as part of its detection and base defense plan.

3. Disrupt the Enemy

The BSB defensive plan is structured to disrupt an attacking enemy's progress after detection and warning to allow time for the base defense forces to react. Disruption also allows time for mobilization of a tactical combat force for reinforcement if necessary. Defending commanders use all available means to disrupt enemy forces. Effective small arms fire will disrupt enemy activity and delay its actions. Commanders disrupt attackers and isolate them from mutual support to defeat them. Repositioning forces, aggressive local protection measures, and employment of obstacles, indirect fires, and ambushes combine to disrupt the threat of an attack.

4. Enemy Destruction

Once the base defenses have detected and disrupted the enemy force, they must destroy the enemy. Well-disciplined, well-aimed, and concentrated small arms and grenade fire can destroy an enemy. Preplanned or adjusted indirect artillery or mortar fire is also extremely effective. Commanders and leaders are conscious of the proximity of the enemy and notify the fire direction center if the fires will be danger close. If the BSB cannot destroy the threat with its internal capabilities and quick reaction force, the BSB coordinates with the BCT for external support. If the threat exceeds the available BSA defensive assets, the BSB's preplanned defensive measures must delay the enemy force until reinforcements from the BCT can destroy the enemy.

V. Displacement of the BSA

The BSB commander must position BSB units close enough to its supported brigade to maintain responsive sustainment. As such, the BSB will have to displace its operating base or base cluster frequently. This is particularly true during execution of large-scale combat. The forward movement of the BCT battalions and BCT commander's movement of the BCT rear boundary will largely dictate when the BSB must displace. The BSB commander may move the base within the BSA boundaries even if the BCT rear boundary or BSA boundaries do not move. This keeps an effective proximity with the BCT units and increases survivability by minimizing time in a single location. The BSB plans base displacement immediately after occupying a new site. Units follow the displacing procedures in their unit's tactical SOP.

Considerations as to why a BSB would displace and reestablish the BSA—

- Lines of communication stretched where FSCs could no longer support their maneuver battalions with a local haul.
- Lack of ability to execute planned and emergency LOGPAC from BSA to combat trains and return in one trip.
- Distance away from its supporting DSSB where DSSB could not support the BSB with a local haul LOGPAC.
- Inability for the BSB to provide uninterrupted support to the BCT and changing operational requirements.
- Persistent chemical threat to the BSA.
- Continued degradation of logistics capability due to enemy indirect fire.
- Inability of the field maintenance company to provide additional field-level maintenance and recovery support to the FSCs.
- Degradation of FSC combat power, necessitating additional throughput from the BSB distribution company.
- Ground lines of communication make ambulance exchange point operations difficult to execute in a timely fashion.

The BSB commander, with close coordination from the BCT commander and BCT S-3, determines when to displace the BSA. Both commanders and staff understand how the displacement of the base will disrupt BCT support operations and medical care. The disruption may be in terms of time, capability, or a combination of both.

The BSB S-3 ensures the displacement of the BSB is coordinated with all supported units, subordinate units, and supporting units. The BSB SPO may coordinate for the DSSB to provide reinforcing support to BCT units during the BSB displacement. The BSB makes all units, and most importantly the supported BCT, aware of when support operations will cease at the existing BSB base location, the location of the new BSB base, when operations will begin at the new BSB base site, and location of a forward logistics element (FLE) supporting the displacement.

The BSB does not have sufficient organic transportation assets to move 100% of its personnel and organic equipment in one lift. Other elements located in the BSA also lack sufficient transportation. This lack of organic BSB transportation may necessitate the further echeloning of sustainment capabilities during displacement. Additionally, downloaded supplies at BSB supply points and disabled equipment at the field maintenance company's MCP place additional demands on transportation. Disabled equipment the field maintenance company cannot repair immediately may need to wait for evacuation to the new BSB site. The BSB maintains supplies uploaded for rapid mobility. For all transportation requirements beyond the BSB's capability, the SPO requests additional support from the sustainment brigade.

Chap 5

III. Echeloned Sustainment

Ref: ATP 4-90, Brigade Support Battalion (Jun '20), chap. 5.

Supporting the BCT is a carefully planned and executed process. The BSB must plan for and synchronize echelon support. Echeloned sustainment is an array of capabilities placed at critical locations to link and facilitate support between echelons in an area of operation. At the lowest level, echeloned sustainment is executed at the company and battalion level. This chapter provides an overview of how the BSB arrays forces and echelons sustainment across the battlefield to support the brigade through the use of field, combat, and unit trains.

I. Brigade Echelon

BSB unit positioning, including those integrated into maneuver battalions, is critical to timely sustainment support. Using echeloned sustainment is a technique to ensure sustainment support is executed effectively. The purpose is to ensure a clear line of communication between echelons and that adequate support is positioned to allow timely support. Sustainment planners echelon assets to weight the effort supporting mission requirements. Echeloned sustainment provides relatively seamless support, extends operational reach, and allows commanders to maintain momentum.

Echeloned sustainment is a deliberate, collaborative effort of the BCT headquarters and its battalions based upon thorough mission analysis. The leadership and staffs at the brigade, battalion, and company levels determine how to echelon the forces. It is not a unilateral decision by a single commander since the decision will affect the effectiveness of support across the entire brigade operation. Leaders at all levels thoroughly understand the capabilities of each support organization and element. This is necessary to ensure utilization and placement of each. Commanders of all brigade organizations understand the manner in which support is echeloned will vary widely based upon mission variables. It may even vary for each battalion during a single operation based on the commander's priority and mission variables.

The BSB is organized to facilitate echeloned support with the FSCs serving as a key element. The FSC is specifically designed and dedicated to link the BSB support to the maneuver battalion and have the requisite capability to execute the support. The FSC eliminates the need for the BSB to develop ad hoc support elements or arrangements. The FSCs are an integral part of the BCT trains.

II. Trains

The BCT uses the trains approach to establish an echeloned sustainment support structure; field trains, combat trains, and company/troop trains. Trains are a grouping of personnel, vehicles, and equipment that provide sustainment to the battalions and subordinate companies of the BCT. The trains are arrayed across the area of operations and link supported units to the BSB and EAB units for support. The BCT S-4, BSB and maneuver battalion commanders and staff, FSC and maneuver company commanders and XOs collaborate to determine the best method of trains employment appropriate for the BCT's sustainment concept of support and to meet the brigade commander's intent. The commanders and staffs allocate personnel to each train and command posts within the trains. Echeloning of support can include elements of the battalion or squadron aid station, S-1 section, S-4 section, and the FSC.

See following pages (pp. 5-22 to 5-23) for an overview and further discussion.

A. Field, Combat, and Company Trains

Ref: ATP 4-90, Brigade Support Battalion (Jun '20), pp. 5-2 to 5-4.

Figure 5-1 depicts an example of echeloned sustainment using field, combat, and company trains including templated distances between locations. The distances in the diagram are not prescribed distances but are depicted to display a typical range of distances commanders may expect and are for planning purposes only. Distances will vary from operation to operation and units must be located for maximum effectiveness. Sustainment and BCT planners and leaders consider operational and mission variables when locating units and the impact the distances have on sustainment support. Distribution platform capability and convoy security are considerations when determining distances.

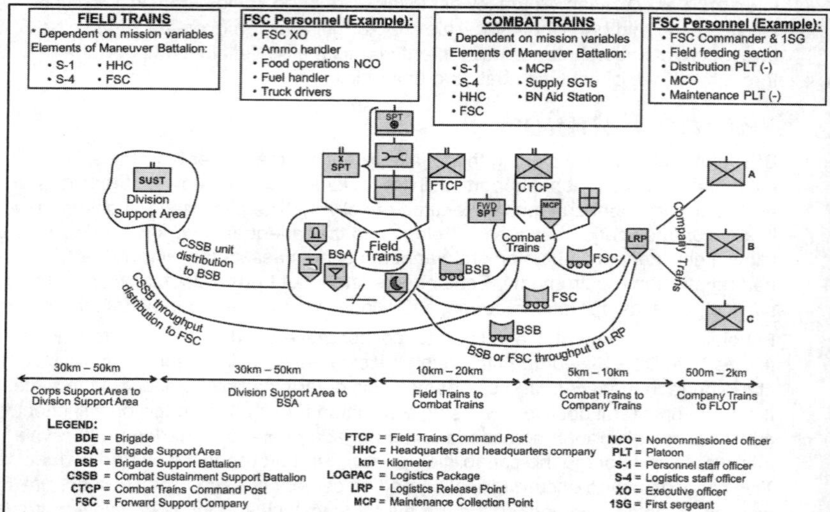

Ref: ATP 4-90, fig. 5-1. Example of echeloned sustainment using field, combat, and company trains.

Field Trains

Each maneuver battalion/squadron normally task organizes a field trains element to provide a centralized location for controlling battalion sustainment support. The field trains consists of a field trains command post (FTCP), battalion sustainment personnel, battalion sustainment vehicles, and supporting FSC personnel located near the BSB headquarters. The FTCP serves as the battalion or squadron commander's primary direct coordination element with the supporting BSB in the BSA. The battalion HHC commander resides in the FTCP to provide command presence. FTCP may also contain a maneuver battalion/squadron S-4 representative, a maneuver battalion/squadron S-1 representative, and maneuver battalion/squadron supply sergeant or representative. Field trains provide direct coordination between the maneuver battalion and the BSB. The field trains are positioned based on mission variable considerations but are normally collocated on the BSB base. The maneuver battalion's HHC (company, battery, or troop) commander or designated representative (such as an XO or first sergeant) can control the field trains, which include battalion or squadron sustainment assets not located with the combat trains.

The FSC places personnel in the field trains that can facilitate the distribution integration functions to ensure timely distribution of all supply classes to the supported battalion.

The FSC commander resides at the FTCP during periods when the battalion HHC commander cannot be present. If the FSC commander is in the FTCP, the FSC XO must move to the FSC location at the combat trains command post (CTCP). Only a small number of FSC personnel work in the FTCP. Too many FSC personnel in the FTCP detracts from the FSC capability. Other FSC personnel may consist of the FSC XO and/or first sergeant, and supply personnel. Their primary function is to communicate maneuver battalion support requirements to the BSB SPO officer and to coordinate for transportation to distribute supplies. The majority of the FSC support capability should remain in the combat trains, closer to the supported battalion when possible.

Combat Trains

Each maneuver battalion task organizes a combat trains element in the maneuver battalion operations area. The combat trains provide a centralized location for battalion and FSC sustainment personnel. Combat trains consist of a CTCP, the battalion or squadron's HHC first sergeant, S-1, S-4, battalion aid station, MCP, company supply sergeants or representatives. The FSC typically positions its commander, first sergeant, field feeding section, the distribution platoon, MCO, and the maintenance platoon in the combat trains. Units consider the mission variables when locating combat trains for the battalion or squadron.

The CTCP plans and coordinates sustainment operations to support tactical operations. The CTCP coordinates and executes sustainment operations between the FTCP and company trains. The CTCP serves as the focal point for all administrative and logistical functions for the battalion or squadron. The CTCP may serve as an alternate command post for the battalion or squadron main command post when necessary. The battalion or squadron S-4 usually serves as the CTCP officer in charge, and the MCO usually serves as the MCP officer in charge.

The FSC and HHC commanders for the maneuver battalions may split locations and responsibility. The HHC commander has primary responsibility for the FTCP with the battalion S-4's assistance for example. The FSC commander is responsible for the CTCP. The FSC commander provides guidance to the battalion/squadron S-4 officer in charge. The HHC and FSC commanders change locations as required to execute support to the maneuver unit. One commander will be present in each location at all times. The BCT and BSB commanders will place assets and leaders in trains locations that best support their mission, sustainment concept of support, and concept of the operation.

The FSC positions the MCP in the combat trains where recovery vehicles have access to easy ingress and egress routes and a large enough area for maintenance operations. The combat trains remains mobile enough to support frequent changes in location to maintain pace with the maneuver battalion/squadron. The CTCP may also change location, time and terrain permitting, when heavy use or traffic in the area may cause detection or security is compromised.

The FSC commander considers dispersion when arraying forces and capabilities in the combat trains. The FSC commander considers the additional security and communication requirements that dispersion requires in the combat trains. Elements located outside the CTCP base require adequate protection and security capability.

Company, Battery, and Troop Trains

Maneuver companies, batteries, and troops establish trains that, for simplicity in the remaining sections, will be referred to as company trains. As with field and combat trains, company trains consolidate support personnel in a single location to facilitate unit resupply and medical support. The commander determines the composition of the company trains that may include the first sergeant, supply sergeant, unit medic, and assorted vehicles. The FSC commanders may also include FSC field maintenance teams (FMTs) and field feeding teams in the company trains. BCT companies have very limited transportation assets for supply distribution. Companies, batteries, and troops have limited transportation assets, usually a single 5 -ton truck. They rely upon the FSC for bulk distribution of class III and class V.

B. Forward Support Company (FSC) Employment

An FSC supports each of the BCT's battalions. The FSC is a critical component in echeloned sustainment and allows the BSB and BCT commanders to weight the logistics effort for the BCT. This section provides additional considerations for FSC employment.

See pp. 5-10 to 5-11 for discussion of the role and organization of the FSC.

The BSB commander retains the ability to surge, mass, and reallocate logistics capabilities in accordance with the BCT commander's intent and the mission variables. The primary asset that allows the BSB commander to accomplish this is the FSC. FSCs are organic to the BSB. However, the FSC may be attached to their respective supported battalion for limited duration missions or phases. The BSB commander advises the brigade commander on establishment of these types of command relationships after thorough mission analysis. The attachment of FSC to maneuver battalions limits the BSB commander's flexibility and affects the entire BCT's operations. Regardless of the command relationship, the FSC is positioned where it provides responsive support to the maneuver battalion.

BSB and BCT commanders and staffs must also consider the FSCs' force structure when weighting BCT's operations. For example, in the absence of a maintenance surge team, all specialized mechanics for the M1 Abrams, M2/M3 Bradley, Stryker vehicles, and M109A6 Paladin reside solely in the FSCs. If a maintenance surge team is attached to the BSB, M1, M2/M3, and Stryker maintenance is available outside the FSC. If a maintenance surge team is attached to the BSB, it can be used to reinforce the FSC to weight the main effort. M109A6 Paladin maintenance capability still solely resides in the FSC even with a maintenance surge team attached. No maintenance capability for these systems resides in the field maintenance company or at echelons above the BCT. The maintenance surge team may be positioned within the BSA or at any of the trains CP locations as required. If a maintenance surge team is not available, the brigade and BSB commander consider placing FSC mechanics to repair main weapon systems with long repair times.

If a maintenance surge team is not available, BSB and BCT staffs may cross-level FSC M1, M2/3, M109A6, and Stryker mechanics from one FSC to another to weight the main effort and to ensure adequate maintenance support for task-organized units where additional main weapon systems are attached.

The FSC position a MCP in the combat trains. Although the FSC can establish a temporary MCP near the company trains to provide a quick consolidation point for damaged equipment, hasty repairs, or to perform BDAR, the FMT will evacuate non-mission capable equipment to the FSC's MCP in the combat trains. Elements of the FSC typically also remain in the field trains to coordinate logistics requirements with the BSB and configure logistical packages for the maneuver battalion.

The FSC commander assists the supported battalion S-4 and XO in developing logistics plans for the maneuver battalion. However, the FSC commander is not responsible for developing the battalion sustainment concept of support or operation order. The S-4, as the primary logistics planner for the battalion, is responsible for determining support requirements and creating the battalion's sustainment concept of support based on commander's intent, mission analysis, and the battalion's MDMP. The battalion S-4 ensures it nests with the brigade's concept of support. The FSC commander and the FSC are the executors of the S-4's plan. Therefore, the FSC commander and company leaders have a vested interest in ensuring the battalion's concept of support created by the S-4 is feasible, suitable, follows the principles of sustainment,and adheres to logistics capabilities and doctrine.

The FSC commander understands the capabilities and limitations of the company's personnel and equipment. The FSC commander must adapt to changing operational and mission variables. The commander thoroughly understands the different types of Army command and support relationships and how they affect the company. This knowledge, coupled with logistic expertise, enables the FSC commander to make credible recommendations to both the supported maneuver commander and the BSB commander on the employment of the FSC. See chapter 2 for a further discussion on command and support relationships.

The maneuver battalion commander and staff, the BSB commander and staff, and the FSC commander collaborate to determine the best method of employment for the FSC appropriate with the brigade's concept of support. FSC employment considerations include:

- FSC location in relation to the supported battalion.
- Decision to separate elements of the FSC by platoon or other sub-elements into multiple locations.
- C2 as well as decision-making necessity by location.
- Benefits and drawbacks of locating FSC elements in the BSA.
- Benefits of collocating battalion staff sections with the FSC.
- Benefits of collocating battalion medical elements with the FSC.
- Security of the FSC locations.
- Establishment and location of a MCP.

The FSC commander can be located in a variety of locations, field trains, combat trains, or, if required, the BSB base. The FSC commanders do not solely position themselves at the BSB base. FSC commanders establish their location where they can best command the FSC and maintain communication with appropriate BSB and supported battalion staffs. Ideally, the FSC commander will be in the combat trains as far forward as possible to command their unit and provide responsive logistics support to their maneuver battalion. The FSC first sergeant, field feeding section, distribution platoon, MCO, and maintenance platoon may also be located in the combat trains.

III. Forward Logistics Element (FLE)

The BSB CDR task organizes a FLE to support high tempo operations that exceed the capacity of existing echelonment of sustainment operation, or to temporarily support the brigade when the BSB displaces to a new location. The intent for employing an FLE is to minimize tactical pauses to the plan and enable the commander's momentum by shortening the lines of communication. The FLE operates out of a forward logistics base or support area and is typically commanded and controlled by the BSB SPO and assistant S-3.

The BSB establishes an FLE based on tactical requirements. The BSB is the lowest echelon unit that can establish an FLE. The FLE's composition is tailored to the supported unit's composition and task. While mission variables dictate an FLE's composition, the BSB often establishes an FLE with fuel assets, ammunition assets, water and class I supplies, recovery assets, and medical personnel. FSC maintenance elements continue normal operations so, typically, there is limited requirement for maintenance capability in an FLE while the BSA displaces. The FSCs continue to provide direct support to their assigned battalions. Due to its fixed location and extended period of operation, security if a significant concern for the FLE and requires a greater degree of security planning.

When displacing the BSA, an FLE enables the BSB mitigate disruptions in support to the BCT. The BSB's capability to support the brigade is diminished while moving. Brigade planners should carefully consider the timing of BSB displacement, employment of an FLE, Role 2 capability and coverage, and composition of the FLE to ensure adequate support is maintained throughout the transition. During fires brigade displacement, the majority of support to fires brigade units will come from CSSB throughput.

IV. Echelons Above Brigade (EAB)

Ref: ATP 4-90, Brigade Support Battalion (Jun '20), pp. 5-6 to 5-8.

Support organizations at EAB that provide support to a BCT normally consist of the DSB, DSSB, brigade logistics support element, and area medical support organizations. Sustainment brigades and CSSBs may reinforce the DSB and the DSSB. Each of these units contribute to the echeloning of sustainment to a BCT. Identifying, planning for, and requesting support from the DSB, DSSB, and medical support organizations (such as the medical brigade or multifunctional medical battalion) is critical before the start of an operation for the efficient echeloning of sustainment. Planners coordinate with echelons above brigade units in order to establish a support relationship if no command relationship exists.

Sustainment Brigade

Sustainment brigades are normally attached to a corps expeditionary sustainment command, known as an ESC, or theater sustainment command. The sustainment brigade attached to a corps expeditionary sustainment command commands and controls all assigned and attached units in an area of operations as directed by the corps commander providing logistics, finance and comptroller, and personnel services to forces operating in the corps AO. The corps commander determines the task organization for the sustainment brigade attached to a corps expeditionary sustainment command. A task-organized sustainment brigade attached to a corps expeditionary sustainment command normally includes attached CSSBs, a petroleum battalion, and a motor transportation battalion to support tactical-level sustainment operations. The sustainment brigade coordinates and synchronizes tactical-level sustainment operations to meet the current and future operations.

See pp. 6-11 to 6-22 for discussion of the sustainment brigade from ATP 4-93.

Combat Sustainment Support Battalion (CSSB)

The CSSB can be attached to sustainment brigades supporting a corps. The CSSB commands and controls all assigned and attached units in an area of operations as directed by the sustainment brigade commander. The unit performs maintenance, transportation, supply, field services, and distribution. The corps commander determines the task organization for the CSSBs. Task-organized CSSBs, normally include a composite supply company, support maintenance company, modular ammunition company, palletized load system truck company, inland cargo transfer company and a field feeding company. The CSSB synchronizes and executes logistics support to functional brigades and multifunctional support brigades attached to the corps.

See pp. 6-16 to 6-17 for further discussion. Refer to ATP 4-93.1, Combat Sustainment Support Battalion for more information on the unit.

Division Sustainment Brigade (DSB)

The DSB is assigned to a division. The commander is responsible for the integration, synchronization, and execution of sustainment operations. The DSB employs sustainment capabilities to create desired effects in the support of the division commander's objectives.

Depending upon operational and mission variables, the DSB can command up to seven battalions. The DSB and its subordinate units provide support to all assigned and attached units in an area of operations as directed by the division commander. The DSB provides logistics, personnel services, and finance and comptroller support to forces operating in the division AO. A DSB assigned to a division includes an organic division sustainment troops battalion (DSTB) and an organic DSSB to support the BSBs. The DSB coordinates and synchronizes tactical-level sustainment operations to meet current and future operations. The DSB is dependent on the division staff for long-range planning capability. The DSB and its subordinate units must be able to move and displace to maintain pace with the division headquarters. Additional modular CSSBs and companies may be attached to the DSB to sustain large-scale combat operations.

See p. 6-6 for further discussion.

Division Sustainment Troops Battalion (DTSB)

The DSTB is organic to DSBs. The battalion's role is to exercise C2 for all units assigned, attached, and OPCON to the DSB headquarters. The battalion plans, prepares, executes, and assesses the internal support requirements for the DSB headquarters. Its core competencies are to establish a battalion command post, execute the operations process, and synchronize internal support operations to support mission requirements.

The DSTB consists of a command group, UMB, and coordinating staff. It serves as a battalion headquarters organized to provide administrative support, life support, and communications for the DSB headquarters. Capable of operating at the tactical level throughout an area of operations, it can command up to seven organizations.

Organic to the DSTB is a headquarters company that includes a maintenance section, medical treatment team, and medical evacuation team. Assigned to the DSTB are a signal company, human resources company, field feeding company, and a financial management support unit, also known as an FMSU. These units all support the DSTB, DSB headquarters, and all the organic, assigned and attached units.

See p. 6-10 for further discussion.

Division Sustainment Support Battalion (DSSB)

The DSSB is organic to DSBs. The DSSB and its subordinate units move and displace to maintain pace with the DSB and division headquarters. The DSSB commands and controls all organic, assigned, and attached units. As directed by the DSB commander, the DSSB performs maintenance, transportation, supply, and distribution. DSSBs have an organic composite supply company, composite truck company, and support maintenance company. Other capabilities are task organized by the division commander based on support requirements. The DSSB synchronizes and executes logistics support to BCTs and multifunctional support brigades attached to the division and non-divisional units operating in the division AO.

Medical Company (Area Support)

The medical company (area support) provides Role 1 and Role 2 medical support to units supporting operations. The medical company (area support) is attached to a multifunctional medical battalion and executes area medical support operations in an area identified in the operation order issued by the multifunctional medical battalion commander. The medical company (area support) can task-organize and is tailorable to the OE and mission requirements dictated by defensive operations.

Sustainment units and maneuver forces evacuate casualties from the point of injury to the next appropriate higher role of care. The brigade support medical company may medically evacuate patients to a Role 3 medical treatment facility or to another Role 2 medical treatment facility until further evacuation to a combat support hospital or Role 4 medical treatment facility is available.

The evacuation of casualties or patients and the role of a medical company (area support) and Role 4 medical treatment facilities are discussed further in ATP 4-90, chap. 8.

The BSB commander may need to coordinate with echelons above brigade sustainment units to provide augmentation to the FLE. This includes identifying and the positioning of EAB sustainment assets in close proximity to geographically dispersed forces to extend operational reach and prolong endurance.

V. Area Support

Area support is not a support relationship and it is not synonymous with general support. Area support is necessary to support units which are not in the original support concept and for which the order has not established a support relationship. Area support is provided as capabilities allow and is secondary to the priority effort and existing support relationships. Normally, BSB units provide area support to units upon request unless supporting that unit will jeopardize the priority effort. If requested support is beyond supporting unit's capability or not authorized by supporting unit's headquarters, the requesting unit will coordinate with its parent unit for support.

The BCT may task the BSB to provide area support to division or corps units in, around, and traversing the BCT's area of operations. The BSB may need to provide area support for small, units such as—a special forces operational detachment A, air defense batteries, or civil affairs.

BSB commanders will determine procedures for supporting units with no established support relationship. Alternatively, the BCT commander may require units to verify authority to support and priority for all unplanned units. In that instance, the supporting unit would notify the BSB SPO, who would contact the brigade S-4 and XO for guidance. When providing area support to a unit, the BSB SPO coordinates with the supported unit's commander, XO, or S-4, to integrate their requirements into the sustainment concept of support and to synchronize support operations.

Chap 6

Division, Corps, and Field Army Support

Ref: FM 4-0, Sustainment (Jul '19), pp. 2-43 to 2-54. When published, also refer to projected new references ATP 4-91, Division Sustainment and ATP 4-92, Corps & Field Army Sustainment (replacing current ATPs).

The effectiveness of the sustainment warfighting function is dependent upon actions of units and staffs at the operational and tactical levels. Understanding unit and staff roles and functions is essential to conducting sustainment operations.

I. Field Army

The Army constitutes a field army in theaters where large-scale combat is possible. Its primary purpose is to prevent and if necessary, prevail in large-scale ground combat against peer or near-peer adversaries. It also enables effective competition against such threats below the threshold of armed conflict. Field armies are organized, manned, trained, and equipped to command and control multiple corps-sized formations including U.S. Army and multinational corps, or a Marine expeditionary force operating within an area. The field army is best suited to serve as the JFLCC for large-scale ground conflict, but requires augmentation from the joint and multinational force to perform the JFLCC role successfully. It may also assume the ARFOR role depending on the JFC's command arrangements, and the situation.

For more information, refer to FM 3-94.

The field army focuses on the threat to successfully compete, deter, and if necessary, prepare for and transition to combat operations as a land component command. The field army is tailored in its capability and capacity as determined by mission and operational variables. To enable continuous shaping and the ability to transition quickly to prevention of conflict, a subordinate field army enables the theater Army to focus on its broad Title 10, Army support to other Services, and DOD EA responsibilities across the entire theater. The ASCC conducts administrative and select operational activities (theater opening, RSOI, Army support to other Services and CUL for tasks that include transportation, fuel distribution, intra-theater aeromedical evacuation, EOD, and logistics management) to allow the field army to focus on tactical operations.

A field army specifically tailored to mission requirements may be assigned to a JFC with an enduring operational requirement. Typically, a sub-unified command is established instead of a JTF when the military operation is anticipated to be enduring or protracted. In such cases, a field army would be appropriate as the Army component or ARFOR to the sub-unified command. The theater Army exercises ADCON over the field army and its subordinate Army forces, and it provides the field army and its joint force command with all Army Service functions. See FM 3-94 for additional information. Fundamentally the field army headquarters is staffed and equipped to perform three roles:

- Army component and ARFOR for a subordinate unified commander.
- Joint force land component headquarters (with augmentation) for large-scale combat operations.
- JTF headquarters (with augmentation) for crisis response and limited contingency operations.

An ESC may be attached to the field army. The ESC is the controlling headquarters for the integration and synchronization of sustainment operations at echelon. The

ESC staff supports the field army staff planning while conducting its own parallel planning for support requirements. The ESC advises the field army staff on issues regarding task organization, sustainment capabilities, and risk. In coordination with the field army G-4, it maintains the sustainment running estimate and take actions to mitigate shortfalls. The field army surgeon and the surgeon cells at each echelon work with their staffs to conduct planning, coordination, synchronization, and execution of AHS support in coordination with the MEDCOM (DS) as the theater medical command. The ESC and its subordinate task organized sustainment units normally have a GS relationship with units in their geographic area.

See p. 7-6 for more information on the ESC from ATP 4-94.

II. Corps

A corps is normally the senior Army headquarters deployed to a JOA. It is optimized to serve as a tactical command during large-scale combat. However, it may perform other roles under different conditions. The corps forms the nucleus for a JTF or joint force land component command to respond to situations exceeding a division's capability or one which requires joint force augmentation. The corps may serve as the ARFOR when it is the only U.S. Army corps assigned to an area. The ARFOR is the Army component and senior Army headquarters of all Army forces assigned or attached to a CCMD, subordinate joint force command, joint functional command, or multinational command (FM 3-94). Regardless of its role, the corps executes both operational and administrative responsibilities.

See facing page (fig. 2-27) for a notional task-organized corps.

The corps commands Army and multinational forces in campaigns and major operations. Its headquarters is organized, trained, and equipped to control the operations of two to five divisions, together with supporting theater-level organizations. In the event conflict escalates, large-scale combat operations may require the corps headquarters to operate under the command of a multinational force land component or subordinate to a field army equivalent established as part of a coalition.

For additional information, refer to FM 3-94.

The corps also has administrative responsibilities when it is designated an ARFOR. When it serves as the ARFOR, the corps is responsible through the theater Army commander for the Service specific support of all Army forces in the JOA, as well as for providing any Army support to other Services with forces deployed in the JOA. The theater Army commander specifies the ADCON responsibilities of the ARFOR. CONUS corps have an AFSB in direct support to integrate and synchronize the delivery of all USAMC support. The AFSB provides a CLSE OPCON to the corps when it deploys for large-scale combat operations. This delivers USAMC capabilities at echelon in support of multi-domain operations to provide responsive support to the corps. Figure 2-27 provides a notional task organization of a corps.

The corps G-4 is responsible for developing the corps sustainment support concept that ensures Army forces are sustained throughout all phases of an operation. The G4 integrates the corps' sustainment staff estimates and annexes of the G-1, G-8, and Surgeon sections in the operations process. The G-4 is responsible for integrating the subordinate personnel services, logistics, financial management, and AHS capabilities. Examples of the corps sustainment cell coordination tasks are:

- Coordinates with the G-9 for HNS.
- Coordinates with the assistant chief of staff, intelligence (G-2), G-3, and engineer officer to requisition cataloged topographic foundation data and existing mission-specific data sets from the DLA.
- Synchronizes and integrates AHS support operations within the corps.
- Coordinates with the G-1 that develops and implements human resources policies and procedures.

Notional Task-Organized Corps

Ref: FM 4-0, Sustainment (Jul '19), fig. 2-27. Notional task-organized corps.

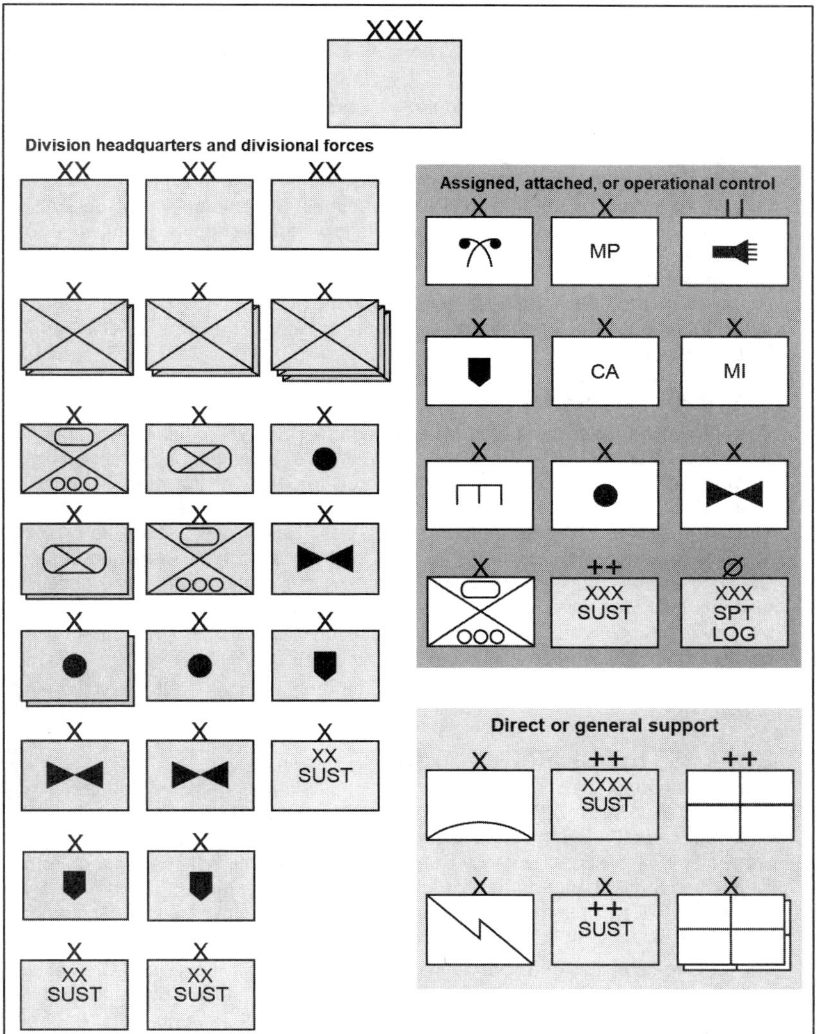

Legend

CA	= civil affairs	MP	= military police
ESC	= expeditionary sustainment command	SPT	= support
LOG	= logistics	SUST	= sustainment
MI	= military intelligence	TSC	= theater sustainment command

A. Expeditionary Support Command (ESC)

An ESC is assigned to the corps. The ESC is the corps' command for the integration and synchronization of sustainment in an operational area. The corps' ESC and its subordinate task-organized functional and multifunctional sustainment units provide GS for all units in corps area of operations as directed by the corps commander. The ESC commander may perform the duties of deputy commanding general (support) if directed by the corps commander. The ESC assists the corps sustainment cell with planning and coordinating sustainment.

The corps' ESC commands and controls all assigned and attached units in an operational area as directed by the corps commander. A task-organized ESC assigned to a corps normally includes enablers that include a corps logistics support element, petroleum group, movement control battalion, and a sustainment brigade task organized with CSSBs to support sustainment operations. The corps' EAB sustainment is dependent on the corps units for medical support, signal support, intelligence, surveillance and reconnaissance, fires, protection (engineer support & route security), and strategic partner planning capability for field maintenance support. The ESC plans for near term operations and synchronizes operational-level sustainment operations to meet the current operations. The corps' ESC is dependent on the corps staff for long-range planning capability.

B. Sustainment Brigade

Sustainment brigades can be attached to a corps ESC. The sustainment brigade attached to a corps ESC commands and controls all assigned and attached units in an operational area as directed by the corps commander providing GS logistics, financial management, and personnel services to forces operating in the corps AO. The corps commander determines the task organization for the sustainment brigade attached to a corps ESC. A task-organized sustainment brigade attached to a corps ESC normally includes attached CSSBs, a petroleum battalion and motor transportation battalion to support tactical-level sustainment operations. The sustainment brigade coordinates and synchronizes tactical-level sustainment operations to meet the current and future operations.

See pp. 6-11 to 6-22. Refer to ATP 4-93 for more information on the sustainment brigade.

Special Troops Battalion (STB)

The special troops battalion is organic to the sustainment brigade. The special troops battalion's role is to exercise command and control for all units assigned, attached, and OPCON to the sustainment brigade headquarters. The special troops battalion plans, prepares, executes, and assesses the internal support requirements for the sustainment brigade headquarters. Its core competencies are to establish a battalion CP, execute the operations process, and synchronize internal support operations in support of mission requirements.

The battalion consists of a command group, unit ministry team, and coordinating staff. It is a battalion headquarters organized to provide administrative support, life support, and communications for the sustainment brigade headquarters. Capable of operating at the tactical level throughout an operational area, it can command up to seven organizations.

Organic to the special troops battalion is a headquarters company which includes a maintenance section, medical treatment team, and medical evacuation team. Assigned to the special troops battalion are a signal company, human resources company and a FMSU. These units all support the special troops battalion, sustainment brigade headquarters, and all the organic, assigned and attached units.

See pp. 6-14 to 6-15 for and overview and further discussion of the STB.

Combat Sustainment Support Battalion (CSSB)

The CSSB can be attached to sustainment brigades supporting a corps. The CSSB attached to sustainment brigades supporting the corps commands and controls all assigned and attached units in an operational area as directed by the sustainment brigade commander and conducts maintenance, transportation, supply, field services, and distribution. The corps commander determines the task organization for the CSSBs attached to sustainment brigades supporting the corps. Task-organized CSSBs attached to sustainment brigades supporting the corps normally include a composite supply company, support maintenance company, modular ammunition company, palletized load system truck company and inland cargo transfer company and a field feeding company. The CSSB synchronizes and executes logistics support to functional brigades and multifunctional support brigades attached to the corps.

See pp. 6-16 to 6-17 for and overview and further discussion of the CSSB.

III. Division

The division is a formation that employs organic and assigned units for operations. It is optimized to serve as a tactical unit of execution for a corps during large-scale combat. It is organized, manned, trained, and equipped for decisive action. Figure 2-28 depicts a notional task-organized division.

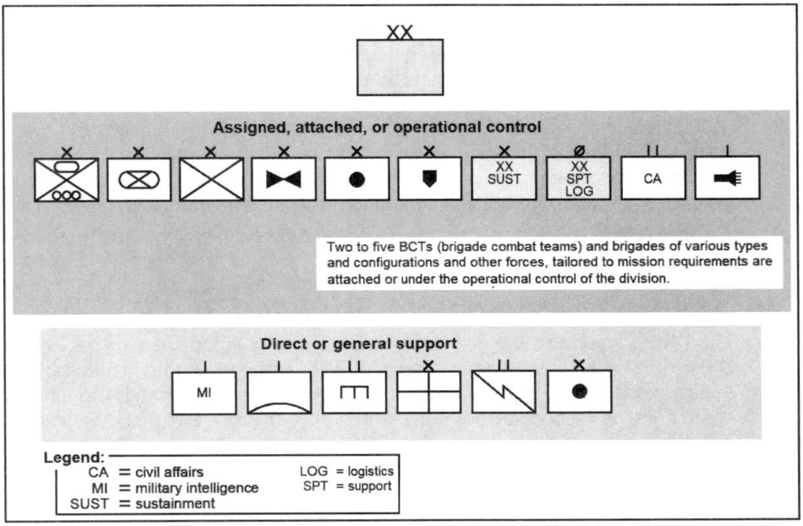

Figure 2-28. Notional task-organized division

During competition, the division may fulfill the ARFOR role for a JFC or theater Army commander. It can also form the nucleus for a very small-scale JTF or joint force land component command, although it would require joint force augmentation to fulfill either role successfully. The division primarily focuses on operational responsibilities. Unless the division is serving as the ARFOR, a higher echelon normally retains ADCON for all but the division's organic, assigned, and attached units. However, a higher echelon commander serving as the ARFOR may designate a division commander as the deputy ARFOR with prescribed responsibilities when the situation warrants.

A division conducting large-scale combat operations combines offensive, defensive, and stability tasks in an AO assigned by its higher headquarters. The division task organizes its subordinate forces to accomplish its mission. The division and its subordinate units must be able to move and displace at the pace of large-scale combat operations. A division will conduct operations with their assigned DSB, and the organic DSSB of the DSB. The DSB provides materiel management capability to the division. Divisions may have additional CSSBs attached to meet operational requirements.

For additional information, refer to ATP 4-93, ATP 4-93.1, and FM 3-94.

Each division has a Direct-Support Readiness AFSBn to integrate and synchronize the delivery of all USAMC support. The AFSBn provides a division logistics support element OPCON to the division when it deploys. This delivers USAMC capabilities at echelon in support of large-scale combat operations to provide responsive support to the division.

A division receives capabilities and units from its corps to conduct operations. The corps commander determines the number and types of BCTs necessary for the divisions to accomplish their respective missions. A division headquarters is organized, trained, and equipped to command the operations of up to five BCTs. Divisions are typically task organized with a combination of armored, infantry, and Stryker BCTs. In addition to the BCTs, a division controls several different types of multifunctional and functional brigades, including the DSB. The division's assigned, task-organized DSB provides GS for all units in or passing through their geographic area.

The human resources operations center is assigned to the corps and division G-1. The human resources operations center provides strength management capability, HR technical expertise, analysis, assessment, training, and assistance in support of the execution and management of operational and tactical HR support, for organizations assigned or attached within the corps and division AO. The human resources operations center uses HR planning considerations to develop performance indicators to measure the accomplishment of HR support. The human resources operations center assesses and analyzes HR systems, processes and procedures, and executes HR training in support of organizations assigned or attached within the corps and division AO. The human resources operations center provides the G-1 and supported units, visibility of the employment and integration of Army HR information and systems.

Refer to ATP 1-0.1 for specific duties and responsibilities of the human resources operations center.

A. Division Sustainment Brigade (DSB)

The DSB is assigned to a division. The DSB is a renamed sustainment brigade. The DSB commander is the primary senior advisor to the division commander and the deputy commanding general (support) for the sustainment warfighting function. The commander is responsible for the integration, synchronization, and execution of sustainment operations at echelon. The DSB employs sustainment capabilities to create desired effects in support of the division commander's objectives.

Depending upon operational and mission variables, the DSB can command up to seven battalions. Figure 2-29 and figure 2-30 *(facing page)* depict notional task-organized DSBs in support of an armored division and an infantry division respectively. The DSB and its subordinate units assigned to a division provides direct support to all assigned and attached units in an operational area as directed by the division commander. The DSB provides GS logistics, personnel services, and financial management to non-divisional forces operating in the division AO. A task-organized DSB assigned to a division includes an organic division sustainment troops battalion (DSTB) and an organic DSSB to support tactical-level sustainment operations. The DSB coordinates and synchronizes tactical-level sustainment operations to meet current and future operations. The DSB is dependent on the division staff for long-range planning capability. The DSB and its subordinate units must be able to move and displace at the pace of large-scale combat operations. Additional modular CSSBs and companies may be attached to the DSB to sustain large-scale combat operations.

(DSB discussion continued on p. 6-10, see DSTB and DSSB.)

Notional Task-Organized DSBs

Ref: FM 4-0, Sustainment (Jul '19), pp. 2-48 to 2-49

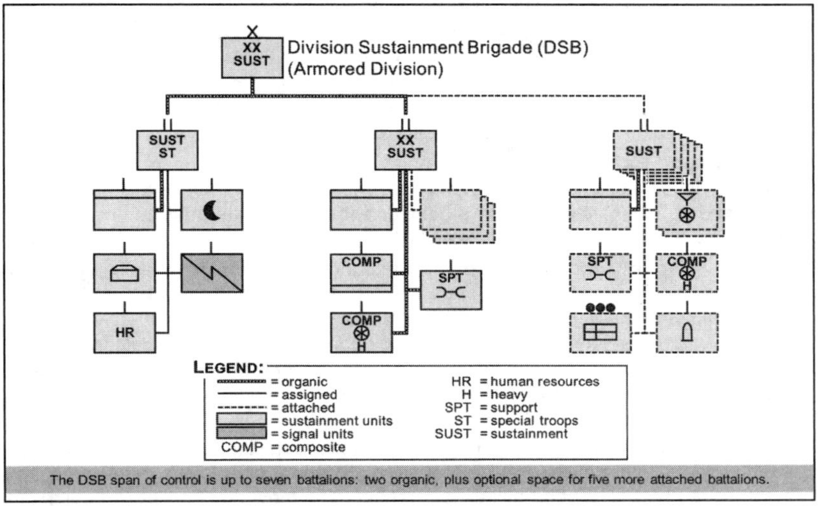

FM 4-0, fig. 2-29. Notional task organized division sustainment brigade for an armored division.

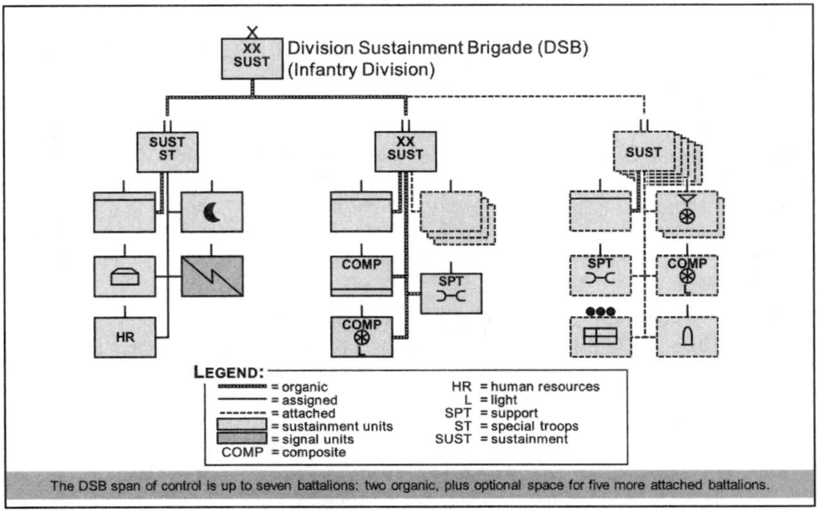

FM 4-0, fig. 2-30. Notional task organized division sustainment brigade for an infantry division.

Division, Corps & Field Army

B. Echelon Above Brigade & Additional Support

Ref: FM 4-0, Sustainment (Jul '19), pp. 2-50 to 2-54.

Echelons Above Brigade Battalions

The battalions below provide logistics support to **division and corps** formations **as well as enabling units operating throughout the AOR**. DSSBs, CSSBs, and functional logistics battalions provide critical logistics capabilities in an operational area. The commanders, staffs, and subordinate units have unique technical skills and specialized training that enables them to provide critical capabilities.

Motor Transport Battalion. The motor transport battalion commander and staff commands and directs assigned or attached units. The battalion synchronizes motor transport operations. The battalion is task organized with subordinate truck companies and is attached to a sustainment brigade. *Refer to ATP 4-13.*

Petroleum Support Battalion. The petroleum support battalion provides command and control for theater bulk petroleum storage and distribution, port terminal operations, and petroleum pipeline operations. The battalion operates a central dispatching and scheduling agency to ensure direct flow of bulk petroleum through the multiproduct military pipeline. The battalion normally attached to a sustainment brigade or the quartermaster petroleum group, also operates bulk and retail supply point distribution, and executes over-the-shore POL storage requirements. *Refer to ATP 4-43.*

Movement Control Battalion. The movement control battalion coordinates and synchronizes the execution of movements and transportation operation to ensure effective and efficient movements to support military operations. The movement control battalion is attached to an ESC or a sustainment brigade and assists with the planning and execution of deployment, redeployment, and distribution operations. *Refer to ATP 4-16.*

Explosive Ordnance Disposal Battalion. The EOD battalion and its subordinate companies perform three essential tasks when not engaged in combat operations: provide EOD support to military and civilian authorities, provide EOD support to the United States Secret Service, and train for wartime missions. *Refer to ATP 4-32.1.*

Terminal Battalion. The terminal battalion synchronizes terminal and watercraft operations to include fixed port or joint logistics over-the-shore operations. Task organized terminal battalions operate at a seaport and may also support amphibious and riverine operations. Terminal battalions provide control of watercraft supporting terminal and sea basing operations. Terminal battalions are task organized with subordinate watercraft companies, seaport operations companies, or causeway companies. The battalion may be attached to a transportation brigade expeditionary, a sustainment brigade, or a SDDC transportation brigade. *Refer to ATP 4-13.*

Echelons Above Brigade Companies

DSSBs, CSSBs and functional battalions are task organized with composite and functional sustainment companies, teams, and detachments. The following paragraphs are a summary of sustainment capabilities that may be task organized to a logistics battalion. Many of these units have specific capabilities and execute complex operations. See the references provided for more details about the unit's capabilities.

Modular Ordnance Company (Ammunition). The modular ammunition company role is to receive, store, and issue munitions. The organization of the modular ordnance company permits assignment of modular platoons tailored for specific functions to support forces or other munitions units as required. Modular ammunition companies operate ammunition supply areas at theater, corps, and other levels as required by the tactical situation. *Refer to ATP 4-35 and ATP 4-35.1.*

Support Maintenance Company. The support maintenance company is organic to a DSSB and attached to a CSSB. Its role is to provide field maintenance support. The support maintenance company is organic to a DSSB. The support maintenance company will normally be attached to a CSSB, but may be attached to any logistics battalion. Support maintenance companies operate at any echelon above the BCT level and may be located at any point in the theater from port to division support area. *Refer to ATP 4-33.*

Composite/Functional Transportation Companies. The transportation truck company is organic to a DSSB and attached to a CSSB. Its role is to provide transportation support. It provides distribution for general supplies, ammunition, fuel, water, unit equipment, and personnel. The 5K medium truck company (POL) is a critical enabler in providing class III (bulk and package) distribution in theater.

Watercraft Companies and Detachments. Watercraft companies and detachments provide transportation support required in the maritime domain, including intra-theater ocean, coastal, littoral, and river transit. Watercraft companies and detachments units are typically assigned to a transportation brigade (expeditionary) for theater opening, closing or distribution. *Refer to ATP 4-15.*

Composite/Functional Supply Companies. The supply company is organic to a DSSB and attached to a CSSB. Its role is to provide supply support. It provides, class I perishable and semi-perishable supply, petroleum supply, water purification and supply support, and shower and laundry services. *Refer to ATP 4-41, ATP 4-42, ATP 4-43, ATP 4-44, ATP 4-46, and ATP 4-48 for additional information.*

Quartermaster Mortuary Affairs Company. The quartermaster mortuary affairs company's role is to provide the full range of mortuary affairs operations across the range of military operations. *Refer to FM 4-0, app. A and ATP 4-46.*

Field Feeding Company. The field feeding company is responsible for providing the Army Field Feeding System to all Soldiers in EAB units not supported by internal field feeding assets. The Army Field Feeding System provides flexibility by presenting various feeding methods. The field feeding company performs sustained tactical feeding to meet the commander's needs as determined by mission variables and logistical support available on the battlefield and afloat. *Refer to ATP 4-41.*

Signal Network Support Company. The signal network support company is assigned to a sustainment brigade special troops battalion and a DSB sustainment troops battalion. The company consists of a headquarters and network support platoon, and two network extension platoons.

Financial Management Support Unit

The FMSU is assigned to a sustainment brigade special troops battalion and a DSB sustainment troops battalion. Its role is to provide financial management. The FMSU provides command and control for up to seven financial management support detachments. *Refer to FM 4-0, app. A and FM 1-06 for more details.*

Human Resources Company

The HR company is assigned to a sustainment brigade special troops. It provides command and control for HR and postal platoons providing casualty, personnel accountability, and postal operations support. *Refer to FM 1-0.*

Theater Gateway Personnel Accountability Team. The theater gateway PAT is attached to a DSTB of a DSB or sustainment brigade. It may also be attached to a DSSB or CSSB. It provides personnel accountability support to the AOR by coordinating and providing personnel accountability operations using the deployed theater accountability system as personnel enter, transit, and depart the theater at the inter-theater APOD.

Military Mail Terminal. The military mail terminal is attached to a sustainment brigade special troops battalion. It provides postal support by establishing a military mail terminal team which coordinates, receives, and processes pro-grade mail, and dispatches retro-grade mail to destinations worldwide. *Refer to FM 4-0, app. A and ATP 1-0.2 for additional information.*

Division Sustainment Troops Battalion (DSTB)

The DSTB is organic to DSBs. The battalion's role is to exercise command and control for all units assigned, attached, and OPCON to the DSB headquarters as shown in Figures 2-29 and 2-30 *(see p. 6-7)*. The battalion plans, prepares, executes, and assesses the internal support requirements for the DSB headquarters. Its core competencies are to establish a battalion CP, execute the operations process, and synchronize internal support operations in support of mission requirements.

The DSTB consists of a command group, unit ministry team, and coordinating staff. It is a battalion headquarters organized to provide administrative support, life support, and communications for the DSB headquarters. Capable of operating at the tactical level throughout an operational area, it can command up to seven organizations.

Organic to the DSTB is a headquarters company which includes a maintenance section, medical treatment team, and medical evacuation team. Assigned to the DSTB are a signal company, human resources company, field feeding company, and a FMSU. These units all support the DSTB, DSB headquarters, and all the organic, assigned and attached units.

Division Sustainment Support Battalion (DSSB)

The DSSB is employed using various task-organizations as shown in Figure 2-31. The DSSB is a renamed CSSB. The DSSB is organic to DSBs assigned to divisions. The DSSB and its subordinate units must be able to move and displace at the pace of large-scale combat operations. The DSSB commands and controls all organic, assigned, and attached units. As directed by the DSB commander, the DSSB conducts maintenance, transportation, supply, and distribution. DSSBs organic to DSBs supporting divisions have an organic composite supply company, composite truck company, and support maintenance company. Other capabilities are task organized by the division commander in accordance with requirements. The DSSB synchronizes and executes logistics support to BCTs and multifunctional support brigades attached to the division and non-divisional units operating in the division AO.

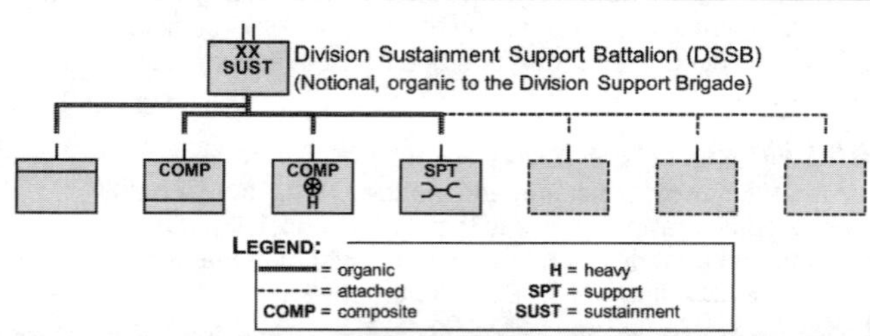

The DSSB span of control is up to seven companies: four organic plus optional space for three more attached companies.

FM 4-0, fig. 2-31. Notional division sustainment support battalion

The human resources operations branch (HROB) is an important planning and coordinating element on the battlefield for the delivery of personnel accountability, casualty operations, and postal support. HROBs are assigned to the ESC DMC and DSB support operations section with the mission to plan current and future personnel accountability, postal, and casualty operations, and coordinate, synchronize, integrate, and assess the emplacement and operations of HR companies, platoons, and teams executing those functions. As replacements flow into theater, the HROB passes inbound replacement information to the supported G-1. The G-1 provides the unit of assignment for inbounds to the HROB so the SPO officer can coordinate delivery to the unit or pickup by the unit.

For more details, refer to FM 4-0, appendix A and FM 1-0.

Chap 6

I. Sustainment Brigade

Ref: FM 4-0, Sustainment (Jul '19), pp. 2-43 to 2-54 and ATP 4-93, Sustainment Brigade (Apr '16), chap. 1.

The sustainment brigade is a multifunctional headquarters integrating and employing all assigned and attached units while planning and synchronizing sustainment operations. It is the Army's primary brigade level sustainment headquarters. Sustainment brigades are usually assigned or attached to a sustainment command. The sustainment brigade and its attached units will normally have a general support relationship with supported organizations.

The sustainment brigade executes logistics and personnel services functions associated with theater opening, sustainment, distribution, and theater closing missions. The sustainment brigade headquarters plans, coordinates, synchronizes, monitors, and controls sustainment operations within its support area.

I. Sustainment Brigade Capabilities

The sustainment brigade supports Army forces at the tactical and operational levels, providing support to brigade combat teams (BCTs), multifunctional and functional support brigades, deployable, self-contained division and corps headquarters, and other units operating in its assigned support area. Depending upon operational and mission variables, the sustainment brigade commands between three and seven battalions. Sustainment brigades are usually assigned or attached to a sustainment command. The sustainment brigade and its attached units will normally have a general support relationship with supported organizations.

The sustainment brigade is expeditionary, inter-operable and agile. These characteristics describe the attributes that the organization requires to be effective. The sustainment brigade is expeditionary as it can deploy task organized forces on short notice to austere locations and conduct sustainment operations immediately upon arrival. The sustainment brigade is inter-operable as it can task organize rapidly and integrate joint, inter-organizational and multinational requirements and capabilities. The sustainment brigade is agile as it can transition sustainment support across all decisive action tasks.

The sustainment brigade is task organized with units required to execute logistics and personnel services. Logistics includes; supply, maintenance, transportation, field services, distribution, and operational contract support. Personnel services are sustainment functions that fund and man the force.

The combat sustainment support battalion (CSSB) is the building block upon which the sustainment brigade capabilities are developed. The organization and operations of most functional logistics battalions are addressed in specific functional Army techniques publications. Organizational information about functional logistics battalions is available in unit authorization documents and from force design resources located at the Combined Arms Support Command Sustainment Unit One Stop website.

A financial management support unit and a human resources company may be attached or assigned to the sustainment brigade.

The sustainment brigade headquarters is designed to operate as a single command element without the ability to conduct split based operations. The sustainment brigade cannot create or operate a tactical command post (CP) without accepting risk in other areas.

The sustainment brigade headquarters plans and conducts base security and protection against level I threats. Level II and III threats require coordination with designated combat reaction forces. The sustainment brigade cannot be assigned an area of operations or manage terrain.

A task organized sustainment brigade is dependent on the following organizations:

- Sustainment brigade signal network support company for signal support.
- Area support medical company for Role 2 medical support.

II. Role and Functions

The role of a sustainment brigade commander and staff is to exercise mission command for task organized sustainment brigades. Mission command is the exercise of authority and direction by the commander using mission orders to enable disciplined initiative within the commander's intent to empower agile and adaptive leaders in the conduct of unified land operations (ADP 6-0).

The sustainment brigade executes logistics and personnel services functions associated with theater opening, sustainment, distribution, and theater closing missions. A function is a practical grouping of tasks and systems (people, organizations, information, and processes) united by a common purpose (ADP 1-01). Properly task organized, a sustainment brigade could be conducting theater opening tasks, sustainment and theater distribution tasks during the early phases of an operation or if it is the only sustainment brigade in the joint operations area (JOA). This same sustainment brigade, with a different task organization, can transition to conducting a theater distribution mission or sustainment mission.

III. Relationships

Commanders task organize the force to provide specific capabilities in support of mission requirements. They task organize the force by establishing command and support relationships. These relationships establish clear responsibilities and authorities between subordinate and supporting units. For every operation, the sustainment brigade commander and subordinate commanders must make every effort to ensure command and support relationships are clearly expressed in orders; their own and those of their higher headquarters and supported organizations. Doctrine sets general guidelines; mission orders will determine the details of the relationships.

Doctrinal relationships are defined and explained in ADRP 5-0, The Operations Process, and FM 6-0, Commander and Staff Organization and Operations.

Sustainment brigade commanders closely evaluate the outcome they wish to achieve and then decide which combination of command and support relationships to assign subordinate units. The relationships must accommodate the known situation and empower subordinate leaders to respond to the unknown. Changes in command relationships do not necessarily require changes in support relationships, especially if the nature of the support does not change. Simple command and support relationships increase the likelihood of success.

The sustainment brigade commander also establishes informal relationships. The informal relationship between the sustainment brigade and the division G-4 (assistant chief of staff, logistics) provides another source of information for the sustainment brigade commander to consider when determining appropriate command and support relationships and internal task organization. A description of the relationship between the division G-4 and sustainment brigade support operations (SPO) is in the organization discussion later in this chapter.

A. Command Relationships

Command relationships define command responsibility and authority. Army command relationships are: organic, assigned, attached, operational control, and tactical

IV. Sustainment Brigade Organization

Ref: ATP 4-93, Sustainment Brigade (Apr '16), pp. 1-9 to 1-16.

The sustainment brigade is a headquarters organization comprised of a command group, staff and a special troops battalion.

A. Staff Organization

The sustainment brigade headquarters plans, coordinates, synchronizes, monitors, and controls sustainment operations within its support area. Figure 1-1, depicts the recommended sustainment brigade staff organization.

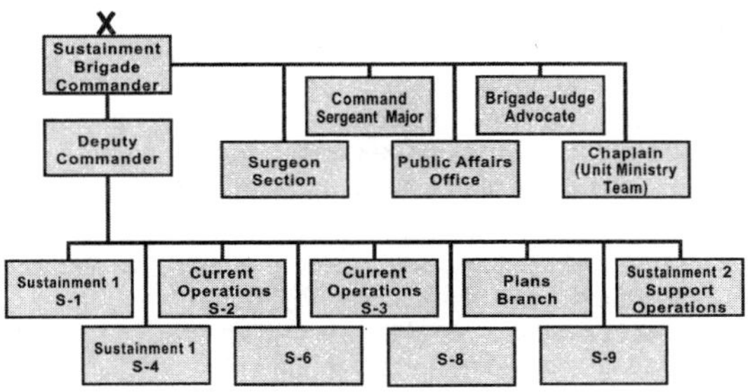

ATP 4-93, fig. 1-1. Sustainment brigade staff organization.

B. Support Operations (SPO)

The brigade support operations plans and coordinates support operations. Support Operations is the staff function of planning, coordinating, and synchronizing sustainment in support of units conducting decisive action in an area of operations. It is performed by support operations coordinating staff of a sustainment unit. The brigade support operations balances external sustainment support requirements with sustainment capabilities. The SPO conducts distribution operations, maintenance management, operational contract support and commodity management of general supplies, ammunition, fuel and water.

The sustainment brigade SPO executes materiel priorities established by the sustainment command's distribution management center in accordance with the ASCC policy and priorities. The SPO manages internal supplies and stocks as well as supervise distribution, maintenance, and materiel management functions within the brigade's geographic support area.

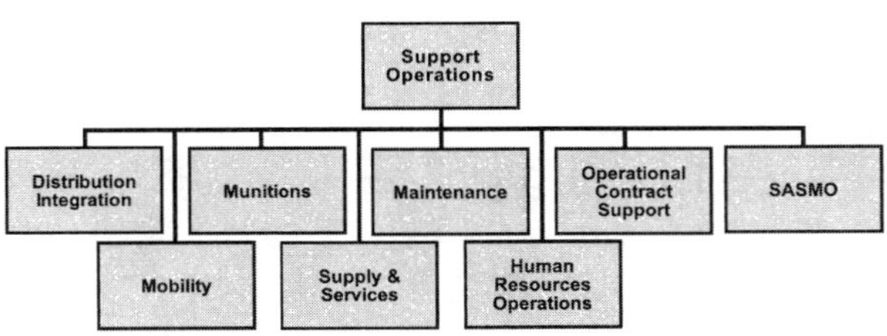

ATP 4-93, fig. 1-2. Sustainment brigade support operations.

Special Troops Battalion (STB) and CSSB

See following pages for discussion of the STB (pp. 6-14 to 6-15) and CSSB (pp. 6-16 to 6-17).

C. Special Troops Battalion (STB)

Ref: ATP 4-93, Sustainment Brigade (Apr '16), chap. 2 and FM 4-0, Sustainment (Jul '19), p. 2-46.

The special troops battalion is organic to the sustainment brigade. The special troops battalion's role is to exercise command and control for all units assigned, attached, and OPCON to the sustainment brigade headquarters. The special troops battalion plans, prepares, executes, and assesses the internal support requirements for the sustainment brigade headquarters. Its core competencies are to establish a battalion CP, execute the operations process, and synchronize internal support operations in support of mission requirements.

STB Organization

The battalion consists of a command group, unit ministry team, and coordinating staff. It is a battalion headquarters organized to provide administrative support, life support, and communications for the sustainment brigade headquarters. Capable of operating at the tactical level throughout an operational area, it can command up to seven organizations.

Organic to the special troops battalion is a headquarters company which includes a maintenance section, medical treatment team, and medical evacuation team. Assigned to the special troops battalion are a signal company, human resources company and a FMSU. These units all support the special troops battalion, sustainment brigade headquarters, and all the organic, assigned and attached units.

The STB headquarters comprises a command group, a coordinating staff and an HHC. When financial management and human resources organizations are attached to the sustainment brigade, they are usually further attached to the STB. The STB headquarters has a limited staff capability and depends upon the brigade staff for support.

Special Troops Battalion (Notional)

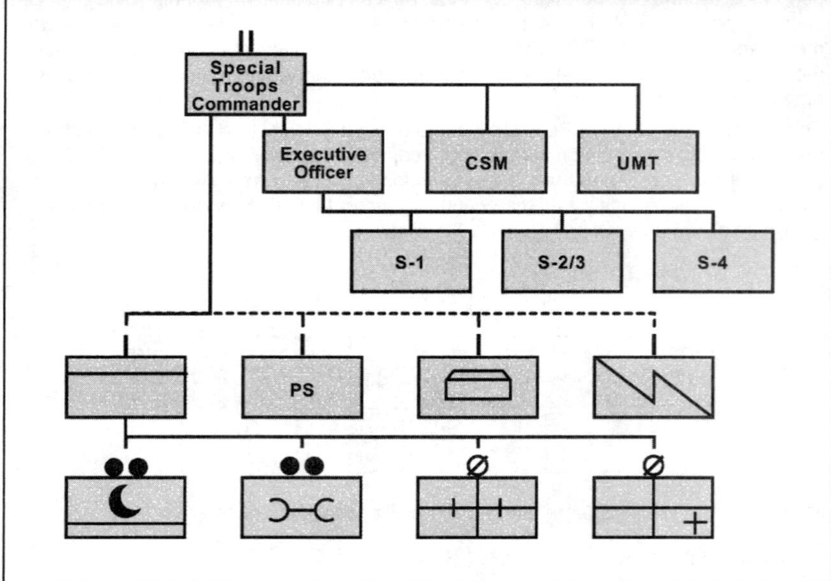

Ref: ATP 4-93, fig. 2-1. Notional special troops battalion.

STB Capabilities

Ref: ATP 4-93, Sustainment Brigade (Apr '16), p. 2-1.

The STB is task organized with companies and detachments that provide capabilities from across the warfighting functions. The STB task organization changes depending on the sustainment brigade's specific operational environment and may include competencies with which the staff is not familiar. There are five to seven organizations assigned or attached to the STB. One of the challenges for the STB commander and staff is adapting to the quantity and complexity of attached units and assigned missions. A clear commander's intent, a well-informed professional staff, teamwork among the units, and cooperation will help to minimize these challenges and foster a positive climate. The STB commanders and staffs contribute to this positive climate during training and sustain it during operations.

The STB commander and staff integrate and control assigned and attached units. The STB provides all administrative support, life support, and communications for the sustainment brigade headquarters. It is a battalion level headquarters available to the sustainment brigade to control assigned and attached mission enabling units.

The STB's organic elements include a command group, coordinating staff, a unit ministry team, and the headquarters company. The staff provides information and advice to the headquarters and headquarters company (HHC) and attached unit commanders. The headquarters company has a field feeding section, a maintenance section, a medical treatment team, and a medical evacuation team. The teams provide Role 1 care for the sustainment brigade HHC and the STB's assigned and attached units.

The STB can control a variety of units based on the operational situation. The task organization may include units that perform functions which are not represented on the staff. Examples include aerial delivery, mortuary affairs, theater gateway teams and military mail terminals. The STB commander and staff must be adaptable and innovative to meet mission requirements.

The STB will have attached companies, detachments and teams with differing command and support relationships. Human resources, Finance and Signal units are the most common attachments to the STB. The command relationship of any unit in the STB task organization will depend on the expected longevity of the relationship between the headquarters involved and additional mission variables. Sustainment units attached to the STB usually have a general support relationship with units in the sustainment brigade's assigned support area.

The STB does not normally have a direct interface with strategic partners. However, the STB may have to provide life support to elements of combat support agencies that are in direct support to sustainment brigade operations. The most likely strategic interface will be with organizations overseeing contracts and contractors. This includes elements of the Army field support battalion such as logistics assistance representatives or a logistics support element designated to oversee LOGCAP execution.

D. Combat Sustainment Support Battalion (CSSB)

Ref: ATP 4-93, Sustainment Brigade (Apr '16), chap. 3 and FM 4-0, Sustainment (Jul '19), p. 2-46.

The combat sustainment support battalion is a flexible and versatile headquarters that controls execution and synchronizes logistics support in a designated area of operations. The CSSB can be task organized with functional companies, teams and detachments that execute transportation (mode, terminal and movement control) operations, maintenance operations, ammunition operations, supply support activity operations, water operations, petroleum operations, aerial delivery operations and mortuary affairs. The CSSB is the building block upon which the sustainment brigade capabilities are developed.

Combat Sustainment Support Battalion

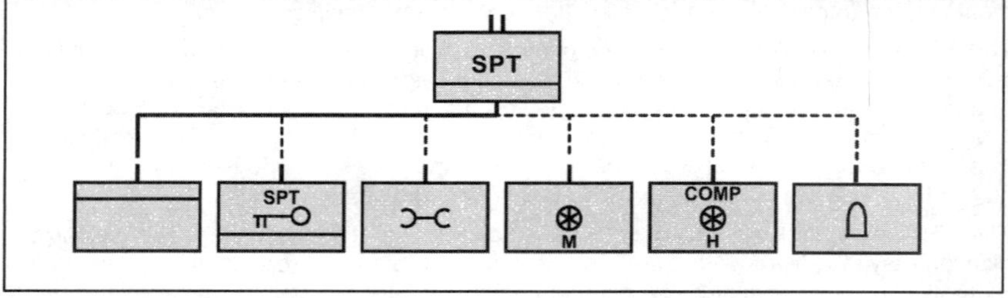

Ref: ATP 4-93, fig. 3-3. Notional combat sustainment support battalion.

The combat sustainment support battalion employs and controls up to seven company-sized assigned and attached units conducting logistics operations and support. The CSSB staff establishes a command post, executes the operations process and synchronizes logistics operations in support of mission requirements. The CSSB supports brigade combat teams, multifunctional support brigades, and other units operating in its assigned support area. The CSSB is task organized with units required to support logistics requirements. A task organized CSSB is dependent on the following organizations:

- The sustainment brigade for administrative support.
- Support maintenance company for field maintenance and recovery support.
- Area support medical company for Role 2 medical support.

The CSSB executes and synchronizes logistics functions as required to support units in its assigned support area. It is task organized to provide specific types of logistics functions support depending on its assigned mission. The CSSB usually has a general support relationship with its supported organization.

The CSSB is a logistics headquarters with a command group, coordinating staff and a headquarters company. The CSSB is task organized with logistics capabilities to support specific requirements. These logistics companies, platoons or detachments include maintenance, supply, transportation mode, terminal and movement control, mortuary affairs and field services.

The concept of support developed by the sustainment brigade may require CSSBs to change task organization frequently to meet mission requirements. Cascading concepts carry the top commander's intentions to the lowest levels, and the nesting of those concepts traces the critical path of concentration and priorities. When requirements change, the supporting unit may be released from its direct support relationship and the support or command relationship reassigned in orders.

Division, Corps & Field Army

CSSB Subordinate Organizations

Ref: ATP 4-93, Sustainment Brigade (Apr '16), pp. 3-5 to 3-6.

The CSSB headquarters and headquarters company is the only organic unit in the CSSB. The CSSB is task organized to meet mission requirements.

Ammunition

The CSSB may be task organized with ammunition units to support decisive action tasks. Ammunition capability includes the functions of receiving, storing, issuing and reconfiguring ammunition packages. An ammunition storage area is operated by one or more modular ammunition platoons with or without an ammunition company headquarters. Ammunition support is fully modular and platoons may be added or reduced from the organization based upon ammunition support requirements and mission variables. The CSSB ammunition staff coordinates receipt and issue of ammunition from the ammunition supply point. The ammunition staff has no materiel management responsibility for class V.

Refer to ATP 4-35, Munitions Operations and Distribution Techniques, for more information about ammunition organizations.

Maintenance

The CSSB may be task organized with support maintenance companies to support decisive action tasks. Maintenance capability includes the functions of wheeled vehicle repair, armament repair, allied trades, radio repair, computer and electronic equipment maintenance, ground support equipment repair and recovery assistance to units within its assigned support area. The SPO maintenance section provides priorities of effort to the support maintenance company, coordinates evacuation of equipment to sustainment maintenance level activities and provides COR support of monitoring the contract execution.

Refer to ATP 4-33, Maintenance Operations, for more information about field maintenance capabilities.

General Supply and Field Services

The CSSB may be task organized with supply and service units to support decisive action tasks. Supply capability include subsistence support, fuel storage and issue, water purification and storage and supply support activity (SSA) support to units within its assigned support area. The type and quantity of supply companies attached to the CSSB will vary based upon mission requirements and units supported. Quartermaster companies and/or platoons may be assigned to a CSSB to provide supported units with field services such as: aerial delivery, mortuary affairs, field feeding, laundry, shower, and water purification.

More information about quartermaster capabilities is in ATP 4-42, General Supply and Field Services Operations, ATP 4-43, Petroleum Supply Operations, ATP 4-44, Water Support Operations, ATP 4-45, Force Provider Operations and ATP 4-48, Aerial Delivery.

Transportation

The CSSB may be task organized with transportation units to support decisive action tasks. Transportation capability includes terminal, mode and movement control support to units within its assigned support area. The assigned mission will determine the number and type of truck companies attached to the CSSB.

Refer to ATP 4-11, Army Motor Transport Operations, for more information.

control. Command relationships unify effort and enable commanders to use subordinate forces with maximum flexibility. The type of command relationship often relates to the expected longevity of the relationship between the headquarters involved and quickly identifies the degree of support that the gaining and losing Army commanders provide. Leaders and Soldiers must understand the different kinds of command relationships and the impact those relationships have on providing and receiving sustainment support.

The sustainment brigade has different command relationships depending on many factors including, mission, priorities of support and transitioning task organization. Sustainment brigades are usually assigned or attached to a sustainment command. The sustainment brigade's command relationship and task organization changes based on changing mission requirements. Subordinate battalions may have different command relationships than the parent sustainment brigade.

The command relationship provides the authority to control unit mission. If a CSSB, or functional logistics battalion, has a command relationship with a unit they do not also have a support relationship with that unit. If the CSSB is attached to a sustainment brigade, the sustainment brigade has the authority to establish priorities and impose further command or support relationships. This relationship enables the sustainment brigade to maximize the capacity of all the subordinate CSSBs. Mission command doctrine describes the intended relationship, not a prescribed relationship.

Army Service Component Command (ASCC)

An Army Service Component Command (ASCC) assigned to a geographic combatant command is organized, manned, and equipped to perform three roles:

- Theater Army for the geographic combatant command to which it is assigned.
- Joint task force headquarters (with augmentation) for a limited contingency operation in that area of responsibility (AOR).
- Joint force land component (with augmentation) for a limited contingency operation in that AOR.

The ASCC is the primary vehicle for Army support to joint, interagency, intergovernmental, and multinational forces. The ASCC headquarters directs functions that include theater opening, theater distribution, reception, staging, onward movement and integration (RSOI), joint logistics over-the-shore operations; and sustainment and security coordination. A theater sustainment command (TSC) assigned to the ASCC is task organized with expeditionary sustainment commands and sustainment brigades to support mission requirements. The sustainment brigade has a command relationship with a sustainment command. The sustainment command has a command relationship with the ASCC.

Refer to FM 3-94, Theater Army, Corps, and Division Operations, for more information about the ASCC.

Corps

The corps headquarters is organized, trained, and equipped to serve as the ARFOR in campaigns and major operations, with command of two or more Army divisions, together with supporting theater-level organizations, across the range of military operations. When required, a corps may become an intermediate tactical headquarters under the land component command, with operational control of multiple divisions (including multinational or Marine Corps formations) or other large tactical formations. The corps headquarters has the capability to provide the nucleus of a joint task force or joint force land component headquarters. The corps normally has one expeditionary sustainment command (ESC) and one medical brigade in direct support. The sustainment brigade normally has a command relationship with an expeditionary sustainment command.

See pp. 6-2 to 6-5 for further discussion.

Sustainment Commands (TSC/ESC)

The TSC synchronizes current and future sustainment operations for an ASCC headquarters. The TSC deploys an expeditionary sustainment command when the TSC determines that a forward command presence is required.

The expeditionary sustainment command is a headquarters which deploys to an area of operations (AO) or joint operations area (JOA). The ESC provides command capabilities when multiple sustainment brigades are employed or when the TSC determines that a forward command presence is required.

The significant difference between TSC and ESC capabilities is scale and scope. The TSC looks across the area of responsibility and shapes sustainment operations. It sets the conditions for successful sustainment operations. The TSC provides guidance to the strategic partners when priority conflicts exist between JOAs. The ESC is focused on the JOA and executing the joint task force or Army forces commander's priorities. The ESC also manages the sustainment mission in the JOA. The TSC maintains oversight of sustainment operations within the operational area with direct coordination with the ESC and its sustainment information systems. This capability provides the TSC commander with the regional focus necessary to provide effective operational-level support to Army or joint task force missions. The TSC may employ multiple ESCs within the theater.

The sustainment command, either the theater sustainment command or the expeditionary sustainment command, is the senior Army sustainment headquarters (less medical) in an area of responsibility in support of the ASCC, Corps or joint task force. The sustainment command plans and coordinates the sustainment functions supporting theater opening and theater closing. They also plan and coordinate theater distribution and sustainment operations in support of Army, joint, interagency, and multinational forces as required. The sustainment command communicates sustainment priorities, as determined by the geographic combatant commander (GCC) and ASCC, to the sustainment brigade commander.

See chap. 7, Theater Support, for further discussion.

Division

The division commands multiple Army brigades and is the Army's primary tactical headquarters for decisive action. When required it may serve as a joint task force or joint force land component headquarters in a limited contingency operation. As required, the division may be the Army component and the joint force land component within a joint task force. Their principal task is directing subordinate brigade operations. Divisions are not fixed formations. They may control more than one type of brigade combat team (BCT). A division can control up to six BCTs with additional appropriate multifunctional supporting brigades. In most cases, deployed sustainment brigades will have a command relationship with a sustainment command and a support relationship with a division.

See pp. 6-5 to 6-10 for further discussion.

B. Support Relationships

Support relationships define the desired purpose, scope, and effect when one capability supports another. Army support relationships are not command authorities and are more specific than joint support relationships.

There are four support relationships in Army doctrine: direct support, reinforcing, general support reinforcing and general support. In the past, these relationships were referred to as field artillery tactical missions. These tactical missions are now referred to as support relationships and are used by the rest of the Army to employ unit capabilities to achieve results required by supported commanders. The sustainment brigade's support relationship is general support unless otherwise ordered. This support is executed through the area support method.

C. Strategic Interface

Ref: ATP 4-93, Sustainment Brigade (Apr '16), pp. 1-5 to 1-9.

The ASCC staff and the sustainment command staff provide the strategic interface for the sustainment brigade. However, there are instances when the sustainment brigade communicates and coordinates directly with unified action partners' representatives to synchronize and integrate support. Unified action partners are those military forces, governmental and nongovernmental organizations, and elements of the private sector with whom Army forces plan, coordinate, synchronize, and integrate during the conduct of operations (ADRP 3-0). The level or degree of interaction will depend on how mature the theater is, what phase of operations and the sustainment brigade's mission. This coordination will be more common for a sustainment brigade supporting operational level forces, a theater opening mission or a largely contracted mission. Sustainment brigade commanders and staff must be familiar with U.S. governmental partners and understand what each partner provides to support ASCC objectives. There are more details about the strategic partners and their actions in support of retrograde of materiel in Appendix A.

Defense Logistics Agency (DLA)

The Defense Logistics Agency provides the Services, other federal agencies, and combined and allied forces with the full spectrum of logistics, acquisition and technical services, including reutilization of military equipment management. The sustainment brigade interfaces with representatives of the DLA organizations which are providing support to them. DLA publishes a Customer Assistance Handbook which includes a description of subordinate DLA organizations and points of contact.

The Defense Logistics Agency supports each geographic combatant commander with a DLA regional commander. This commander is the focal point for coordinating all DLA activities throughout the theater and can provide flexible support on demand. Once the Services identify capability required, the requirement goes through the joint task force commander, the combatant command and to the joint staff for approval. Following approval, DLA organizes its capabilities to meet requirements. The DLA regional commander will stand up a Defense Logistics Agency Support Team, commonly known as DST, to provide direct support to the area of operations.

The DLA support team provides logistical support to conflicts, disasters (both natural and man-made), emergencies, mobilizations and other contingency operations around the world. DLA support teams are responsible to the combatant commander. The DLA support team works directly with the sustainment command and integrates materiel management support of DLA common commodities such as subsistence, protective clothing, general supplies, and bulk petroleum. They also provide disposal support as appropriate including the disposal of hazardous wastes. The DLA support teams require protection, life support, use of common-user land transport assets and may need terrain prepared for their operations.

U.S. Army Materiel Command

The United States Army Materiel Command (USAMC) provides technology, acquisition support, materiel development, logistics power projection, and sustainment (less medical) to the Army. Three of the USAMC's major subordinate commands that have important roles providing national-level support to the sustainment brigade are the U.S. Army Military Surface Deployment and Distribution Command, the Army Contracting Command and the Army Sustainment Command.

Civil Affairs Operations

Civilian organizations, such as other government agencies, intergovernmental organizations, and nongovernmental organizations, bring resources and capabilities that can help establish host-nation civil authority and capabilities. Sustainment brigades may be required to support stability and foreign humanitarian operations that are often sustainment intensive. In these operations, sustainment brigades often work closely with or directly support intergovernmental, non-governmental and other agencies and organizations. This support may include ground transportation, provision of equipment and supplies, port operations and must be specifically authorized by the Secretary of Defense. Sustainment brigade commanders and staff must be familiar with the legal authorizations to provide support to the interagency and inter/non-governmental organizations or indigenous populations and institutions.

Contractors

Commanders can expect that contractors will be involved in operations. The management and control of contractors differs from that of Soldiers and Department of the Army civilians. During military operations, Soldiers and Army civilians are under the control of the military chain of command. Commanders can direct Soldier and Army civilian tasks, special recognition, and disciplinary action. However, they do not have the same control over contractors. The terms and conditions of the contract establish relationships between the military and the contractor.

Multinational

Sustainment of forces is a national responsibility. However, certain efficiencies and effects can be obtained through sharing, supporting and/or receiving support from allied or coalition forces. Chapter 138 of Title 10 United States Code (USC) authorizes exchanging support between U.S. services and those of other countries. Depending on the extent or scope of the multinational support agreement, sustainment brigades may be required to coordinate, control and support these types of operations.

Host Nation

Host-nation support is civil and military assistance rendered by a nation to foreign forces within its territory during peacetime, crises or emergencies, or war based on agreements mutually concluded between nations. Host-nation support may include the use of sea and aerial ports of debarkation, warehousing for storage, transportation assets, personnel such as stevedores and other distribution related or supported capabilities. Many host nation support (HNS) agreements have already been negotiated between existing allies. There are certain sustainment efficiencies that can be achieved to facilitate a unity of effort through the use of host nation (HN), allied and intergovernmental organization agreements. These can be pre-existing agreements or agreements that are generated after deployment to a theater. A comprehensive analysis of HN capabilities and plans for incorporating these resources provides sustainment commanders with an array of options. Note that HNS is different than contracted support.

Other

In addition to doctrinally defined relationships, there are critically important professional relationships that cannot be categorized as either a command or a support relationship. The sustainment brigade should develop and participate in partnership activities to build the capacity of partners to secure populations, protect infrastructure, and strengthen institutions. This engagement builds a strong relationship with the indigenous populations and institutions which also benefits the sustainment brigade's efforts to obtain contracted transport and goods.

Army National Guard sustainment brigades have official relationships that are not doctrinal. The State Governor commands Army National Guard sustainment brigades until they are federalized. They exercise command through the state adjutants general. When a National Guard sustainment brigade is activated by the governor of the state, they establish doctrinal command and support relationships with other military units, if there are any. The sustainment brigade will have non-doctrinal relationships with other organizations or state agencies that they are supporting or supported by.

Joint Forces

The Services are responsible for operational logistics support systems, platforms, and their execution to support the force. However, the sustainment brigade may provide common user support, common-user logistics, and common-land transportation support to the joint force. If this is the case, the joint force commander (JFC) will annotate the details in orders. A sustainment brigade providing support to joint forces is not a command or center for joint logistics, it is a brigade supporting a joint force, as per the operations order

See chap. 8, Joint Logistics, for further discussion. Refer to FM 4-95, Logistics Operations, for more information on the Army's responsibilities as executive agent to other Services.

Special Operations Forces

The sustainment brigade special operations (SO) airborne (ABN) is a unique Army sustainment brigade because it maintains global situational awareness of deployed Army special operations forces logistics support structures. The sustainment brigade (SO) (ABN) sets the operational-level logistics conditions in order to enable Army special operations forces missions. It is assigned to United States (U.S.) Army Special Operations Command and focuses on operational to tactical sustainment support. During periods where only special operations forces are operating in a theater, support may be executed under the sustainment brigade (SO) (ABN).

ATP 3-05.40, Special Operations Sustainment, provides more details on special operations sustainment.

When deployed, the sustainment brigade (SO) (ABN) acts as the single logistics headquarters for a joint special operations task force. The sustainment brigade (SO) (ABN) integrates Army special operations forces support requirements into the ASCC support plan and ensures a timely response to Army special operations forces requirements. The sustainment brigade (SO) (ABN) may also serve as an early entry control element for one CSSB in support of a conventional force expansion in the theater of operation until relieved by a conventional sustainment brigade. The sustainment brigade provides general support to special operations forces.

Transportation Brigade Expeditionary

The transportation brigade expeditionary is a transportation headquarters controlling all assigned and attached units while managing and conducting seaport operations. The transportation brigade expeditionary is task organized with battalions engaged in non-permissive port opening, water terminal, and watercraft operations to meet mission requirements. The transportation brigade expeditionary is attached to a sustainment command. The sustainment brigade provides support to the transportation brigade expeditionary and coordinates with the staff to synchronize onward movement as part of RSOI.

Brigade Support Battalion (BSB)

Brigade support battalions (BSB) are organic to BCTs and are tailored to support its brigade. Aviation brigades have a similar support unit, the aviation support battalion. The BSB provides supply, field maintenance, transportation, and Role 2 medical support to the supported brigade. The aviation support battalion provides aviation and ground field maintenance, brigade-wide satellite signal support, resupply of all commodities, and Role 1 medical support. The sustainment brigade provides the support these support battalions are lacking, such as water purification and storage, petroleum storage and transportation support.

The sustainment brigade commander may recommend a direct support relationship if mission and operational variables indicate. Support relationships designated by appropriate orders will be used to specify the details of the support relationship, including the designated priorities of support.

See chap. 5, Brigade Support, for further discussion of the BSB.

II. Sustainment Brigade Employment

Ref: ATP 4-93, Sustainment Brigade (Apr '16), chap. 5.

The sustainment brigade plays an important role in providing maneuver forces with operational reach, prolonged endurance, and freedom of action. It is task organized to conduct sustainment functions associated with theater opening, theater distribution, sustainment, and theater closing missions. The sustainment brigade will have multiple CSSBs to provide distribution and sustainment support to forces operating in or transiting its assigned support area. There are examples of sustainment brigade task organization for deployed sustainment brigade missions.

I. Joint Operations

A joint force is one composed of elements of two or more military departments operating under a single JFC. Army forces operate as an interdependent joint force; there is a purposeful reliance by one service on another service' capabilities. Sustainment brigade commanders and staffs operate as part of an interdependent joint force and may be required to provide common-user logistics support to another service or multinational force. Commanders and staff must be aware of which logistics capabilities are shared by the joint forces so that the sustainment brigade provides the required support. Regardless of command and support relationships established by the JFC, Service component commanders retain responsibility for administrative and logistics support of their forces.

The U.S. Army has a major role in joint operations by conducting theater opening tasks and setting logistics conditions for conducting military operations in a theater. The Army may also provide common-user logistics for the joint force to include class I, class III, class IX and transportation.

The GCC may designate the senior logistics headquarters of a service as a joint command for logistics. The joint command for logistics is responsible for coordinating and synchronizing logistics support across the joint force. If designated, the GCC would augment the selected unit with joint augmentation.

The sustainment brigade may be required to establish or operate an intermediate staging base. The intermediate staging base is a secure temporary theater staging facility. They may be used to sustain forces in the area of operations. Intermediate staging base required capabilities are contingent on the operational situation and are located where they can best support the force.

Units establishing the intermediate staging base must deploy early to be prepared to receive deploying forces and to operate the nodes inherent in the theater distribution plan. Intermediate staging bases serve as a principal staging base in order to secure a lodgment to project the force for the rapid delivery of combat power to an AOR.

Units staging through an intermediate staging base will require life support, including housing, sanitation and health care. This life support may be as simple as improvised shelters and unit field sanitation or as complex as a force provider module. Planners must identify force provider requirements early in the planning phase. All the force provider companies are in the reserve component and the modules are in prepositioned stock.

Refer to ATP 4-45, Force Provider Operations, for more information.

The theater sustainment command is responsible for sustainment support to an area of responsibility. An area of responsibility is the geographical area associated with a combatant command within which a geographic combatant commander has authority to plan and conduct operations (JP 1). Theater of operations, area of operations or joint area of operations may all be included in an AOR.

II. Theater Opening

Ref: ATP 4-93, Sustainment Brigade (Apr '16), pp. 5-2 to 5-3.

Theater opening is the ability to establish and operate ports of debarkation (air, sea, and rail), to establish a distribution system, and to facilitate throughput for the reception, staging, and onward movement of forces within a theater of operations (ADP 4-0). Theater opening is a complex joint process involving the JFC and strategic and joint partners such as U. S. Transportation Command, DLA and the joint deployment and distribution operations center. The joint deployment and distribution operations center supports the geographic combatant commander's operational objectives by synchronizing and optimizing the interface of inter-theater and intra-theater distribution.

The sustainment command is responsible for planning and executing sustainment tasks enabling theater opening and RSOI. They also develop the theater sustainment concept of support.

When given the mission to support theater opening, the sustainment brigade is task organized with CSSB or functional logistics battalions. It performs the following tasks: establish in-transit visibility; conduct transportation management; support theater RSOI; conduct distribution and distribution management; support movement control, support expeditionary contracting efforts and establish initial theater sustainment.

Theater Opening

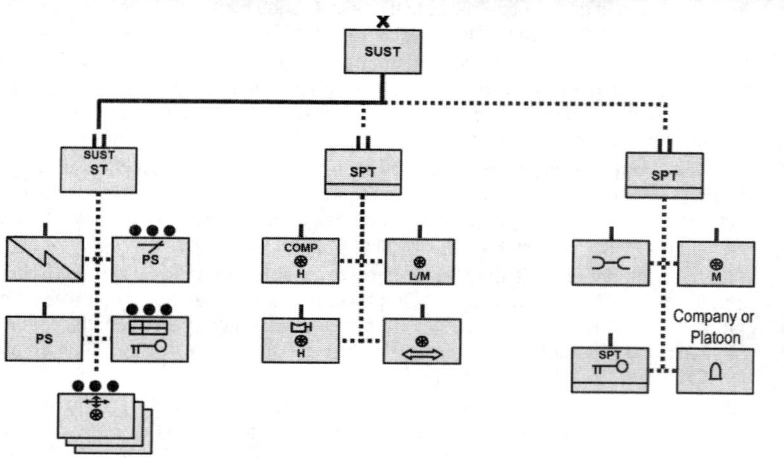

ATP 4-93, fig. 5-1. Notional task organized sustainment brigade conducting theater opening tasks.

In addition to the task organized sustainment brigade, a transportation brigade expeditionary is also in the JOA directing logistics over the shore or port opening operations. The sustainment brigade will become larger as the operational area and missions mature. The longer the operation lasts the more likely reserve component units are activated.

The sustainment command identifies initial theater opening requirements for contract and host-nation support. The deployed sustainment brigade works with the supporting contingency contracting team or battalion which assesses and acquires available HN infrastructure capabilities and contracted support. Working together, units conducting theater opening functions set the conditions for effective support and lay the groundwork for subsequent expansion of the theater distribution system.

Port Opening

Port opening is a subordinate function of theater opening. Port opening is the ability to establish, initially operate and facilitate throughput for ports of debarkation to support unified land operations (ADRP 4-0). Throughput refers to the quantity of cargo and passengers that can pass through a port or a transportation terminal on a daily basis. This is not to be confused with throughput distribution, which is a method of distribution explained in ATP 4-11, Army Motor Transport Operations. The port opening process is complete when the ports of debarkation and supporting infrastructure are established to meet the desired operating capacity for that node. The port is the first node in the theater distribution system. Seaport opening requires units designed to support seaport of debarkation operations; such as harbor master and watercraft detachments.

Reception, Staging, Onward Movement, and Integration (RSOI)

RSOI is a joint task force operation heavily enabled by the sustainment community. The sustainment command controls the theater distribution system which includes transportation modes and transshipment and supply nodes. Nodes are locations within the distribution system where a movement requirement is originated and processed for onward movement.

- **Reception** is the initial step in RSOI and is defined as receiving units into the area of operation. Reception planning and execution is the responsibility of the commander assigned the responsibility for RSOI. Synchronizing transportation reception activities are critical to facilitating throughput at the ports of debarkation which involves mission command, movement control, and port operations. The sustainment brigade establishes theater gateway personnel accountability, performs departure/arrival airfield control group functions at the aerial port of debarkation, conducts motor transport operations and provides sustainment support for units supporting reception tasks.
- **Staging** is that part of the RSOI operation that reassembles and reunites unit personnel with their equipment and schedules unit movement to the tactical assembly area, secures or uploads unit basic loads, and provides life support to personnel. A staging base is a controlled area where unit reassembly may occur. There will be at least one staging base per seaport of debarkation /aerial port of debarkation pairing. The sustainment brigade provides supply, maintenance support and human resources support for units executing staging tasks. Units staging through a staging base will require some life support.
- **Onward movement** involves unit movement from ports to theater staging bases or forward to the tactical assembly area. The primary factors affecting onward movement are transportation capability and capacity, physical infrastructure, and protection. The sustainment brigade coordinates motor transport support for units without sufficient transportation lift to move themselves or mitigate degradation of tracked assets over long distances while conducting onward movement. Personnel and equipment reassembled as combat-ready units are moved to the tactical assembly area based on the JFC priorities.
- **Integration** is the synchronized transfer of capabilities into an operational commander's force prior to mission execution. The transfer may require interaction and familiarization among units and that arriving units meet certain standards before being completely integrated into the combat plan.

See chapter 9, Deployment & Redeployment, for further discussion of RSOI.

III. Support to Decisive Action

The sustainment brigade will support decisive action operations which are characterized by units conducting offense, defense and stability tasks simultaneously. The sustainment brigade plays an important role in providing maneuver forces with operational reach, prolonged endurance, and freedom of action. Sustainment brigade units will operate in and traverse through areas wherein the threat includes a variable mix of conventional military units, special operations forces, well-armed and highly motivated paramilitary forces, loosely organized irregular forces, and armed, organized criminal organizations, including terrorists. Supported units are confronting anti-access and area-denial threats which possess sophisticated and asymmetric capabilities.

Leveraging host-nation support can be a critical enabler to operations if there is a constrained flow of forces into theater resulting in a delay of sustainment forces to the joint operational area. The degree to which a host nation can assist with logistics will depend on the operation's circumstances and impacts how much the sustainment brigade may leverage local markets and work force. This information will be known when a thorough sustainment preparation of the operational environment is completed.

See chap. 2, Sustainment Operations, and chap. 3, Sustainment Execution, for related discussion.

A. Protection

Sustainment brigade and CSSB commanders ensure the protection tasks are integrated into all aspects of operations to safeguard personnel, systems, and physical assets. Personnel includes combatants and noncombatants (contractors, host nation support and refugees). Commanders and staffs synchronize, integrate, and organize capabilities and resources throughout the operations process in order to preserve combat power and mitigate the effects of threats and hazards.

See pp. 6-31 to 6-38 for further discussion of protection considerations. Refer to JP 3-0, Joint Operations, and ADRP 3-37, Protection, for more information about protection.

B. Supporting the Force

Sustainment enables the tactical commander to maintain combat power. The sustainment brigade supports Army forces at the tactical and operational levels. Its focus is the continuous management and distribution of supplies, human resources support, financial management support, and allocation of maintenance to prolong supported commander's operational endurance. These functions are generally categorized in to two missions; sustainment and theater distribution. A small scale or short duration operation may have a sustainment brigade task organized for sustainment and theater distribution. A sustainment brigade may be task organized to conduct only theater distribution operations.

See the following pages (pp. 6-28 to 6-29) for and overview and further discussion of supporting the force.

C. Redeployment

Redeployment is the transfer or rotation of forces and materiel to support another joint force commander's operational requirements, or to return personnel, equipment, and materiel to the home and/or demobilization stations for reintegration and/or out processing (JP 3-35). Planning for redeployment should be considered early and continue throughout the operation. It is best accomplished in the same time-phased process in which deployment was accomplished. Redeployment processes are the same whether it is a company sized unit, BCT or redeployment as a result of theater closing.

See chapter 9, Deployment & Redeployment, for further discussion.

D. Emplacing the Sustainment Brigade

Ref: ATP 4-93, Sustainment Brigade (Apr '16), pp. 5-5 to 5-6.

The sustainment brigade will typically establish a base within its assigned support area to provide centralized control of operations. Decentralized execution of operations may be conducted within the base or at designated locations within its support area. Unless a sustainment brigade is direct support to a division, the area in which a sustainment brigade operates is different than divisional boundaries. For example, one sustainment brigade may provide support to more than one division (or major combat force) or, more than one sustainment brigade may provide support to a single division. The sustainment command remains a part of the coordination link for all support.

The maneuver enhancement brigade may be responsible for the terrain assignment and establishing secure movement corridors. The sustainment brigade base will be integrated into area terrain management and protection plans will be based on established command and support relationships and the physical space occupied. Within the support area, the sustainment brigade answers to the maneuver enhancement brigade, or the identified terrain owner, for protection, security, and related matters. The sustainment brigade's assigned support area is determined by the physical array of forces, support requirements during progressive phases, proximity of sustainment capabilities, natural land boundaries, available road networks, and other considerations.

Throughout operations, the sustainment brigade commander and staff consider the proximity of the sustainment brigade assets in relation to supported units. Proper proximity to the supported unit affects timely support and must be considered in terms of time and distance. There are many aspects of the operational environment that affect decisions on where to position units. Mission analysis is required to ensure prudent decisions are made on both the number and placement of sustainment units in relation to the supported organizations.

Proximity of the sustainment brigade assets to supported unit is very important since they most often use area support as the method of support. Area support is a method of logistics, medical support, and personnel services in which support relationships are determined by the location of the units requiring support (ATP 4-90). It is a method in which general support is provided to units operating within an established geographic boundary. Sustainment units use this method to support units located in or passing through a specified geographic location.

As supported units move about the area of operations, the CSSB or its subordinate units may have to relocate to remain an appropriate distance from supported units. The time it takes for support to reach the supported unit given the operational conditions at that time is the primary consideration for placement. Generally, the CSSB should be able to distribute commodities to supported units and return the same day.

Consider the following factors when determining proximity of supported units:

- Threat order of battle and activity
- Available road network and condition
- Fuel consumption and en route fuel capability
- Frequency of support missions
- Close air support
- Medical evacuation support
- Route clearance support
- Route security support
- Weather

Supporting the Force

Ref: ATP 4-93, Sustainment Brigade (Apr '16), pp. 5-6 to 5-14.

Sustainment enables the tactical commander to maintain combat power. The sustainment brigade supports Army forces at the tactical and operational levels. Its focus is the continuous management and distribution of supplies, human resources support, financial management support, and allocation of maintenance to prolong supported commander's operational endurance. These functions are generally categorized in to two missions; sustainment and theater distribution. A small scale or short duration operation may have a sustainment brigade task organized for sustainment and theater distribution. A sustainment brigade may be task organized to conduct only theater distribution operations.

Sustainment brigades conducting sustainment are task organized with CSSBs that are further task organized with supply, services, transportation, and maintenance units. The brigade operates multi-class SSAs in accordance with sustainment command's directives; provides field level maintenance support and field services; and replenishes the brigade support battalions and echelons above brigade units. Figure 5-3 is an example of a sustainment brigade task organized for sustainment tasks.

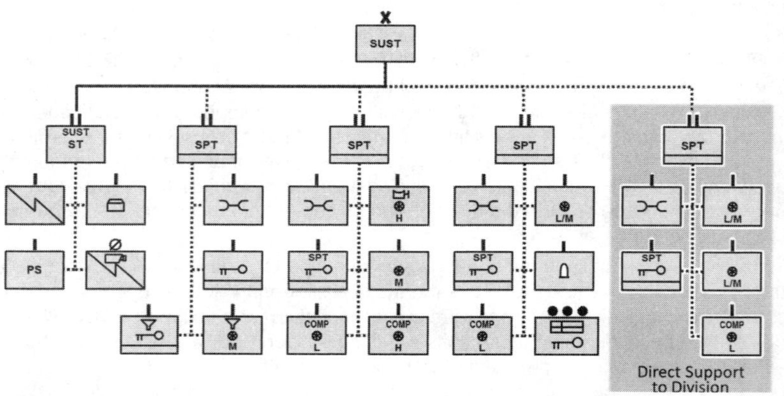

ATP 4-93, fig. 5-3. Notional task organized sustainment brigade conducting sustainment operations.

The notional sustainment brigade in figure 5-3 is task organized to support multiple BCTs and echelon above brigade units with replenishment of all classes of supply, field services and maintenance. This sustainment brigade also maintains visibility of the distribution system (including readiness of Army air and ground transport assets), reallocates resources to maintain optimal system performance and controls the common-user transportation assets as required.

The CSSB in figure 5-3 are task organized to provide support to units operating in its assigned support area. One CSSB is task organized to provide direct support to a division. It is specifically organized with a multi-capable quartermaster supply company and a composite truck company. These companies provide water purification, petroleum storage, and troop transport to the BCT. This task organization complements the BSB capabilities and provides capabilities the BSB doesn't have.

There may be a requirement to conduct throughput distribution to a specific task force or small organization. The sustainment brigade may consider establishing a forward logistics element to support a unique requirement. A forward logistics element is comprised of task-organized multifunctional logistics assets designed to support fast-moving offensive operations in the early phases of decisive action (ATP 4-90).

A sustainment brigade conducting theater distribution tasks is organized with CSSB that are further task organized with functional quartermaster and transportation units. The brigade is focused on receiving and distributing materiel to and from one node to another for further distribution. It is supply point to supply point with no direct replenishment to units. Figure 5-4 is an example of a sustainment brigade task organized for theater distribution tasks.

The notional sustainment brigade in figure 5-4 is task organized to establish and operate multimodal distribution hubs, synchronize multiple node operations (an SSA, an ammunition supply point, a centralized receiving and shipping point, and a convoy support center), maintain visibility of the distribution system (including readiness of Army air and ground transport assets), reallocate resources to maintain optimal system performance and control the common-user transportation assets as required. One CSSB is task organized to transship cargo, operate regional distribution hubs, including centralized receiving and shipping points.

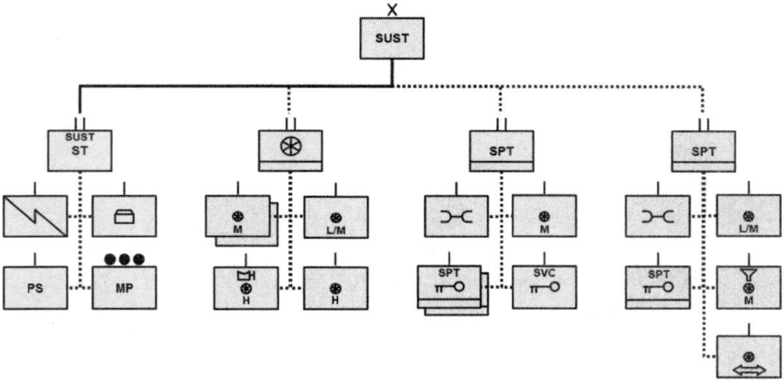

Ref: ATP 4-93, fig. 5-4. Notional task organized sustainment brigade conducting theater distribution operations.

As mission requirements change, the sustainment command could determine functional battalion headquarters are required. Functional Transportation and Quartermaster battalions are limited in their ability to control multi-functional sustainment units but may be appropriate in some operational environments. Planners must consider the decisive action tasks and projected duration of operations that equate to requirements over time and space to determine the appropriate unit mix. Commanders may augment a battalion staff with Soldiers who have functional expertise if required.

Support Operations

The sustainment brigade support operations plans, coordinates and synchronizes sustainment operations. The sustainment brigade SPO conducts:

- Materiel management
- Sustainment automation support
- Distribution planning and integration
- Transportation operations
- Human resources operations
- Operational contract support

See chap. 2, Sustainment Operations, and chap. 3, Sustainment Execution, for related discussion.

IV. Theater Closing

Theater closing is the process of redeploying Army forces and equipment from a theater, the drawdown and removal or disposition of Army non-unit equipment and materiel, and the transition of materiel and facilities back to host nation or civil authorities (ADP 4-0). Theater closing begins with the termination of operations.

Termination criteria accounts for a wide variety of operational tasks that the joint force may need to accomplish, including disengagement, protection (including force health protection support to conduct retrograde cargo inspections and pest management operations), transition to post-conflict operations, and redeployment. Planning for the transition from sustained combat operations to the termination of operations, and then a complete handover to civil authority, begins during plan development and continues throughout all phases of a campaign or major operation.

During theater closing the sustainment command coordinates with the JFC's planning team. Sustainment decisions should always consider the eventuality of redeploying Army forces and equipment from a theater. The execution of theater closing tasks is synchronized with tactical commanders, base commanders and strategic partners, including supporting contractors.

If the sustainment brigade is designated as the headquarters responsible for overseeing theater closing tasks, it will focus on redeployment, drawdown of non-unit materiel, and transitioning of materiel, facilities and capabilities to HN or civil authorities. Considerations include:

- What were the property accountability policies during the operation?
- How long was the operation?
- What classes of materiel are part of the retrograde of materiel?
- Does retrograde of materiel include containers? How many?
- Who else is conducting theater closing tasks?
- Which strategic partners are identified (USAMC, DLA or U.S. Transportation Command [USTRANSCOM])?
- Are there clear lines of property accountability?
- Is the sustainment brigade responsible for theater closing tasks in their assigned support area,throughout the entire JOA, or some other geographic area?
- What is designated for foreign military sales?
- Has the GCC staff identified which real property transfers to local government?

The CSSB and its subordinate units, execute the redeployment plan as directed by the sustainment brigade. Planners in the sustainment brigade ensure there are dedicated units redeploying forces, equipment, and retrograde of materiel while another unit is focused on continued area support. Key functions during this phase are transportation support, staging and upload of strategic lift, movement control, maintenance and recovery support, and field services. Examples of teams organized to accomplish base closure and transfer, redistribution, retrograde, and disposal of materiel are outlined in appendix A.

Transitioning facilities to HN or civil authorities is another aspect of theater closing. Although basing is an engineer responsibility, a sustainment brigade could be assigned as the headquarters responsible for closing bases in a joint operational area.

Refer to ATP 3-37.10, Base Camps, for more information.

Strategic partners are critical to theater closing. The USAMC is responsible for pre-positioned materiel and theater provided equipment (TPE). The AFSB and the defense logistics agency are responsible for identifying where equipment requiring sustainment maintenance is shipped. The USAMC representative provides disposal codes and a DLA representative identifies locations where equipment and materiel are disposed.

III. Protection Considerations

Ref: ATP 4-93, Sustainment Brigade (Apr '16), chap. 5 (protection responsibilities). The remainder of material in this section is retained from previous SMFLS editions (adaptations from FMI 4-93.2, Sustainment Brigade, Feb '09, app. B and JP 3-11).

Protection consists of those actions taken to prevent or mitigate hostile actions against DOD personnel (to include family members), resources, facilities, and critical information. Additionally, counter proliferation and consequence management actions associated with chemical, biological, radiological, nuclear, and high yield explosive weapons, which includes toxic industrial material and improvised explosive devices (IED) should be addressed.

I. Responsibilities of the Sustainment Brigade

Sustainment brigade and CSSB commanders ensure the protection tasks are integrated into all aspects of operations to safeguard personnel, systems, and physical assets. Personnel includes combatants and noncombatants (contractors, host nation support and refugees). Commanders and staffs synchronize, integrate, and organize capabilities and resources throughout the operations process in order to preserve combat power and mitigate the effects of threats and hazards.

Sustainment commanders plan for all the supporting tasks of the protection warfighting function but often focus on coordinating security operations conducted to protect friendly forces, installations, and routes in their assigned support area. Sustainment brigade and CSSB commanders dedicate assets to protection tasks and systems based on an analysis of the operational environment, the likelihood of threat action, and the relative value of friendly resources and populations. Criticality, vulnerability, and recoverability are some of the most significant considerations in determining protection priorities. The list below includes some of the activities sustainment commanders consider as protection priorities:

- Base and base camp defense
- Critical asset security
- Node protection
- Response force operations. The sustainment brigade establishes a response force to protect the base it is occupying and coordinates for additional security.
- Lines of communication security. The sustainment brigade coordinates this with the terrain owner.
- Convoy security

Base camps may evolve from unit locations established during major combat or other military operations. These base camps may start out as a single unit or very small number of units that are capable of providing protection for their assets with organic and attached supporting capabilities. In these cases, the senior commander

The protection warfighting function is the related tasks and systems that preserve the force so the commander can apply maximum combat power to accomplish the mission. Commanders incorporate protection when they understand and visualize threats and hazards in an operational environment. Refer to AODS6-1 (w/ SMARTupdate 1): The Army Operations & Doctrine SMARTbook (Guide to FM/ADP 3-0 Operations & the Elements of Combat).

on site executes all base camp responsibilities in a dual-hatted role. As the base camp grows in size and complexity, specifically the number and variety of units and missions being supported, the assumed additional responsibilities can detract from the senior commander's focus on the primary mission.

As a base camp grows in size, a specialized base camp commander and staff may be required to minimize the demand on the senior tenant unit and free that commander from the details involved with managing a base camp. Units that will potentially perform this role include regional support groups, construction battalion maintenance units, and maneuver enhancement brigades.

Although primarily aligned with the protection warfighting function, base defense is enabled by all of the warfighting functions. Base camp defense includes the activities needed to defeat Level I and Level II threats to a base camp or base cluster, and shape or delay Level III threats until they can be defeated by a tactical combat force or other available response forces that is part of the higher commander's area security efforts.

Refer to ATP 3-37.10, Base Camps, for more information on base camp operations, security and defense, and transfer and closure.

In the sustainment brigade, the S-3 plans and SPO participate in the detailed planning and coordination for support operations. They ensure the logistics support plan and the protection plans are coordinated and nested within the command and with higher headquarters plans. The published rules of engagement also shape how we protect the force.

During execution, the S-3 current operations, the S-2 and the SPO mobility branch staff integrate their efforts to conduct cross boundary coordination for lines of communication. Convoys may include contracted drivers or contracted trucks with drivers.

Sustainment brigade commanders consider how evolving relevant operational or mission variables affect force employment concepts and tactical actions that contribute to the brigade's mission. Awareness of potential ground threats is especially critical in the area of sustainment operations in countering improvised explosive devices, military grade land mines and explosively formed penetrators. Lethal threats likewise are of concern to the convoy planner. Successful sustainment operations result from the fusion of current intelligence, Soldiers familiar with current tactics, techniques and procedures, standard operating procedures, and deliberate and careful use of counter improvised explosive devices enablers.

Refer to ATP 3-90.37, Countering Improvised Explosive Devices, for more information about countering improvised explosive devices.

Sustainment brigade commanders and staffs should continue to refine their understanding of protection and their ability to prepare and execute protection plans.

FM 3-21.8, The Infantry Rifle Platoon and Squad, includes fundamental of tactics and essentials for developing site or protection plans. This manual also includes a chapter on defensive operations and an explanation of engagement area development.

II. Fire Support Considerations

The sustainment brigade, and all subordinate commanders and staffs, must have a thorough understanding of fire support procedures, organizations, and assets that are available within their AO. Normally, the ESC will have a dedicated fires battalion assigned for fire support that a sustainment brigade may call upon. However, this unit might not be available until the theater of operations is mature.

The sustainment brigade S-3 must include fire support considerations into their planning and the fire support information must be disseminated to all subordinate elements. All fire support assets need to be assessed (artillery, mortar, UAS, CAS, and naval) to determine availability and coordination requirements. Planning should focus on close coordination with the fires units and detailed procedures to request fires support. It should be routine for all personnel to receive training on call for fire procedures and to rehearse the procedures.

III. Protective Measures

Ref: Adapted from FMI 4-93.2, Sustainment Brigade (Feb '09), pp. B-8 to B-10.

Sustainment brigade units take several measures to reduce their vulnerability to enemy operations. These measures include dispersion, cover, concealment, camouflage, intelligence gathering, obstacles, and air and missile defense.

A. Dispersion

Sustainment brigade organizations disperse as much as possible throughout the assigned AO. Dispersion as a protective measure is balanced against the potential deficits to support operations and the base cluster defense system. Dispersion helps avoid catastrophic damage from air and mass destruction weapons. Even if a sustainment brigade unit is not the primary target, it may be attacked as a target of opportunity. The dispersion required depends on the following:

- Type of threat
- Terrain
- Defensibility

B. Engineer Support

Engineer survivability support will be important for the elements of the sustainment brigade. When available, it may be used for a variety of protection hardening measures in support of the sustainment brigade to include survivability (see FM 5-103, Survivability Operations) support and support to camouflage, concealment, and decoys. Engineers may also provide geospatial support to the sustainment brigade's protection efforts.

C. Cover, Concealment, and Camouflage

The enemy cannot target sustainment brigade resources that it cannot detect. Cover, concealment, and camouflage remain critical to protecting sustainment units, facilities, and supplies from enemy detection and attack. Cover includes natural and artificial protection from enemy observation and fire. When selecting sites, advance parties consider the type of cover available. Concealment includes natural or artificial protection from enemy detection. Sustainment brigade units use concealed ingress and egress points and halt locations within support locations. Camouflage consists of using natural or artificial objects or tactical positions to confuse, mislead, or evade the enemy.

D. Obstacles

Obstacles slow, impede, or channel enemy movement and incursion. They buy time until reaction forces can deploy or a response force can arrive. Effective use of obstacles involves sound counter-mobility planning and early warning. Obstacles in urban environments are as important as in the field. Strategically placed obstacles provide protection against terrorist access to buildings.

E. Air and Missile Defense

Air defense artillery (ADA) forces cannot provide dedicated air and missile defense (AMD) for all sustainment brigade assets in the AO. The commander positions brigade organizations to take advantage of coverage that available AMD forces provide. Using base clusters makes it possible for AMD units to cover more Brigade assets than if units disperse throughout the AO, but reduces the benefits of dispersion. Brigade assets identified AMD priorities that do not receive dedicated support are positioned to take advantage of the coverage provided by AMD units protecting higher-priority assets.

Passive air defense operations include the means a unit uses to avoid enemy detection, along with measures to minimize damage when attacked. Sustainment brigade units use OPSEC to conceal their location from enemy visual and electronic surveillance. Elements within base clusters disperse as much as possible. Dispersal along with field fortifications and obstacles significantly reduce casualties and damage from air and missile attack.

IV. Bases and Base Clusters

Ref: Adapted from FMI 4-93.2, Sustainment Brigade (Feb '09), pp. B-14 to B-16.

The sustainment brigade must integrate its HQ and subordinate elements FP plans into the base and base cluster defense plans. These plans are integrated with the defense plans of the MEB or BCT having the overall responsibility for the AO, and are copied to the higher sustainment commander. This defense method protects elements from level I threats in their assigned areas. Commanders ensure all bases and base clusters in their AO's train and prepare for their roles. Cooperation and coordination elements are critical.

Bases and base clusters form the basic building block for planning, coordinating, and executing base defense operations. The sustainment brigade S-3, with input from the sustainment brigade SPO, organizes units occupying the support HQ AO into base clusters. The sustainment brigade SPO recommends appointments of base or base cluster commanders from units in the cluster to the sustainment brigade S-3. The base cluster commander is usually the senior commander in the base cluster. The base cluster commander forms a base cluster defense operations center (BCOC) from the staff and available base assets.

A base may be a single-service or a joint-service base. The base cluster commander appoints the base commanders. Base commanders form base defense operating centers (BDOCs).

Within the base cluster, three commanders have distinct responsibilities. These three — the individual unit commander, the base commander, and the base cluster commander — are discussed below. Non-sustainment units residing within a base which a sustainment brigade commands will have a command relationship and responsibilities established by order to the sustainment brigade for protection.

Individual Unit Commanders

The commanders of units in a base are responsible for the following:

- Participating in base defense planning
- Providing, staffing, and operating base defense facilities in accordance with base defense plans
- Conducting individual and unit training to ensure their forces' readiness to perform their assigned tasks in defense of the base
- Providing appropriate facilities and essential personnel for the BDOC and the base commander
- Providing liaison personnel to advise the base commander on matters peculiar to their units
- Providing internal security of the base
- Providing communications systems, including common-user communications, within the command

Sustainment brigade units use observation posts, listening posts, or unattended sensors on likely avenues of approach to collect intelligence on threat activity. In areas where the populace is friendly, local law enforcement or government agencies can provide information on threats in the area. BCOCs implement an integrated warning plan within their cluster and with adjacent bases or base clusters.

Base Commander

The base commander is responsible for base security and defense. All forces assigned to the base are under OPCON for base defense purposes. The base commander's responsibilities for base defense include:

- Establishing a BDOC from available base assets to serve as the base's tactical operations center (TOC) and focal point for security and defense. The BDOC assists with planning, directing, coordinating, integrating, and controlling base defense efforts.
- Establishing an alternate BDOC from base resources or, if base assets are not available, designating a HQ element from units dedicated to the base for its local defense
- Planning for including transient units by ensuring that base defense plans include provisions for augmenting the regularly assigned base defense forces with units present at the base during periods of threat

Base Cluster Commander

The base cluster commander is responsible for securing the base, coordinating the defense of bases within the base cluster, and integrating base defense plans into a base cluster defense plan. Specific responsibilities include:

- Establishing a BCOC from the staff and available base or base cluster assets to serve as the base cluster's TOC and focal point for planning, directing, coordinating, integrating, and controlling base cluster defense activities
- Providing appropriate facilities, housing, and services for necessary liaison personnel from bases from within the cluster

Base and Base Cluster Defense Plan

Base and base cluster commanders develop and implement comprehensive defense plans to protect their support capability. The defense plan includes measures to detect, minimize, or defeat level I and defend level II threats. To maximize mutual support and prevent fratricide, the base and base cluster commanders coordinate defense plans with adjacent base and base clusters and joint, multinational, and HN forces. The sustainment brigade S-3 ensures that all plans conform to the overall TSC or BCT security plans.

Area Damage Control (ADC)

Commanders of bases and installations within the JSA coordinate requirements for area damage control (ADC) with the JSC through their respective chains of command. Commanders establish priorities for ADC missions as part of their planning process at the base or installation level. All units are responsible for providing ADC within their base or installation to the extent of their capabilities.

The sustainment brigade takes ADC measures before, during, and after hostile action or natural disasters to reduce the probability of damage, to minimize its effects, and to reestablish normal operations. Necessary repair begins after the damage is contained.

Other forces and assets that contribute to the ADC mission include—

- Ordnance
- MP
- Chemical
- CA
- Maintenance
- Medical
- Signal
- Supply
- Transportation
- Transiting units

HNS can be a vital resource for ADC in the AO. Early HNS identification and coordination are essential to supplement ADC efforts. Responsibilities and support from HN assets are negotiated at theater level and are part of the status-of-forces agreements and treaties.

V. Convoy Security

The most serious threat faced by the deployed sustainment brigade elements occurs when moving. The sustainment brigade must coordinate with maneuver, military police, and medical units operating in the AO for additional security capability, medical support, and/or route security assessments.

Movement control always includes convoy defense considerations. Supply routes are assumed not to be secure on a high threat area. Therefore, convoy movements between supported unit areas are combat operations. The TSC and supported unit commanders and staffs work together to integrate defensive capabilities into convoys. Adequate convoy security depends on two critical components. These are thorough staff planning to counter enemy plans and capabilities and individual Soldier training to counteract enemy action. All convoy personnel must be familiar with published rules of engagement and local tactics, techniques, and procedures developed for the current situation on the convoy route.

The convoy commander ensures that troops are trained in convoy defense techniques. The damage a convoy prevents or incurs when attacked often depends on the adequacy of convoy defense training. It also depends on the route and timing of the convoy in relation to the enemy situation and the adequacy of the intelligence and information convoy leaders receive in advance of the operation. The following paragraphs discuss in the broadest of terms the considerations of convoy protection.

There is no such thing as an administrative move on the noncontiguous operational area. Once outside the base perimeter you have a tactical convoy. A tactical convoy is a deliberately planned combat operation. Its mission is to move personnel and/or cargo via a group of ground transportation assets in a secure manner to or from a target destination. Tactical convoys operate under the control of a single commander in a permissive, uncertain, or hostile environment. Tactical convoys should always have access to the COP and be characterized by an aggressive posture, agility, and unpredictably. There should never be less than two personnel in the cab, one to drive and one for protection.

Each tactical convoy must be prepared to take appropriate action in the face of ambush and once contact is made based on rules of engagement and TTPs. Training, experience, and unit SOPs will accelerate tactical convoy preparation and prepare unit personnel to take appropriate actions on contact.

Standard troop leading procedures (modified to reflect convoy operations) are included in Multi-Service Tactics, Techniques and Procedures for tactical convoy operations and should be used to ensure all planning elements are considered when preparing to conduct convoy operations.

A. Movement Corridors

The Maneuver Enhancement Brigade (MEB) is a multifunctional headquarters, task organized according to METT-TC that provides security and protection within its assigned areas. The sustainment brigade is likely to operate within a MEB AO and it will rely on the protection MEB elements provide throughout designated movement corridors.

A movement corridor (MC) is part of a layered and integrated security approach to LOC security. Layered security constitutes concentric rings that increase in survivability and response measures. The first ring (the center ring) being the ability of every convoy to defeat a Level I threat and to delay a Level II threat. The next ring (middle ring) provides the increased security/protection capability in support of the center ring activities, capable of defeating Level I and Level II threats, and supports the defeat of Level III threats. The middle ring is also capable of integrating fires, CAS, MEDEVAC, safe havens, vehicle removal/recovery operations, and so forth in support of the center ring and central effort that is the MC concept. The final ring (the

B. Main Supply Routes (MSRs)/ Alternate Supply Routes (ASRs)

Ref: Adapted from FMI 4-93.2, Sustainment Brigade (Feb '09), pp. B-12 to B-13.

MSRs are routes designated within the MEBs or higher headquarters AO upon which the bulk of sustainment traffic flows in support of operations. An MSR is selected based on the terrain, friendly disposition, enemy situation, and scheme of maneuver. Supply routes are selected by the MEB S-4 in coordination with the BSB support operations officer and MEB S-3. They also plan ASRs for use if a MSR is interdicted by the enemy or becomes too congested. In the event of CBRN contamination, either the primary or alternate MSR may be designated as the dirty MSR to handle contaminated traffic. All ASRs must meet the same criteria as the MSR. MPs assist with regulating traffic and engineer units maintain routes. Security of supply routes in a noncontiguous AO may require the MEB commander to commit non-logistics resources.

Some route considerations are:

- Location and planned scheme of maneuver for subordinate forces to include combined arms forces, artillery units, and other forces moving through the MEB's AO
- Route characteristics such as route classification, width, obstructions, steep slopes, sharp curves, and type roadway surface
- Two-way, all-weather trafficability
- Weight classification of bridges and culverts
- Requirements for traffic control such as at choke points, congested areas, confusing intersections, or along built-up areas
- Number and locations of crossover routes from the MSR to ASRs
- Requirements for repair, upgrade, or maintenance of the route, fording sites, and bridges
- Route vulnerabilities that must be protected. This may include bridges, fords, built-up areas, and choke points.
- Enemy threats such as air attack, conventional and unconventional tactics, explosive hazards, ambushes, and chemical strikes
- Known or likely locations of enemy penetrations, attacks, chemical strikes, or obstacles
- Known or potential civilian/refugee movements that must be controlled or monitored

outer most ring) is the final ring of protection and brings with it the ability to defeat all level of threats through the integration of all joint capabilities.

An MC is a protected LOC that connects two support areas. Within the MC are main and alternate roads, railways, and/or inland waterway supply routes used to support operations. Within an AO, there is an MC network that consists of multiple MCs that connect inter-theater APOD and SPOD; intra-theater Aerial Port of Embarkation and APOD's; operational level support, distribution, and storage areas; and brigade support areas. The MEB is responsible for tactical LOC operations and security within its AO and, as assigned, within its higher headquarters AO. The width and depth of an MC will be dependent on METT-TC factors and the commanders' guidance.

The establishment of an MC network is the result of applying multiple functions and establishing required command, control, and support relationships. The collective integration and synchronization of units, capabilities, and facilities will provide a comprehensive three-dimensional protection capability for the designated LOC's, the unit and convoy movements on the LOC's, and the units supporting LOC and movement operations. A fully developed MC will consist of military police units providing route regulation and enforcement, straggler and dislocated civilian control, area and route security, convoy escort, response force operations, and logistical units conducting and managing movement control. Supporting functions include units and capabilities for vehicle recovery and storage, cargo transfer, refueling, road maintenance and repair, MC safe haven support facility construction and repair, CBRN detection and response, aerial reconnaissance, and medical treatment and evacuation.

The sustainment brigade S-3 coordinates with the MEB when planning tactical convoys through movement corridors. This may be done through movement control elements collocated with the MEB or the division transportation officer.

C. Danger Areas

Intersections, ramps, traffic circles, over and underpasses, rest halts, or halts to recover disabled vehicles can all be danger areas for convoys.

D. Battle Drills

Battle drills are the pre-planned, rehearsed responses to contact. FM 4-01.45 provides a simple decision matrix for reaction to contact and examples of how convoy participants might respond. Battle drills on how to react to convoy ambushes must be developed and rehearsed.

E. Improvised Explosive Devices (IEDs) and Vehicle Borne IED's (VBIEDs)

IEDs are one of the greatest threats to convoys and are often used to initiate an ambush. Convoy personnel should always expect an ambush immediately following an IED detonation. Convoy commanders should brief convoy personnel on the latest IED threat: what types of IEDs are being used and where they have previously been emplaced along the route.

IEDs and VBIEDs represent an attractive attack weapon for asymmetrical threat forces. They can inflict substantial damage to convoys (equipment and personnel) while providing the threat with very limited exposure counter-attack by US forces.

IEDs are rigged from any explosive materiel available to threat forces and detonated by timing device, trip wire or pressure trigger, or are command detonated. IEDs are most easily emplaced during periods of limited visibility making morning periods exceptionally dangerous for exposure to IEDs. IEDs are often used as the opening weapon in an ambush. Convoy participants must be prepared to counter an assault by direct fire from hidden roadside positions immediately following an IED encounter.

Chap 7

Theater Support

Ref: FM 4-0, Sustainment (Jul '19), pp. 2-19 to 2-40. When published, also refer to projected new reference ATP 4-93, Theater Sustainment (replacing current ATPs).

I. Theater Army

The theater Army is the senior Army headquarters in an AOR, it consists of the commander, staff, and all Army forces assigned to a CCMD. Each theater Army (U.S. Army Africa, U.S. Army Central, U.S. Army Europe, U.S. Army North, U.S. Army Indo-Pacific Command, and U.S. Army South) has operational and administrative responsibilities. Its operational responsibilities include command of forces, direction of operations, and control of assigned operational areas. Its administrative responsibilities encompass the Service-specific Title 10 requirements for equipping, sustaining, and training forces; unit readiness, discipline, and personnel matters.

The theater Army serves as the ASCC of the GCC. It is organized, manned, and equipped to perform that role. The ASCC is the command responsible for recommendations to the JFC on the allocation and employment of Army forces within a CCDR's AOR.

Refer to FM 3-94 for additional information.

Theater Army tasks and functions include—

- Executing the CCDR's daily operational requirements.
- Opening the operational area (for example-JOA, area of operations, and theater of operations).
- Serving as a JTF or joint force land component for crisis response and limited contingency operations.
- Serving as the primary interface between the Department of the Army, Army commands, and other ASCCs.
- Exercising OPCON of deployed Army forces not subordinated to a JFC.
- Exercising ADCON of all Army forces operating within the AOR.
- Exercising OPCON of all joint forces attached to it as either a joint force land component headquarters or JTF headquarters, as required by the CCDR.

The theater Army main command post (CP) monitors current operations and conducts contingency planning and crisis action planning. The main CP conducts current operations as well and maintains a capability to employ an operational-level CP. As an ASCC, the theater Army prepares support estimates and outlines the responsibilities and requirements for maintaining access to and setting the theater where U.S. military presence is forward stationed or deployed. The theater Army executes many of these responsibilities through the TSC. The TSC is the Army's command for the integration and synchronization of sustainment in the AOR. The MEDCOM (DS) is also assigned to the ASCC. It is the theater medical command responsible for command and control, integration, synchronization, and execution of AHS support within the AOR. The MEDCOM (DS) commander coordinates with the ASCC surgeon (as the staff proponent with execution through the assistant chief of staff, operations (G-3) channels under the authority of the ASCC commander) to provide AHS support within the AOR. Key tasks and functions associated with the theater Army's role in sustainment include—

Theater Army (Primary Staff) Roles

Ref: FM 4-0, Sustainment (Jul '19), pp. 2-20 to 2-24.

Assistant Chief of Staff, G-1/Adjutant General-Personnel

The theater Army G-1 is the principal staff officer for all matters concerning human resources support (military and civilian). The ASCC G-1's primary function is to plan and prioritize human resources support in conjunction with the human resources sustainment center (HRSC) to maximize the readiness and operational capabilities of forces within the theater. Specific responsibilities of the G-1 center around the human resources support core competencies of manning the force to build and sustain combat power through replacement operations, providing human resources services focused on essential personnel services, casualty and postal operations, coordinating personnel support such as morale, welfare and recreation and band support, and coordinating human resources planning and operations. The G-1 also serves as the senior adjutant general (AG) officer in the command. The G-1 is ultimately responsible for planning the personnel readiness of the command. As a member of the staff the G-1 participates in the theater Army commander's operations process and develops the personnel support portion of Annex F (Sustainment) to the operation order or operation plan in coordination with the HRSC.

Theater Army G-1 and TSC Cooperation

The theater Army G-1 and the TSC HRSC have clearly different roles. Both are focused on ensuring the theater Army has adequate HR sustainment support and some actions by both may be similar or complementary. The HRSC is a TSC staff element that provides theater-level HR casualty, personnel accountability, and postal support to Army forces within theater IAW the policies, procedures, and priorities established by the ASCC. As a staff element of the TSC, the HRSC is the primary participant in the planning, integration, and execution of theater-wide HR support.

The role of the ASCC G-1 is to direct Army force HR policy IAW CCMD and Army policy.

Assistant Chief of Staff, G-4/Logistics

The assistant chief of staff, logistics, (G-4) is the principal staff officer for sustainment plans and operations, supply, maintenance, transportation, services, and OCS. The G-4 helps the supported unit commander maintain logistics visibility with the commander and the rest of the staff. The G-4, in coordination with the TSC and/or ESC support operations (SPO) officer, prepares annex F (Sustainment), annex P (Host-Nation Support) and annex W support (Operational Contract Support which must include appendices 1, 2, and 3 to ensure contracting support unit requirements to deploy are identified, contractor personnel, and contractor management are planned and sustained in the expeditionary force AO) to the operation order or operation plan. G-4 staff will identify needed interagency and intra-agency support and facilitate development of memoranda of agreement/understanding (as applicable) in order to unify and integrate theater sustainment. An example of a G-4 staff action is coordinating the selection of main supply routes and logistics support areas (with the engineer officer) and recommending them to the G-3.

The G-4 is the principal staff officer within a command headquarters for sustainment plans and operations. The G-4 sets the sustainment priorities and is ultimately responsible for planning sustainment support for operations. As a member of the staff the G-4 participates in the theater Army commander's operations process and develops the sustainment support concept based on the G-3 operations concept. This support concept is included in the theater Army operation order (OPORD) as annex F and is communicated to the TSC through the order dissemination process.

Within corps and division headquarters the G-4 has coordinating staff responsibility for the G-1, G-8, transportation officer, and the surgeon. Theater Army G-4, brigade and battalion logistics staff officer (S-4)s do not have this responsibility. The G-4 plans and coordinates with the ESC and sustainment brigade staff, provides input to necessary orders, develops and implements the sustainment support plan. The G-4 works closely with the Chief of Staff and G-3 to ensure mutual understanding of all aspects of planning, preparation, execution and assessment of sustainment support for operations.

Theater Army G-4 and TSC Cooperation

The theater Army G-4 and the TSC have clearly different roles. Both are focused on ensuring the theater Army has adequate sustainment support and some actions by both may be similar or complementary. However, their actions are distinctly separate and confusing or shifting responsibilities from the G-4 to the TSC or vice versa can negatively affect the sustainment provided to the command.

Role of the TSC and ESC Commander

The TSC commander uses command and control to execute the theater Army support concept as directed by the OPORD issued by the theater Army commander. The TSC commander accomplishes this through his internal operations process and by issuing his internal OPORD to subordinate units. The concept of operations section of the OPORD describes tasks to subordinate units to accomplish the support mission. The ESC commander uses similar processes to execute the theater Army concept when attached to the TSC. The TSC commander, in the role as the ASCC deputy commanding general (support), is responsible for overseeing the development of the theater Army concept of support that specifies how capabilities will be delivered over time, who is responsible for delivering the capabilities, and defines the critical sustainment tasks necessary to achieve objectives. The TSC staff may assist the theater Army G-4 with information required to complete the theater Army OPORD. The TSC commander may also coordinate with strategic providers for support as directed by the theater Army commander.

Assistant Chief of Staff, G-8/Financial Management

The ASCC financial management (G-8) is the principal staff officer responsible for all resource management support to the theater of operation. The Financial Management Support Center (FMSC) advises the G-8 on finance operations throughout theater. The G-8 manages the planning, programming, budgeting, and execution process and determines sources of funding to support resource requirements.

Engineer Section

Senior staff engineer in the theater Army command responsible for the coordination and synchronization of engineer operations across the AOR. Responsibilities include coordination of efforts within the civil affairs brigades, nongovernmental, intergovernmental organizations, U. S. Army Corps of Engineers, other Services, the HN, and other interested parties.

Theater Army Surgeon's Section

The ASCC surgeon is the theater Army staff proponent responsible for (in coordination with the MEDCOM [DS] commander) the provision of AHS support within the AOR. The ASCC surgeon has staff responsibility for medical planning, coordination, and policy development for AHS support to deployed forces. This officer advises the ASCC commander concerning the health of the command, recommends changes to the theater evacuation policy, and provides input to and personnel in support of the theater patient movement requirements center, as required. Organizations from battalion through Army Service component command level are authorized a surgeon. The Army Medical Department (AMEDD) leverages the surgeon's cells (staff channels) at each echelon and medical command channels (through the MEDCOM [DS], medical brigade [support], and medical battalion [multifunctional]) to provide AHS support to the deployed force. Integration of these two chains and other elements of sustainment occur at command headquarters at echelon and not just between sustainment organizations.

- Developing Army plans to support the theater campaign plan.
- Tailoring Army forces for employment in an AOR.
- Prioritizing personnel replacement fill for the theater.
- Setting the theater for the execution of strategic plans.
- Controlling RSOI for Army forces in an AOR.
- Providing Army support to other Services (transportation, fuel distribution, intra-theater aeromedical evacuation, explosive ordnance disposal (EOD), and logistics management).
- Providing support as directed by the CCDR to other unified action partners.
- Identifying sustainment capability gaps and determining how to fill those gaps, identifying funding sources, and acquiring, and distributing funds.
- Assigning the role of deputy commanding general (support) for the ASCC to the TSC when required.
- Managing theater ammunition.

The ASCC commander has the authority to assign command and support relationships to deployed theater enabling commands. The authority includes integrating and synchronizing capabilities (such as transportation, engineers, EOD, medical, and logistics) until later enabling commands arrive in theater.

II. Theater Sustainment Command (TSC)

The TSC is the Army's command for the integration and synchronization of sustainment in the AOR. The MEDCOM (DS) is also assigned to the ASCC. It is the theater medical command responsible for command and control, integration, synchronization, and execution of AHS support within the AOR. The TSC connects strategic enablers to the tactical formations. It is a theater-committed asset to each ASCC and focuses on Title 10 support of Army forces for theater security cooperation and the CCDR's daily operational requirements. The TSC commands assigned HRSCs and FMSCs. The TSC commander also commands and task organizes attached ESCs, sustainment brigades, and additional sustainment units. The TSC executes the sustainment concept of support for planning and executing sustainment-related support to the AOR for all the Army strategic roles (shape OEs, prevent conflict, prevail in large-scale ground combat, and consolidate gains).

See pp. 7-7 to 7-11 for further discussion of the TSC from ATP 4-94.

TSCs execute sustainment operations through their assigned and attached units. The TSC integrates and synchronizes sustainment operations across an AOR from a home station command and control center or through a deployed CP. The TSC has four operational responsibilities to forces in theater: theater opening, theater distribution, sustainment and theater closing.

The task-organized TSC is tailored to provide operational-level sustainment support within an assigned AOR. It integrates and synchronizes sustainment operations for an ASCC including all Army forces forward-stationed, transiting, or operating within the AOR. The TSC coordinates Title 10, Army support to other Services, DOD EA, and lead service responsibilities across the entire theater.

The TSC organizes forces, establishes command relationships and allocates resources as necessary to support mission requirements, and exercises command and control over attached sustainment forces. The TSC supports the ASCC sustainment cells with planning and coordinating theater-wide sustainment. The execution of sustainment is decentralized, performed by the HRSCs, FMSCs, ESCs, sustainment brigades, and other sustainment organizations. The medical logistics management center (MLMC) forward team collocates with the distribution management center (DMC) of the TSC or ESC to serve as the liaison to the MEDCOM (DS). MEDCOM (DS) is responsible for integrating and executing medical operations.

Theater Support Command (TSC)

Ref: FM 4-0, Sustainment (Jul '19), pp. 2-20 to 2-24.

Fig. 2-11 depicts a representative TSC staff, and figure 2-12 depicts the TSC DMC. The DMC is the principal staff section for coordinating sustainment across an operational area. It is headed by the support operations officer and is a coordinating staff section unique to TSCs and ESCs. The DMC is responsible for sustaining the force in accordance with the theater Army priorities. The staff focuses on detailed planning for operational area opening, distribution, sustainment, and operational area closing operations.

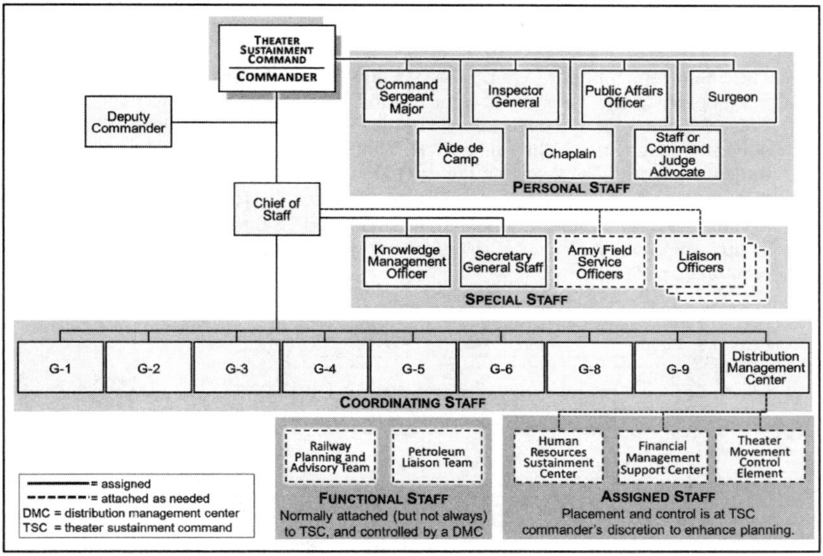

Ref: FM 4-0, fig. 2-11. Theater sustainment command staff with additional staff capabilities.

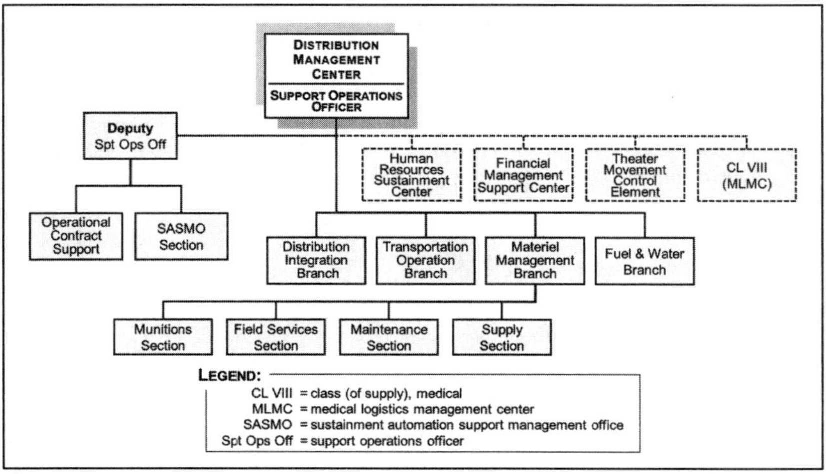

Ref: FM 4-0, fig. 2-12. Theater sustainment command distribution management center.

Theater Support

III. Expeditionary Sustainment Command (ESC)

At the theater echelon, one or more ESC's are attached to a TSC. The ESC attached to a TSC commands and controls all assigned and attached units in an operational area as directed by the TSC commander. A task-organized ESC attached to a TSC normally includes one or more sustainment brigades, a transportation brigade expeditionary, and a movement control battalion to support theater opening, theater distribution, and theater closing operations. The ESC plans for near term operations and synchronizes operational-level sustainment operations to meet requirements for current operations. It may perform as a forward CP for the TSC if directed. The ESC attached to a TSC is dependent on the TSC staff for long-range planning capability and enabling capabilities like signal support.

See pp. 7-12 to 7-14 for further discussion of the ESC from ATP 4-94.

The ESC plans for near term operations and synchronizes operational-level sustainment operations to meet the current and future operational requirements of the TSC. The ESC supports deployed forces in the JOA and/or AO while the TSC maintains a theater-wide focus of the entire AOR. The ESC is task organized with many of the sustainment units discussed under the theater Army echelon. The ESC, usually through an attached sustainment brigade, provides GS to forces in the JOA and/or AO. Figure 2-14 depicts a notional task organized TSC respectively. The ESC and its subordinate units must be able to move and displace at the pace of large-scale combat operations.

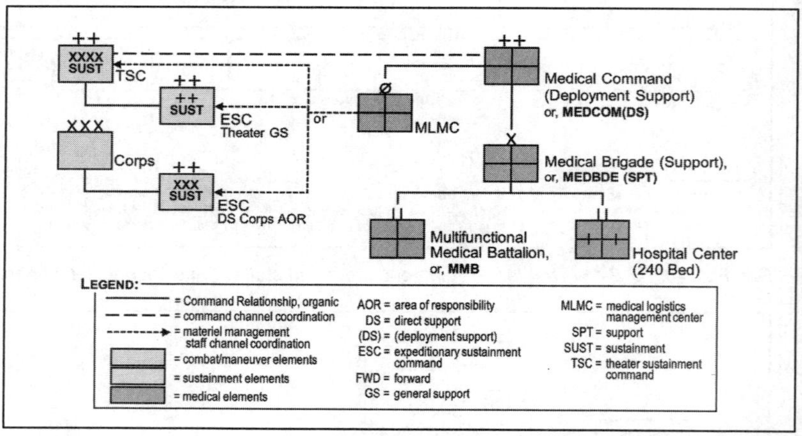

Ref: FM 4-0, fig. 2-13. Notional area of responsibility command and control of sustainment forces.

Chap 7

I. TSC/ESC (Mission, Roles, and Organization)

Ref: ATP 4-94 (FM 4-94), Theater Sustainment Command (Jun '13), chap. 2.

I. Theater Support Command (TSC)

The TSC is responsible for providing sustainment support for an area of responsibility. The ESC is responsible for providing sustainment support for a joint operations area or specified area of operations. The sustainment commands are task organized with sustainment brigades and other modular sustainment forces structure to accomplish the mission. The combination of these capabilities gives the sustainment commander the ability to organize and provide tailored support such as theater opening, theater distribution and sustainment support to forces, and the theater closing within an area of responsibility (AOR).

A. TSC Mission

The sustainment warfighting function is the related tasks and systems that provide support and services to ensure freedom of action, extend operational reach, and prolong endurance. The endurance of Army forces is primarily a function of their sustainment. Sustainment determines the depth and duration of Army operations. It is essential to retaining and exploiting the initiative. Sustainment is the provision of the logistics, personnel services, and health service support necessary to maintain operations until mission accomplishment.

TSC Operational Responsibilities

The TSC has three operational responsibilities to forces in theater:

1. **Theater Opening**
2. **Theater Distribution**
3. **Sustainment**

The mission of the TSC is to provide mission command for operational level logistics within an assigned AOR. The TSC is capable of planning, controlling, and synchronizing operational-level Army deployment and sustainment for the ASCC, joint force commander (JFC), or multi-national joint force commander. It provides a centralized sustainment mission command structure for the ASCC; and supports all phases of operations from phase 0 to phase 5.

The TSC executes its mission through human resource sustainment centers, financial management centers, and the use of modular forces, to include expeditionary sustainment commands (ESC), sustainment brigades, combat sustainment support battalions (CSSB), and other modular sustainment formations. Sustainment brigades, CSSBs, and functional sustainment units serve as the building blocks of the force structure designed to execute TSC missions within the theater.

As required by mission, enemy, terrain and weather, troops and support available, time available, and civil considerations (METT-TC), the TSC may extend its operational reach by deploying a single ESC or multiple ESCs or sustainment brigades into specified areas of operations (AO)/joint operational areas (JOA) in order to more effectively provide responsive support to Army forces. ESCs provide mission command for theater opening, theater distribution, and theater sustainment on an area basis within and between specified AOs/JOAs.

1. Theater Opening

Theater opening is the ability to establish and operate ports of debarkation (air, sea, and rail), establish a distribution system, and to facilitate throughput for reception, staging, and onward movement and integration of forces within a theater of operations (ADP 4-0).

Sustainment brigades and multi-functional CSSBs will attach to the TSC or ESC based on mission requirements for theater opening operations. Sustainment brigade's tasked with theater opening execution may be augmented with additional personnel. The augmentation elements provide the sustainment brigade with additional manpower and expertise to command, control and coordinate TO functions such as managing transportation assets, synchronizing RSOI operations, port operations, throughput, node and mode management, intermodal operations, and movement control. One of the first steps the TSC/ESC must take in theater opening is to establish a port to conduct initial sustainment operations. The transportation brigade (expeditionary) can be attached to the sustainment command to support terminal operations as part of the theater opening mission.

Port opening is a subordinate function of theater opening. Port opening is the ability to establish, operate and throughput forces, equipment, and supplies through a port(s) of debarkation (POD). The port opening process is complete when the POD and supporting infrastructure is established to meet the desired operating capacity for that node (ADRP 4-0).

See pp. 1-24 to 1-25 for further discussion from ADP 4-0 and pp. 6-24 to 6-25 from ATP 4-93.

Port Operations

Port opening requires functional units specializing in port of debarkation operations. Ports of debarkation operations are essential for the maximum throughput of personnel and equipment. They are the first node that sustainment arrives into theater. It is from these locations that sustainment and cargo enter their Service theater distribution system. Units to provide this functional expertise may include (but are not limited to) a Transportation Brigade (Expeditionary), Terminal Transportation Battalion, SDDC Transportation Battalion, Joint Task Force-Port Opening, Air Mobility Command Contingency Response Groups, Navy Cargo Handling Battalions, and Movement Control Teams.

See facing page for further discussion.

2. Theater Distribution

Distribution is the primary means to prolong endurance. Distribution is the operational process of synchronizing all elements of the logistic system to deliver the 'right things' to the 'right place' at the 'right time' (JP 4-0).

Theater distribution is the flow of equipment, personnel, and materiel within theater to meet the commander's mission. Theater distribution extends from the ports of debarkation or source of supply (within theater) to the point of need. It includes transportation management and movement control, warehousing, inventory control, order administration, site and location analysis, packaging, data processing, equipment accountability (materiel management), people, and communications.

Port Operations

Ref: ATP 4-94 (FM 4-94), Theater Sustainment Command (Jun '13), pp. 2-2 to 2-3.

Port operations units may include (but are not limited to) a Transportation Brigade (Expeditionary), Terminal Transportation Battalion, SDDC Transportation Battalion, Joint Task Force-Port Opening, Air Mobility Command Contingency Response Groups, Navy Cargo Handling Battalions, and Movement Control Teams.

Joint Task Force – Port Opening (Aerial Port of Debarkation)

The JTF-PO (APOD) is a joint capability provided by USTRANSCOM that is designed to rapidly establish and initially operate an APOD, establish a distribution node, and facilitate port throughput within an AOR. The JTF-PO (APOD) is not a standing task force, but is a jointly trained, ready set of forces constituted as a joint task force at the time of need. The JTF-PO (APOD) is normally mission commanded by an Air Force AMC Contingency Response Wing or Group commander. Army elements of a JTF-PO (APOD) will normally include a transportation detachment (rapid port opening). The detachment includes limited movement control functions, cargo transfer capabilities, and transportation assets. It does not include heavy equipment truck, Rough Terrain Cargo Handler equipment, or crane assets.

The JDDOC must request JTF-PO from USTRANSCOM. When they are deployed, they serve a direct support role to the GCC, and can serve as the initial theater opening capability for up to 60 days, or when relieved by follow on forces. The TSC or ESC should work closely with the JTF-PO in order to seamlessly transfer robust theater opening capabilities within the AO/JOA.

Joint Task Force – Port Opening (Seaport of Debarkation)

The JTF-PO (SPOD) is a joint capability provided by USTRANSCOM that is designed to rapidly establish and initially operate an SPOD, establish a distribution node, and facilitate port throughput within a theater of operations. Its design and capabilities are similar to those of the JTF-PO (APOD). The JTF-PO (SPOD) is mission commanded by an Army SDDC battalion or Navy Expeditionary Port Unit commander. Like the JTF-PO (APOD), the JTF-PO is a jointly trained ready set of forces constituted in a time of need. They are comprised of Army and Navy elements, and may be augmented by additional port opening enablers such as expeditionary contracting, SDDC subject matter experts, and ship husbandry subject matter experts.

The JTF-PO (SPOD) does not have organic cargo handling capability, and must rely on contracted assets/personnel, or theater opening forces provided by the GCC.

Transportation Brigade (Expeditionary)

The Transportation Brigade (Expeditionary) or TBX, is normally assigned to a FORSCOM unit to provide mission command of assigned and attached port, terminal and watercraft units conducting expeditionary intermodal operations in support of unified land operations. This unit is normally employed in a theater of operation to provide mission command for port opening and operation of inland waterway, bare-beach, degraded and improved seaports.

SDDC Transportation Brigade

SDDC Transportation Brigades are GCC aligned and provide single port management functions in an area of operations. SDDC serves as the Army Service component command of US Transportation Command, and is a major subordinate command to Army Materiel Command. This relationship allows the SDDC Transportation Brigade to establish linkage to the joint deployment and distribution enterprise and Army Materiel Command's Materiel Enterprise. The brigade also works with commercial transportation industry as a coordinating link between DoD surface transportation requirements and the capability industry provides.

As the senior logistics headquarters for the Army, the combatant commander may designate the TSC as a joint command for logistics. When exercising this option the combatant commander must specify the control and tasking authorities bestowed on the TSC as well as the command relationships it will have with the Service components. The primary element of the TSC for theater distribution is within the support operations (SPO). Within the SPO, the distribution management center (DMC) becomes the center of gravity to sustain distribution within the theater.

See pp. 3-38 and 7-29 to 7-30 for further discussion.

3. Sustainment

Unified land operations require commanders to generate and maintain combat power to execute operations. The TSC is responsible for army theater sustainment enabling the GCC/JFC to apply combat power. Sustainment comprises of logistics, personnel services, and health service support.

The TSC and ESC is able to provide the sustainment warfighting function through the means of the support operations.

See pp. 7-32 to 7-33 for further discussion of providing sustainment.

B. TSC Location and Echeloning

The TSC headquarters may operate from outside the AOR and may be collocated with the ASCC in order to most effectively perform its mission command functions. When required, the TSC employs ESCs to provide forward mission command within an AO, and improve the span of control, or mission command a specified function. However, ESC deployment timelines may not meet TSC mission command requirements and the TSC headquarters may be required to relocate and position itself within the AOR.

A primary consideration for this relocation is continuity of mission command. A TSC commander will typically maintain mission command continuity by echeloning elements of the headquarters. Echeloning provides the commander with the capability to place minimum mission command capabilities forward while continuing to support the force. This lead element is commonly referred to as an early entry command post, and its structure is entirely at the discretion of the commander based on mission requirements. Once the early entry command post is in place and communications with all nodes are established, the balance of the TSC headquarters moves forward by echelon.

In addition to continuity of mission command considerations, the TSC commander is concerned with the organizational design of the echeloning elements; specifically establishing functional groupings and establishing functional responsibilities. Identification of responsibilities and authority for each echeloning element provides clarity and direction with respect to the exercise of authority and continuity in the conduct of on-going operations.

TSC mid range planning should also consider the requirement to deploy an ad hoc TSC headquarters mission command element on a long-term basis for split-based operations. The early entry force would be the basis for such an element and the overall structure and manning would be based on existing missions, tasks, and resources available. Commanders must consider mission requirements, organize the force, and allocate resources appropriately while maintaining a balance to support both locations effectively. This ad hoc element would remain in place for the time needed to mobilize and station an ESC in the area.

C. TSC Tasks

Ref: ATP 4-94 (FM 4-94), Theater Sustainment Command (Jun '13), pp. 2-3 to 2-4.

The HQDA Standardized Mission Essential Task Lists (METL) are the Department of the Army's official listing of fundamental tasks that brigade and higher units are designed to perform in any operational environment. These METLs consist of:

- Missions essential tasks (METs) - expressed as Army Tactical Tasks (ARTs)
- Tasks Groups (Higher order Collective Tasks) - expressed as T&EOs
- Supporting Collective Tasks - expressed as T&EOs contains guidance for developing the TSC mission essential task list (METL).

The TSC, with guidance from the ASCC, develops its METL based on tasks published by the HQDA G3. The standardized METL listings that are to be performed in support of major operations are either stated or implied in the ASCC and GCC operational plans. The ASCC commander is the approving authority for the TSC METL.

The following TSCs METs and tasks groups are valid as of 15 May 2012:

MET 1: Conduct Mission Command

- Task Group 1: execute the operations process
- Task Group 2: manage information and data

MET 2: Provide Operational Sustainment

- Task Group 1: conduct logistics support in the operational area
- Task Group 2: direct theater reception, staging, and onward movement
- Task Group 3: provide human resource support
- Task Group 4: provide funding support to financial elements

MET 3: Conduct Operational Area Security

- Task Group 1: conduct base security operations

II. Expeditionary Support Command (ESC)

A. ESC Mission

The ESC provides mission command for attached units in an area of operation as defined by the TSC, ARFOR, or JTF. The ESC provides operational reach and span of control. The ESC plans and executes sustainment, distribution, theater opening, and reception, staging, and onward movement for Army forces. The ESC may serve as the basis for an expeditionary joint sustainment command when directed by the combatant commander or his designated coalition/ joint task force (JTF) commander. When the ESC serves as an expeditionary joint sustainment command, the headquarters should be augmented by personnel and equipment from other Services.

B. ESC Role

The role of the ESC is to deploy to an AO/JOA and provide mission command capabilities when multiple sustainment brigades are employed or when the TSC determines that a forward command is required. The ESC may be employed directly under the mission command of the corps, ARFOR, or JTF as designated by an appropriate order. The TSC maintains oversight of sustainment operations within the AO/JOA with direct coordination with the ESC and its sustainment information systems. This capability provides the TSC commander with the regional focus necessary to provide effective operational-level support to Army or JTF missions. The TSC may employ multiple ESCs within the theater.

The forward deployment of the ESC facilitates agile and responsive support by placing the ESC in relative proximity of the supported force and its operational environment. Positioned to provide a regional focus, the ESC is optimally placed to refine that portion of the TSC logistics preparation of the theater assessment applicable to the JTF area of operations and to array logistics forces accordingly.

Depending on the command structure within the theater, ESCs may be employed to support specific Army forces within a specific AO/JOA; or to support other ESCs or sustainment brigades with theater opening or theater distribution capabilities.

The ESC role in supporting a JTF is less about supply and more about physical distribution and readiness. Its purpose is to build and sustain JTF combat power through agile and responsive JOA-wide support. It achieves its purpose through the effective synchronization and execution of TSC plans and directives in support of JTF operational requirements; executing distribution management responsibilities for its specified AO/JOA; establishing a command climate where close coordination and collaboration with the JTF enables decisive action when unanticipated events rapidly occur in a specific operational environment.

The organizational design of the TSC and ESC are similar, however there are several differences in capabilities. The TSC has regional and organic trial defense support capability and a G5 Plans section.

The ESC is focused on synchronizing operational-level sustainment operations to meet the day-to¬day and projected operational requirements of the JTF or supported force. It accomplishes this, in part, by establishing mid range and short range planning horizons that are derived from the JTF OPLAN, commander's intent, CCIR, operational tempo, and distribution system capacity.

C. ESC Functions

By design, ESC operations are limited in scale and scope when compared to the TSC. The ESC employs reach capabilities to provide the entire spectrum of support in accordance with TSC plans, policies, programs, and mission guidance.

When supporting a JTF, the ESC establishes control of decentralized operations conducted in support of JTF operations. It executes operational control of TSC

ESC Command and Staff Roles

Ref: ATP 4-94 (FM 4-94), Theater Sustainment Command (Jun '13), pp. 2-20 to 2-21.

Role of the ESC Commander

The ESC commander's role is to establish a positive command climate, prepare the command for operations, direct it during operations, and continually assess subordinates. ESC commanders visualize the nature and design of operations through running estimates and input from subordinates. They describe operations in terms of time, space, resources, purpose, and action; employing intent, commander's critical information requirements, and mission orders to direct planning, preparation, and mission execution. The commander mission commands their forces to do that centralized planning and decentralized execution ensures the best practice to sustain the AO/JOA.

- **Commander's Critical Information Requirements (CCIR).** ESC commanders use CCIR to focus information collection on what they need to support critical decisions. CCIR enable commanders to make informed decisions during planning and course of action selection. During preparation and execution, CCIR's address information that commanders require in order to make informed decisions associated with decision points.
- **Mission Orders.** ESC commanders direct with mission orders. Mission orders are directives that emphasize to subordinates the results to be attained, not how they are to achieve them. They begins the mission command process which is the Army's preferred method for exercising mission command. The TSC/ESC commander's intent and concept of operations set guidelines that provide unity of effort while allowing subordinate commanders to exercise initiative in planning, preparing, and executing deployment and sustainment operations. Mission orders stress not only the tasks required of subordinates but also understanding their context and purpose.

Role of the ESC Deputy Commander

The ESC deputy commander serves as the second in command to the ESC commander. His role, responsibilities, and authority vary, based on the commander's desires, the ESC mission, and the scope and complexity of operations. The relationship between the deputy commander and the staff is unique to each ESC.

The ESC deputy commander may assume duties, to include command duties, as delegated by the commander, either explicitly or by standard operating procedures, when the commander is temporarily absent from the command.

Because ESC deputy commanders must be able to assume command at any time, they always keep abreast of the situation. Commanders inform their deputy commanders of any changes in the commander's visualization or commander's intent. The chief of staff keeps the deputy commander informed of staff actions.

ESC commanders typically delegate authority to their deputy commanders to act in their name for specific fields of interest and responsibility. Doing this reduces the burden of commanders' responsibilities and allows them to focus on particular areas or concerns while their deputy commanders concentrate on others.

Role of the ESC Staff

ESC staffs provide commanders with relevant information in usable forms that help commanders achieve accurate situational understanding and make decisions. Situational understanding enables commanders to make well informed and timely decisions and allows staffs to rapidly synchronize and integrate actions in accordance with the commander's intent. The ESC staff synchronizes sustainment functions effectively and efficiently to sustain unified land operations.

forces and orchestrates the effective and efficient flow of logistics and deploying units to, and retrograde and redeploying units from, the JTF; synchronizing operational-level inter-modal distribution operations with the maneuver force's battle rhythm. The ESC maintains multiple means of communication with the TSC DMC to facilitate effective coordination for intertheater and intratheater deliveries to the JTF and materiel management functions not performed by the ESC.

D. ESC Organization

The ESC organizational structure is very similar to the TSC structure. The ESC is organized with a personal, special, and a coordinating staff comprised of a G-1-G-4, SPO, G-6, and G-8. In addition to their common staff functions, the ESC staff develops policies and plans for their respective areas of responsibility and provides guidance, priorities, and allocations to subordinate commands/units. They also review the plans of counterpart staff elements and those of subordinate units.

For the most part, the significant difference between TSC and ESC capabilities is scale and scope. Although they are structured following the same organizational design, the ESC lacks specific support functions such as trial defense capability, and a special troops battalion. The ESC scope is also different. The TSC is concerned with supporting ASCC/joint requirements from a theater-wide perspective, whereas the ESC is concerned with supporting ASCC/joint requirements from a regionally focused theater of war, theater of operations, AO or JOA perspective. There are other differences in capabilities such as long range planning and materiel management. METT-TC considerations determine the number of ESCs required to mission command TSC deployment and sustainment operations theater-wide.

ESC Command Staff Organization

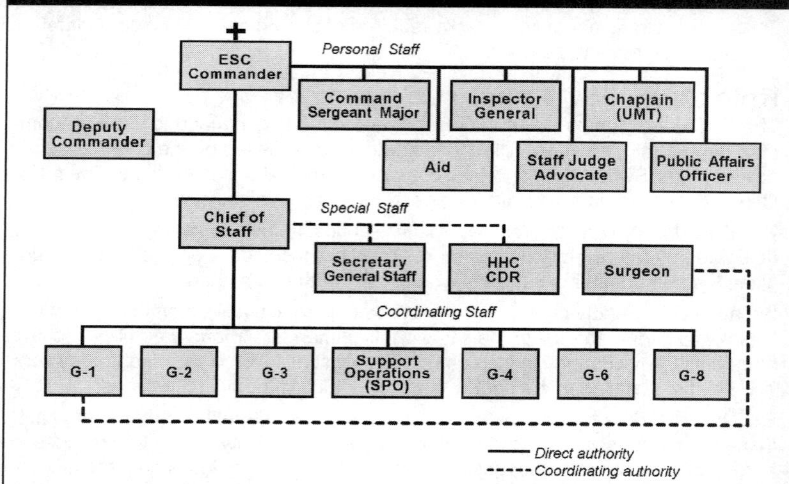

Ref: ATP 4-94, Theater Sustainment Command (Jun '13), fig. 2-4, p. 2-23.

III. TSC and ESC Subordinate Commands

This section discusses the functional and multifunctional organizations routinely assigned to the TSC or ESC as subordinate commands. Subordinate mission command relationships and force tailoring decisions are determined by the TSC or ESC based upon METT-TC considerations.

A. Movement Control Battalion (MCB)

The MCB provides mission command to movement control teams operating within an AO/JOA. It commands between four and ten movement control teams (MCT), plans and schedules movements, and is responsible to the TSC/ESC for the execution of the TSC movement program and performance of the theater transportation system. The MCB provides transportation asset visibility and coordinates the use of common-user transportation assets, intermodal container assets such as international organization for standardization containers, 463L pallets, and flatracks. The MCB also provides in-transit visibility of unit moves and convoy movements.

Movement Control Teams (MCT)

MCTs are attached to the MCB in order to provide decentralized execution of MCB movement responsibilities throughout a specified AO. MCTs may be employed on an area basis or at critical nodes in order to facilitate effective movement control.

Refer to ATP 4-16 for more information on movement control and MCT roles and functions.

Petroleum Pipeline and Terminal Operating Battalion

Petroleum pipeline and terminal operating battalions are normally attached to the TSC. Petroleum pipeline and terminal operating battalions are responsible for the operation and maintenance of a military petroleum distribution system that may include ports of entry, pipelines, tank farms, and tactical marine terminals. Their core capabilities include scheduling and directing the flow of bulk petroleum products through multiproduct military pipelines and coordinating the movement of bulk petroleum products by barge, rail, and truck.

Petroleum pipeline and terminal operating battalions are also responsible for implementing a quality assurance program and may operate a base petroleum products laboratory.

Petroleum Supply Battalion

Petroleum supply battalions are normally attached to the TSC and further attached to a sustainment brigade. In such cases, they serve as the link between the pipeline systems and direct support (DS) supply units in their specified AOs.

Petroleum supply battalions receive bulk petroleum via pipeline, rail, truck, or barge from terminals operated by a petroleum pipeline and terminal operating battalion. They may also receive deliveries from Defense Logistics Agency (DLA), Defense Energy Support Center and commercial sources/ contracts. These battalions receive, store, and transfer bulk petroleum to DS supply units. They operate 5,000- or 7,500-gallon tankers and, when feasible, rail cars or barges to distribute bulk fuels. When required, they can also provide bulk and retail supply point distribution. Petroleum supply battalions provide technical and operational supervision for the storage and distribution of petroleum products within their specified area of operations.

B. Sustainment Brigade

Sustainment brigades provide mission command of theater opening, theater distribution, and sustainment operations. Combat sustainment support battalions (CSSB) are the building blocks of the sustainment brigades. Their headquarters' designs are standardized and they can consist of up to eight companies. CSSBs are modular and task organized to support theater opening, theater distribution, area sustainment, or life support missions.

The sustainment brigade materiel management effort is focused on the management of its supply support activities (SSA) in accordance with TSC plans, programs, policies, and directives. The sustainment brigade may also provide materiel management of bulk supplies through oversight of stockage areas such as bulk fuel and ammunition storage areas. The sustainment brigade coordinates and controls supply functions, including the redistribution of intratheater excess, to meet the operational requirements of the TSC and its supported units, employing near real-time situational awareness of stock records and asset visibility to provide responsive and agile support.

See pp. 6-11 to 6-22 for complete discussion of the sustainment brigade.

Transportation Theater Opening Element (TTOE)

The transportation theater opening element (TTOE) is attached to a sustainment brigade when that brigade is assigned the mission of early entry and establishment of an area of operation's logistics base. The TTOE provides an additional 55 transportation personnel and allows the brigade to function as a seaport operator and distribution manager. With this capability, the brigade establishes the initial surface distribution system for an area of operations.

TTOE capabilities provide a sustainment brigade with the staff augmentation and functional expertise necessary to efficiently and effectively conduct theater opening operations (less health service support) that include RSOI of deploying Army forces. RSOI functions include coordinating, synchronizing, and clearing of aerial ports of debarkation/sea ports of debarkation (APOD/SPOD) holding areas, staging areas, and marshalling areas; personnel and unit equipment integration; life support; and the multi-modal onward movement of units and/or supplies to tactical assembly areas and/or distribution hubs. The execution of RSOI functions require close coordination with supported commanders, the TSC, joint partners, and the HN.

Other TTOE functions include:

- Evaluating and ensuring that the appropriate mode is employed and fully integrated with materiel distribution requirements.
- Providing advice on the use and implementation of assigned, attached, contracted, and HN motor transport assets.
- Providing guidance on positioning of motor transport, air, and rail assets throughout the AO.
- Monitoring and maintaining the status of all modal transportations assets in the AO and ensuring proper tasking.
- Providing advice on the use and implementation of assigned, contracted, and HN terminal and watercraft operations.
- Providing terminal infrastructure assessment.
- Monitoring and coordinating operations and positioning of all terminal operations in the AO, to include motor, rail, inter-modal, air and sea.
- Monitoring and maintaining status of terminal assets in the AO to ensure they are properly employed and not over-tasked.

At some point along the deployment-employment-sustainment continuum the TTOE may be attached to the ESC to facilitate theater-level movements in accordance with the TSC movement program and support ongoing deployment/redeployment operations.

Combat Sustainment Support Battalion (CSSB)

The CSSB is the building block upon which TSC sustainment capabilities are developed. The CSSB is tailored to meet specific mission requirements. Attached capabilities, drawn from the force pool, may include transportation, maintenance, ammunition, supply, mortuary affairs, airdrop, field services, water, and petroleum. Selected CSSB may be organized to provide specific types of support to division aligned brigades lacking full internal logistics capabilities.

Employed on an area basis, the CSSB plans, coordinates, synchronizes, monitors, and controls sustainment operations (less health service support) within a specified AO; supporting units in or passing through its geographic area.

See pp. 6-16 to 6-17 for further discussion of the CSSB.

Support relationships designated by OPORD will normally be used to specify the details of the support to supported relationship, including the designating priorities of support. The CSSB and its attached units will normally have a general support relationship with supported organizations. In certain circumstances, a CSSB could be aligned to provide direct support to a corps or division in an area of operations. Sustainment operations would be coordinated through the G4 to receive operational or strategic level support from higher echelon elements.

Motor Transportation Battalion

Motor transportation battalions are assigned to the TSC and are normally attached to sustainment brigades conducting theater distribution missions. A motor transportation battalion's core capabilities include motor transport operations and terminal operations (less seaport). It mission command's three to seven motor transport or cargo transfer companies.

The motor transportation battalion staff translates mission orders from the sustainment brigade into specific requirements; operationally controlling the operation of truck terminals, trailer transfer points, and trailer relay operations within its assigned AO.

Transportation Terminal Battalion

Transportation terminal battalions are assigned to the TSC and are normally attached to sustainment brigades. A transportation terminal battalion's core capabilities include terminal and inland waterway operations. Terminal operations include truck, rail, air, and marine terminals.

A transportation terminal battalion mission command's three to seven transportation terminal operations or watercraft companies. The transportation terminal battalion staff translates mission orders from the sustainment brigade into specific requirements; enabling the effective and efficient flow of materiel and personnel into and out of the theater.

Ordnance Battalion (Ammunition)

Ordnance battalions (ammunition) are assigned to the TSC and are normally attached to sustainment brigades. An ordnance battalion is typically employed to provide mission command of modular ammunition units operating theater storage areas at the operational level; providing technical supervision of subordinate unit ammunition operations, except for inventory management functions performed by the TSC SPO DMC Munitions Branch.

Subordinate ammunition units receive munitions from the national level; maintain theater stocks; conduct operational-level reconfiguration; and distribute munitions to forward storage areas located throughout the theater.

Financial Management Center (FMC)

The financial management center (FMC) functions as a staff element of the TSC and asserts technical coordination over all Army financial management companies and detachments in theater. The FMC Director, in coordination with the TSC G-8 or Support Operations, is the principal advisor to the ASCC commander and ASCC G-8 on all aspects of financial management operations. The FMC provides technical oversight of all Army financial management operations in the theater to include negotiations with HN banking facilities, advising unit commanders on the use of local currency, and coordination with national providers (US Treasury, Defense Finance and Accounting Service [DFAS], Assistant Secretary of the Army for Financial Management and Comptroller and United States Army Finance Command [USAFINCOM]) to establish financial management support requirements. The FMC sustains Army, joint and multinational operations by providing timely contractual and procurement payments and a theater disbursing capability.

C. Human Resources Sustainment Center (HRSC)

The human resources sustainment center (HRSC) functions as a staff element of the TSC. The HRSC provides theater-level support to the ASCC G-1 and enables the TSC Commander to plan, integrate and execute HR support to the theater. The TSC is the key linkage between the ASCC G-1 who provides the policy, direction, and guidance for HR support to the theater and the HRSC, which executes the HR support mission for postal, casualty, reception, replacement, return to duty, redeployment, rest & recuperation and personnel accounting and strength reporting. The HRSC has a defined role to ensure that the theater HR support plan is developed and then supported with available resources within the TSC. The HRSC is the technical link to HR organizations which execute postal, reception, replacement, redeployment, return to duty rest & recuperation, casualty operations, and personnel accountability support functions.

The HRSC is a multifunctional, modular organization that integrates and ensures execution of HR support throughout the theater as defined by the policies and priorities established by the ASCC G-1 in postal, casualty, reception, replacement, redeployment, return to duty rest & recuperation, and personnel accounting and strength reporting core competencies. The HRSC provides planning and operations technical support to the TSC DMC. The HRSC provides technical guidance to the HR operations branch in sustainment brigades and ESCs, and HR companies and teams. The HRSC's flexible, modular and scalable design increases the HRSC director's ability to recommend HR support based upon the number of units and Soldiers supported and METT-TC. The HRSC's ability to directly coordinate needed sustainment resources with the TSC DMC to support postal and reception, replacement, redeployment, rest & recuperation operations is critical to mission success.

The HRSC provides technical guidance and ensures execution of the personnel accounting, postal, R5, and casualty core competencies performed by HR (Standard Requirements Code 12) elements, including theater gateway teams, military mail terminal, HR Companies, platoons, teams and the HR operations branches in the sustainment brigades and ESCs. The HRSC provides operational planning, and current and future operations management. It coordinates support for the TSC and ASCC G-1, ensures connectivity and resource support for postal, reception, replacement, return to duty, redeployment, rest & recuperation and casualty units; integrates personnel data when necessary, and participates in the TSC distribution management process.

IV. Attachments

The term "Attachments" as used here does not reflect a specified command relationship. The organizations described below reflect capabilities that may be provided to the TSC based upon support agreements with strategic providers and/or METT-TC considerations within the theater or JOA. Appropriate command relationships are attached OPCON, or tactical control (TACON).

Theater Support

A. Medical Logistics Management Center Support Team (MLMC)

A medical logistics management center (MLMC), a subordinate unit of the MEDCOM (DS), provides for the centralized management of medical materiel and maintenance throughout a theater. The MLMC is capable of deploying forward support teams while maintaining base operations within CONUS. MLMC support teams provide centralized management of medical materiel, primary medical items, medical maintenance, and coordination of the distribution of Class VIII materiel throughout a theater, theater of operations, or JOA

MLMC support teams have the capability to deploy an early entry element to support theater opening operations. In most scenarios, the early entry element will deploy and establish initial operations. As the theater base expands, a follow-on element deploys and the two elements merge to form a single MLMC support team that is collocated with the TSC.

The MLMC has two forward support teams. One team is deployed per theater and collocates with the TSC in order to facilitate the integrated and synchronized flow of Class VIII materiel throughout the theater, theater of operations, or JOA. The MLMC support team accomplishes this by:

- Executing liaison functions between forward deployed medical units and CONUS-based strategic assets.
- Coordinating distribution of Class VIII materiel with TSC support operations.
- Developing and executing medical maintenance programs.
- Performing centralized management functions for critical medical items.
- Providing technical expertise in resolving medical maintenance problems within the theater.

When the Army is designated, by the GCC, as the single integrated medical logistics manager for joint operations, the MLMC support team will execute the information management and distribution management portion of the single integrated medical logistics manager mission.

B. Sustainment Brigade (Special Operations) (Airborne)

The Sustainment Brigade (Special Operations) (Airborne), SB (SO) (A), when deployed acts as the single logistics command element for a special operations task force (SOTF) or when necessary a special operations joint task force (SOJTF). The SB (SO) (A) plans, integrates, and coordinates Army common and special operations forces (SOF) peculiar logistics to sustain SOF. The brigade is designed to serve as an early entry element to provide mission command of one to three CSSBs. The brigade monitors and updates the COP; synchronizes and manages sustainment and distribution operations; determines and anticipates sustainment requirements; plans, coordinates, and synchronizes both current and future sustainment operations for deployed SOF units. The brigade integrates ARSOF support requirements into the ASCC support plan and ensures a timely response to ARSOF logistics requirements. It can operate as a standalone SOF logistics command post or as a lateral staff agency that coordinates with a TSC and can provide continuous, 24-hour operations for all SOF sustainment requirements.

If deployed, the SB (SO) (A) is intended to remain an interim capability and any one of the following triggers would cause a transfer of the mission command capabilities to a conventional Army mission command capability: mission command of more than three CSSBs; geographical dispersion beyond the capability of the SB (SO) (A); operations longer than six months in duration or until a theater logistics infrastructure is developed. After being relieved in place, the brigade would then employ an ARSOF support cell embedded with a conventional sustainment brigade, the ESC, or the TSC, to coordinate, monitor, and synchronize logistics support for SOTF/SOJTF operations, other ARSOF operations, and for multinational SOF where the Army is the lead Service for logistics.

The brigade has the capability to provide technical supervision, utilizing its ARSOF liaison element and ARSOF support cell, to assist the ASCC/TSC and theater special operations command in the planning and execution of logistics and HSS/FHP supporting SOF requirements.

Note. The SB (SO) (A)'s ARSOF liaison element, a multifunctional logistics element, is the logistics planning and coordination link between the ASCC, TSC, theater special operations command, SB (SO) (A), and the ARSOF command structure. ARSOF liaison elements are assigned to United States Army Special Operations

Command SB (SO) (A), but habitually attached to both the theater special operations command and ASCC to coordinate logistical support provided by the ASCC to deployed SOF.

C. ARSOF Support Cell

The ARSOF support cell is a task organized deployable team comprised of multi-functional logisticians from within the SB (SO) (A) operations division. Its mission is to coordinate, monitor, and synchronize logistics support for the SOTF/SOJTF operations, other ARSOF operations, and for joint/combined SOF where the Army is the lead Service for logistics.

To facilitate this support, an ARSOF support cell can be employed in the following four scenarios: to reinforce the group support battalion of special forces groups acting as a SOTF/SOJTF by providing mission command of theater opening/theater distribution modules in an austere theater; to serve as the initial command post for a deployed SB (SO) (A) or, with augmentation from the SB (SO) (A), to serve as an interim sustainment mission command capability in support of an ARSOF-led JTF until a theater logistics infrastructure can be developed; to coordinate and monitor Army common and SOF-peculiar sustainment and HSS/FHP in support of ARSOF by collocating with deployed ESCs and TSCs. When not deployed, the ARSOF support cell personnel man the home station operations center and provide reach-back support to the ARSOF liaison element.

II. TSC Support Operations

Ref: ATP 4-94 (FM 4-94), Theater Sustainment Command (Jun '13), chap. 3.

Army forces within an AOR are supported by the ASCC's theater sustainment command. The ASCC normally retains mission command of TSC assets in order to effectively and efficiently integrate and synchronize operations across the entire AOR. There are situations, however, in which TSC assets may be under the tactical control (TACON) or operational control (OPCON) of a supported Army forces.

Actions within an AOR may span the entire range of military operations. When armed conflict does occur, it may involve only a portion of the AOR. In situations such as this, the geographic combatant commander (GCC) may designate the region in conflict as a theater of war and/or theater of operations. When a theater of operations, or subsequent area of operations (AO), is established within a theater, the ASCC commander establishes support priorities in accordance with ARFOR requirements to achieve GCC objectives. A support relationship is established between the ARFOR and the TSC which permits the TSC to employ theater-wide resources to provide timely, responsive support to the ARFOR.

Notional TSC Command Structure

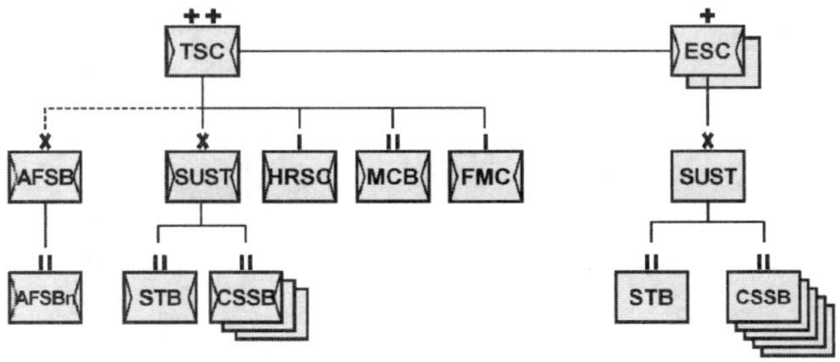

TSC = Theater Sustainment Command
ESC = Expeditionary Sustainment Command
AFSB= Army Field Support Brigade
SUST = Sustainment Brigade
HRSC = Human Resource Sustainment Center
MCB = Movement Control Battalion
AFSBn = Army Field Support Battalion
STB = Special Troops Battalion
CSSB = Combat Sustainment Support Battalion
FMC = Financial Management Center

Ref: ATP 4-94, Theater Sustainment Command (Jun '13), fig. 3-1, p. 3-2.

The TSC support structure is based upon detailed mission analysis and other METT-TC considerations. Required capabilities may vary based upon the situation or by phase of the operation. Modular unit designs provide the means to effectively build required capabilities as operational requirements change. Additionally, TSC planners consider the use of host nation support (HNS), contracted support, and support from other Services as alternatives to adding force structure to the TSC.

A TSC may be required to simultaneously synchronize and integrate deployment and sustainment operations across an AOR that contains multiple theaters of operations. In situations such as this, the TSC may employ multiple expeditionary sustainment commands (ESC) to provide responsive support to Army forces.

The TSC staff's primary role in the mission command process is the development of plans and mission orders. Effective parallel and collaborative planning produces plans and mission orders that:

- Foster mission command by clearly conveying the commander's intent.
- Assign tasks and purposes to subordinates.
- Contain the minimum coordinating measures necessary to synchronize the operation.
- Allocate or reallocate resources.
- Directs preparation activities and establishes times or conditions for execution.

Mission command requires a common understanding of the situation; achieving this ensures seamless strategic to operational sustainment. Situational understanding is accomplished through collaboration and coordination with Army and joint partners at the strategic, operational, and tactical levels. The TSC commander and staff must be cognizant of the logistics situation throughout the AOR. Disciplined logistics status reporting using all logistics information systems and mission command systems available help to achieve situational understanding and a logistics COP.

Notional Theater Construct

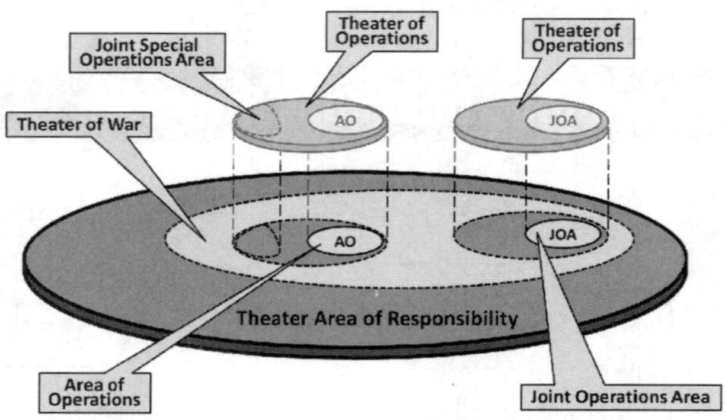

* The combatant commander can also establish a theater of operations or JOA/JSOA or both outside the theater of war to handle a situation.

Ref: ATP 4-94, Theater Sustainment Command (Jun '13), fig. 3-2, p. 3-3.

A key factor in maintaining operational awareness and enabling unity of effort is TSC participation in GCC and subordinate joint force commander (JFC)-level boards, bureaus, centers, cells and working groups. Established on an as required basis, boards and centers set policies and priorities, provide for improved synchronization and integration, and enable the effective flow of resources in support of operational objectives.

Note. Under certain METT-TC conditions, the TSC may be responsible to chair certain boards or centers. Boards and centers are resourced from the internal assets of the participating organizations - a necessity that may cause manning issues for the participating organizations.

Although all TSC staff sections play an important role in supporting mission requirements, it is the support operations (SPO) section that is the TSC center of gravity. It is within the support operations section that materiel and distribution management occurs that enables synchronized and integrated operational-level sustainment support throughout the AOR.

TSC Support Operations Section

Ref: ATP 4-94 (FM 4-94), Theater Sustainment Command (Jun '13), pp. 3-4 to 3-5.

The SPO manages the theater distribution network linking the strategic and operational levels of logistics operations. The SPO, by exercising staff supervision over the DMC, maintains visibility, capacity, and control of the distribution system. Through the coordinated efforts of its internal branches, the DMC, exercises control using current and emerging information technologies that enable the DMC to accurately monitor support from the strategic to tactical level in near real-time.

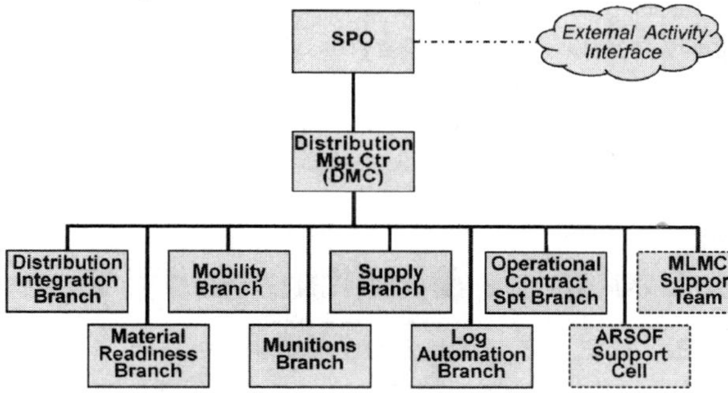

The distribution integration branch (DIB) coordinates and synchronizes the movement of all personnel, equipment, and supplies into and out of the AOR. To do this job effectively, the distribution integration branch relies on coordination and information exchange between the supply and the mobility branches. A complete awareness of the logistics status of subordinate and supported units enables the DMC to optimize resources and task subordinate organizations in support of on-going and future operations.

The materiel readiness branch (MRB) provides staff supervision over maintenance issues impacting force readiness. It determines requirements and manages the maintenance capabilities for supported units of the command. Within the MRB, there are various supporting sections. The ground maintenance section, electronic maintenance section, and aviation maintenance section conduct theater maintenance trend analysis, and identifies equipment maintenance issues. The MRB will then coordinate resolution with appropriate elements of the TSC, ASCC, and Army Materiel Command (AMC).

The mobility branch, comprised of an air, land, and sea section, provide staff supervision of all allocated transportation assets and coordinates directly with the movement control battalion (MCB) assigned to either the TSC or ESC. The MCB coordinates all movement in the theater. The mobility branch also coordinates with joint and strategic partners (i.e. Joint deployment distribution operations center [JDDOC] and the Military Surface Deployment and Distribution Command [SDDC]) in order to synchronize intertheater and intratheater deployment and distribution efforts; and optimize intratheater distribution by employing all transportation modes available in theater.

The munitions branch provides staff supervision and visibility of conventional ammunition. It determines munitions requirements, manages supply capability and conducts materiel management for supported units of the command.

The supply branch provides staff supervision over all supply operations except Class V and Class VIII. It performs materiel management of supply (subsistence, general supplies, construction material, and repair parts) for a designated area of responsibility.

The log automation branch establishes the logistics automation plan, policies, operational functions, system readiness, and maintenance support for logistics automations systems and networks in support of subordinate units.

Theater Support

The support operations section, and in particular the distribution management center (DMC), is augmented as required by METT-TC or in accordance with support agreements in order to synchronize requirements and employ constrained resources more effectively and efficiently. The TSC SPO serves as the focal point for coordination on all matters pertaining to sustainment for:

- Supported units and major commands.
- Other Services or multinational partner forces.
- Other Army major commands (to include specialized commands).
- Strategic-level organizations providing support in the theater.
- Joint boards, centers, and bureaus.

Coordination is essential for the following reasons:

- Ensuring a thorough understanding of the commander's intent.
- Ensuring complete and coherent staff actions.
- Avoiding conflict and duplication by adjusting plans or policies before implementation.
- Considering all factors affecting the situation.

II. Support to Joint & Multinational Operations

A. Lead Service Responsibilities

The GCC assigns lead Service common-user logistics (CUL) responsibilities, normally through the mid range planning process, in order to achieve efficiencies and eliminate redundancies. The GCC usually assigns lead Service responsibilities to the dominant user and/or most capable Service for a particular common supply item or service. In many cases, the lead Service for CUL and other support within a joint or multinational Force is an Army responsibility. If the GCC assigns the ASCC to perform these lead sustainment functions, the TSC SPO will become responsible for managing them within a joint and multinational operational framework

See facing page for further discussion.

B. Multinational Support

First, and foremost, logistics support is a national responsibility. However, to require each nation to perform all logistics functions separately would be inefficient, expensive, and hinder the multinational force commander's (MNFC) ability to influence operations. Accordingly, the US is a member of various alliances and multinational forums that have developed doctrine and procedures, such as implementing agreements and international standardization agreements for combined efforts to support multinational logistics operations. When participating in multinational operations, US forces will conform to previously approved international agreements.

When the military operation requires the involvement of nations that are not members of a formal alliance or multinational forums, logistics planners must be involved in the initial planning to help define the terms for coalition support. The importance of commencing this planning process as soon as possible cannot be overstated.

For multinational operations to be successful, the MNFC must be given sufficient authority over logistics resources to ensure that operational priorities can be effectively supported. The degree of authority the MNFC has will depend upon existing agreements and ad hoc arrangements negotiated with participating nations based on the operational environment and/or as specified in the operations plan.

The main logistics tasks of the MNFC are to develop the overall logistics concept for the operation, including concepts for specific logistics functions; and to mange common logistics support of the multinational force (MNF) within the scope of authorities granted by nations.

A. Lead Service Responsibilities

Ref: ATP 4-94 (FM 4-94), Theater Sustainment Command (Jun '13), p. 3-5.

The GCC assigns lead Service common-user logistics (CUL) responsibilities, normally through the mid range planning process, in order to achieve efficiencies and eliminate redundancies. The GCC usually assigns lead Service responsibilities to the dominant user and/or most capable Service for a particular common supply item or service. In many cases, the lead Service for CUL and other support within a joint or multinational Force is an Army responsibility. If the GCC assigns the ASCC to perform these lead sustainment functions, the TSC SPO will become responsible for managing them within a joint and multinational operational framework. These lead Service support functions may include:

- Supply management for Class I, II (common), III (B), and IV.
- Production, packaging, storage, and distribution of bulk water.
- Receipt, storage, and issue of Class VIII items in theater.
- Common-user land transportation and movement control.
- Rotary aircraft and vehicular medical evacuation.
- Transportation engineering for highway movements.
- Facility construction and repair.
- Financial management support.
- Legal support.
- Explosive ordnance disposal support.
- Airdrop equipment and systems.
- Billeting, medical, and food service support for transient personnel during other than unit moves.
- Environmental management, to include handling hazardous material.
- Mortuary affairs support.
- Postal operations support.
- Casualty liaison.
- Retrograde.
- Reception, staging, and onward movement.

The TSC executes many of the Army's support responsibilities to other Services. The TSC assists the ASCC assistant chief of staff, logistics sections, planners in identifying all lead Service support requirements (to include joint, multinational, and interagency requirements) so that scarce resources can be distributed throughout the force. The TSC synchronizes those support responsibilities falling to other Army theater-level commands with applicable portions of the distribution plan.

For relatively small MNF operations, MNFCs typically form multinational deployment and sustainment staff sections to facilitate coordination and support multinational operations. In the case of larger, more complex MNF operations requiring more coordination and common support, the MNFC may establish a multinational joint logistics center (MJLC) and/or multinational integrated logistics unit to plan and coordinate MNF logistics activities. Essential planning considerations include:

- MNF composition.
- Compatibility.
- HN support limitations.
- Overall infrastructure conditions in the area of operations (AO).
- Budget limitations.
- Contracting.
- Information exchange agreements.

Multinational Joint Logistics Center (MJLC)

The MJLC may consist of various functional coordination centers that provide centralized coordination of common support services, such as movement control, contracting, host nation support, and the provision of bulk fuel and rations. In addition to functional coordination tasks, the MJLC may be assigned the responsibility for coordinating the efforts of logistics units provided by nations to serve at the theater/operational-level for common support of the entire MNF.

In the case of US-led multinational operations, the JFC may opt to expand upon the tools available for managing joint logistics operations and adapt them to the multinational environment. Options include: using a predominant Service organization as the nucleus operational logistics activity to manage common requirements; or using the lead Service to provide CUL support to multinational forces. Regardless of the approach used, MNF augmentation is required to support multinational logistics activities.

Note. MJLC is a North Atlantic Treaty Organization concept/term. For US-led multinational operations, the "MJLC" may be referred to as the Combined/Joint Logistics Center and operate under the staff supervision of the C/J-4.

Although logistics is a national responsibility, varying degrees of CUL support in multinational operations can be expected in order to achieve economy of effort and avoid duplication. Just as for US Services, allies and multinational partners can delineate responsibilities among themselves based on theater requirements and the ability of each country to provide materiel and services. Unity of effort among multinational partners is essential. Selected CUL support, to include limited multinational cooperation, is possible for some logistics functions, such as providing bulk fuel, water, class I, movement and movement control, port arrangements/operations, contingency contracting, engineering, and sharing of facilities such as distribution and warehousing, and organizational options based on METT-TC and force command structure.

In operations where the US is a significant contributing nation, the combatant command may designate the US force as the lead nation for selected CUL support to the entire multinational force and/or as the role specialist nation for a specific common item. A role specialist nation is a nation that has agreed to assume responsibility for providing a particular class of supply or service for all or part of a multinational force. Routinely, the Army component of the US force conducts these CUL-related missions. In which case, the TSC will play a major role in providing CUL support to multinational forces.

Note. Currently, defense reutilization and marketing service is prohibited from providing disposal support, including hazardous waste disposal support, to other nations.

Supporting multinational operations is a major challenge. Potential problem areas include language, cultural differences, differences in doctrine, terminology and definitions, methods for computing requirements, mobility, interoperability, infrastructure, competition among Services and alliance and/or multinational members for common support, environmental considerations, and national resource limitations. In addition, multinational support is subject to US fiscal constraints. An accounting system may be required to ensure the appropriate nation or international agency is properly billed for the logistics it receives. The TSC must recognize these potential problem areas, harmonize them when realistically possible, and account for them during planning and execution.

The relationship between the TSC and the MNF is supporting to supported. In general, the MNFC identifies his support requirements in terms of priority, location, timing, and duration. The TSC commander determines the forces, methods, and procedures to be employed in providing the support. If the TSC commander, subject to his existing capabilities and other assigned tasks, cannot fulfill the MNFC's requirements, then the establishing commander, as established by formal or implementing agreements, is responsible for determining a solution – i.e. a change in overall priorities or allocation of resources.

This command relationship provides the TSC with the control it requires to effectively and efficiently operate the intratheater segment of the distribution system while simultaneously providing responsive support to Army, joint, and multinational forces.

Parallel and collaborative planning between TSC and the MJLC is essential to providing responsive and flexible CUL support to multinational forces.

Many of the same mechanisms that work with joint operations will work in multinational operations; however, one aspect that requires special attention in multinational operations is contractor support. Each nation, and each Service representing each nation, can obtain contracted support. The TSC planners must work with their multinational counterparts to plan for the sharing of theater support and HN contracts to minimize competition for contracted support; and to ensure that contracted support is reasonably available to all participants.

C. Intergovernmental Organizations (IGO), Nongovernmental Organizations (NGO), and International Agency Cooperation

Within the theater, the GCC is the focal point for planning and implementing regional military strategies that require IGO, NGO, and international agency coordination.

Normally, the GCC will form a civil-military operations center to facilitate coordination with other agencies, organizations, and the HN. Achieving unity of effort is essential to mission success and mitigating human suffering. Key civil-military operations center tasks include:

- Carrying out GCC guidance and decisions regarding civil-military operations.
- Providing a partnership forum for military and other participating organizations to meet the needs of the populace.
- Receiving, validating, and coordinating requests for routine and emergency military support from IGOs, NGOs, and international agencies.

Refer to JFODS5-1: The Joint Forces Operations & Doctrine SMARTbook (Guide to Joint, Multinational & Interorganizational Operations). Updated for 2019, topics include joint doctrine fundamentals (JP 1), joint operations (JP 3-0 w/Chg 1), an expanded discussion of joint functions, joint planning (JP 5-0), joint logistics (JP 4-0), joint task forces (JP 3-33), joint force operations (JPs 3-30, 3-31, 3-32 & 3-05), multinational operations (JP 3-16), interorganizational cooperation (JP 3-08), & more!

US agencies, IGOs, NGOs, and international agencies provide for their own logistics support. However, US military logistics capabilities are frequently requested and provided to these organizations. This support may include intertheater and intratheater airlift; ground transportation of personnel, equipment and supplies; airfield control groups; and port and railhead operations groups as authorized by Title 10 US Code.

III. Support Operations Roles

A. Conduct Reception, Staging, Onward Movement, and Integration (RSOI) Operations

RSOI is a set of complex processes involving the GCC and his Service component commands, and strategic and joint partners such as USTRANSCOM. In order to conduct efficient and effective RSOI operations, a seamless strategic-to-theater interface is required.

Balance is central to the relationship between deployment and theater distribution. To achieve balance, the flow of units, equipment, and materiel in the intertheater and intratheater systems must be regulated to allow for a continuous and controlled flow of units, equipment, and materiel. To accomplish this, the supported GCC maintains overall responsibility for planning RSOI operations. Responsibility for execution, however, may be assigned to a joint headquarters or Service component command. In either case, the ASCC is heavily involved in RSOI operations due to dominant user and Army executive agency responsibilities.

The TSC enables efficient and effective RSOI by building a theater infrastructure from a combination of existing and deployable assets capable of supporting the deployment process and rapid force generation. It relies upon subordinate sustainment brigades, augmented by theater opening elements, to conduct port of debarkation support operations; provide life support; and execute theater distribution operations. Army health support is provided by the MEDCOM (DS). Under certain METT-TC conditions, the TSC may also employ one or more ESCs to provide a forward-based mission command presence that provides oversight of RSOI/or theater distribution operations.

Within the TSC, the SPO provides staff oversight of TSC RSOI efforts; coordinating and synchronizing reception, staging, and onward movement activities with subordinate commands and strategic/joint headquarters to maintain a balanced flow of supplies, personnel, equipment, and units consistent with strategic lift capabilities and ASCC/GCC priorities.

Collaborative planning and coordination between the TSC and strategic/joint headquarters is especially critical to the TSC's ability to synchronize and integrate intratheater deployment and distribution operations. This planning and coordination effort provides the TSC with the means to successfully:

- Monitor airlift and sealift flow.
- Provide movement control of arriving supplies, personnel, equipment, and units.
- Establish theater-wide capabilities required to meet anticipated transportation and throughput capacities.
- Provide life support.
- Establish effective liaison among the Service components and strategic providers.
- Identify HNS requirements.

See pp. 9-29 to 9-38 for complete discussion of RSOI operations.

B. Provide Theater Distribution

Theater distribution is enabled by a distribution management system. The Army distribution system is designed to optimize available infrastructure, reduce response time, maximize throughput, and support time-definite delivery. Effective distribution management synchronizes and optimizes the various sub-elements of the distribution system. Methods may include: maximizing containerization, increasing standardized transportation and materiel handling equipment, integrating aerial re-supply as a routine method of delivery, synchronizing and integrating retrograde operations across all available transportation modes, reducing storage, reducing transportation mode transfer handling requirements, and increasing ITV in an AO/joint operations area (JOA).

See pp. 3-38 to 3-39 for related discussion of theater distribution from FM 4-0.

TSC/ESC Distribution Management Roles

The TSC is the distribution manager of the intratheater segment of the global distribution system. The TSC's primary role in distribution management is a seamless flow of supplies, personnel, and equipment throughout the intratheater segment of the distribution system; delivering the right quantity, at the right time, and in the right location. In part, the TSC accomplishes this through close coordination with the JDDOC and supported J-4/G-4s. It is also accomplished, in part, through the effective use of ITV to monitor distribution flow. It executes distribution operations in accordance with priorities identified by the ASCC G-4. If an ESC is deployed, it performs the role of distribution manager for its specified theater of operations, AO or JOA. The ESC and sustainment brigades monitor, track, and execute distribution operations in accordance with TSC guidance.

Note. If the TSC serves as a joint logistics center, the JDDOC may collocate with the TSC DMC and assist the TSC in joint oversight of distribution and deployment.

Under certain METT-TC conditions, an ESC may also have support relationships with other ESCs or sustainment brigades supporting other specified AOs/JOAs.

As the theater distribution manager, the TSC maintains a theater-wide focus; participating in and coordinating with applicable joint logistics boards, centers, bureaus responsible for resolving issues concerning competing priorities and the allocation of constrained resources.

TSC distribution managers conduct parallel and collaborative planning with supporting and supported commands in order to facilitate the effective execution of distribution operations in accordance with ASCC priorities and supported commander requirements.

The ESC performs the same function both from and within its specified area of operations; coordinating with the TSC and supported J-4/G-4 in order to update its time definite delivery schedules and distribution priorities. It also collects and analyzes ITV distribution information to monitor the distribution flow inside the theater of operation/JOA.

TSC and ESC DMCs synchronize operations within the distribution system to maximize throughput and follow-on sustainment. They manage all facets of transportation including the effective use of air, land, and sea transportation assets. DMCs maintain connectivity with supporting and supported headquarters and use all possible measures to establish and maintain a COP.

TSC and ESC roles in physical distribution include maintaining visibility of theater distribution assets within the distribution network. The TSC and ESC can direct cross-leveling of distribution resources to meet requirements and/or optimize the distribution flow.

The TSC will have situational awareness on convoy operations, and will coordinate with the supporting maneuver enhancement brigade, joint security coordination center, or other Army forces to select convoy routes. When developing the ground distribution system, the DMC should account for identifying, evaluating, and comparing factors which tend to facilitate convoy movement and control such as movement restrictions, route classification, traffic flow, choke points, and rest halts. Implicit in this task, is the requirement to develop a thorough understanding of the current enemy situation along the route to include the identification of danger areas and potential ambush sites. Depending on the level of threat, the TSC or ESC should work with supporting maneuver enhancement brigade or local terrain manager to provide convoy escorts or establishing a movement corridor that provides for coordinated responses to Level I and Level II threats.

As required, the TSC DMC may establish distribution management boards or sit in on a joint transportation board to ensure distribution management processes are linked with theater-level processes/boards. The TSC DMC examines current operations to ensure success in achieving the effects the combatant commander desires on the battlefield.

Refer to ATP 4-0.1 for additional information on army theater distribution.

C. Provide Movement Control

The TSC provides mission command for operational-level movement control and multimodal operations in the theater. It is responsible for developing plans, policies, and programs that support the efficient use of Army transportation assets and the efficient flow of supplies, personnel, equipment, and units throughout the intratheater distribution system. The TSC accomplishes this, in part, through effective coordination with the JDDOC to maintain operational awareness of the global distribution system and joint requirements for common user land transportation assets.

The TSC manages intratheater movements through its subordinate MCB(s); implementing priorities established by the ASCC in support of the GCC concept of operations. Critical TSC tasks include:

- Balancing existing transportation capabilities of the distribution system with the day-to-day and projected operational requirements.
- Preparing estimates, plans, policies, and procedures for movement control, mode operations, and terminal operations.
- Managing transportation flow capacity by maintaining visibility of resources that are being transshipped at transshipping nodes.
- Coordinating the movement of major units.
- Developing policies and procedures to control, regulate, and expedite the movement of intermodal assets (i.e. leased containers, flatracks, and 463L pallets) within the theater.
- Maintaining liaison with JDDOC, JTF-PO, and HN transportation agencies, mode operators, and supported units.
- Protecting movement control assets.

In order to more efficiently control movements within the theater, the TSC may decide to divide the theater into transportation movement regions. This approach permits centralized control by the TSC and decentralized execution of movement control functions by subordinate MCBs.

In addition to facilitating the synchronized flow of units, supplies, equipment, and materiel along main and alternate supply routes, the MCB, through its subordinate movement control teams, performs movement control functions at APODs, SPODs, distribution hubs, and other critical nodes to expedite port clearance and provide for the uninterrupted flow of resources and capabilities in support of Army requirements.

Operating in accordance with GCC plans and policies, the MCB is responsible for managing the use of trailers, containers, air pallets, and flatracks located throughout the intratheater distribution system. Included in this responsibility is the requirement to coordinate with users to expedite return of these assets to the distribution system based upon GCC priorities.

Inter-modal operations are critically affected by the manner in which container management policies are enforced and container management is subsequently executed. Therefore, it is vital to TSC distribution operations that visibility and control of containers be maintained IAW GCC policies. Adherence to GCC policies will ensure adequate numbers of containers are available to support intratheater distribution system requirements.

D. Provide Materiel Management

The TSC provides the ASCC with a centralized materiel management capability that provides increased efficiencies and effectiveness by reducing redundant materiel management layers, centralizing materiel management functions, and employing a theater-wide view of resources. The result of which is responsive support to Army requirements and reduced customer wait time.

The TSC performs materiel management for all classes of supplies [less medical – CL VIII materiel management is provided by the MEDCOM (DS)] and maintenance management for those activities for which the TSC has control and responsibility. TSC personnel perform the day-to-day planning for operations; providing the theater interface between strategic and operational-level support. TSC materiel management responsibilities include managing, cataloging, requirements planning, requirements validation and prioritization for procurement, distribution, redistribution of excess, and retrograde of materiel. The TSC performs these functions within the parameters of policies, plans, priorities, and allocations developed in coordination with the ASCC G-4.

Note. Inherent in TSC maintenance management responsibilities is the requirement to integrate sustainment maintenance activities in support of the ASCC to include coordinating with USAMC or other elements that may control or have over sight of these entities.

Army materiel managers are undergoing a fundamental logistics information system streamline process, and in the interim, will recognize materiel management structures designed for our modular units, and structures designed for the Global Combat Support System-Army (GCSS-A). Material Management at the Sustainment Brigade is considered Level I management. Material Management at the Expeditionary Support Command and Theater Support Command are considered Level II & III respectively. The following levels encompass:

- Level I: Sustainment Brigade materiel management has a highly interactive direct management role in GCSS-A as well as a supply support activity and unit level quality surveillance and oversight role.
- Level II: The ESC or equivalent at this level of materiel management has less of an interactive direct management role in GCSS-A, and more of a quality surveillance and oversight role over Level I materiel mangers.
- Level III: The TSC, ASC, National Guard Materiel Management Center, or equivalent at this level of materiel management has more of a quality surveillance and oversight role over both Level I and II materiel managers.

In accordance with ASCC support priorities, the TSC provides direction for receiving, storing, and issuing theater stocks. When the required stocks are not available or stock replenishment is required, the TSC passes requirements to the appropriate CONUS national inventory control point. For requirements being considered for local procurement, the TSC validates the requirement prior to forwarding it to the local procuring activity.

Requisitions flow from the requesting unit directly to systems controlled by TSC materiel managers in accordance with standard operating procedures. The corps/

E. Provide Sustainment

Ref: ATP 4-94 (FM 4-94), Theater Sustainment Command (Jun '13), pp. 3-4 to 3-5.

Supply

The mission determines sustainment requirements, and is further influenced by factors such as deployment timelines, troop density, infrastructure, geography, and theater policies. The TSC staff considers these factors as well as others when developing a concept of support to meet ASCC requirements. Inherent in this is an understanding of the supported commander's priorities and status of available resources.

Typically, during the early stages of an operation, the TSC will prioritize replenishment of certain classes of supplies (I, IIIB, and V) to supported units based upon an analysis of the applicable supported plan, supported commander's priorities, and planning factors. The TSC may rely upon Army pre-positioned stocks to meet initial surge requirements for sustainment. Push systems will most likely be utilized in the early stages of an operation, and as distribution capabilities expand, a pull system, based upon anticipated/actual requirements, is implemented to achieve greater effectiveness and efficiencies.

Services

The TSC is responsible for planning, resourcing, monitoring, and analyzing field services support to deployed Army forces. TSC field services operations include field laundry, showers, light textile/clothing repair, force provider, mortuary affairs, aerial delivery support, and coordination with DLA for hazardous waste removal.

Services, such as shower and laundry support, are provided on an area basis throughout the theater and may be called forward to support brigade combat team sustainment operations. Force provider assets may be employed at major nodes within the theater to provide life support services to units flowing into or out of the theater. Force provider amenities include a dining facility, showers, laundry, and billeting.

The TSC, usually in coordination with the Joint Mortuary Affairs Office (JMAO), determines the requirement for and placement of Army theater MA assets. Requirement and placement considerations include casualty estimates from the ASCC G-3, force structure, and the mortuary affairs concept of support. During Contingency Operations, units evacuate remains through mortuary affairs collection points (MACPs), located throughout the AOR, to the theater mortuary evacuation point.

Maintenance

The TSC is the fleet maintenance manager for Army forces deployed in an AOR, theater of operations, AO, or JOA. It is in this capacity that the TSC collects, analyzes, and monitors readiness data of subordinate and supported units. This enables the TSC to effectively manage maintenance support to units and systems in accordance with ASCC priorities. Systemic issues, beyond TSC capabilities, are passed to the appropriate national level maintenance manager via the Army field support brigade (AFSB) for resolution.

As the fleet maintenance manager for deployed Army forces, the TSC develops plans, policies, programs, and procedures involving the maintenance of ground missile and aviation equipment in accordance with requirements established by the ASCC. The intent of which is to maximize combat readiness through the effective application of field and sustainment maintenance capabilities.

To this end, TSC maintenance managers work closely with the AFSB (OCONUS) to ensure effective sustainment maintenance support to Army forces; analyzing readiness data for both systemic problems and those associated with the unique aspects of the specific operational environment, such as, environmental conditions and usage levels. This coordination and collaboration also provides the basis for the effective employment of USAMC sustainment maintenance capabilities throughout the AOR, theater of operations, AO, or JOA.

TSC field maintenance activities involve the collection and analysis of maintenance data and reports; enabling the TSC to enforce ASCC priorities relating to the repair of specific types of equipment or support of specific units. These same activities provide the means to identify significant trends and deviations from established standards; enabling TSC maintenance managers to take action to ensure the maximum number of combat systems remain fully mission capable. TSC actions may include disseminating technical information and the allocation or reallocation of resources and capabilities to support maintenance requirements.

Operational Energy

Operational energy is the sum of energy and associated systems, information and processes required to train, move, and sustain forces and systems for military operations. Commanders at all levels must reinforce behaviors and employ technical solutions that reduce the amount of energy we waste and make more effective use of the energy consumed. To achieve this goal, energy considerations should be included during mission planning and resourcing and then monitored during mission execution. These conservation practices reduce supply and maintenance requirements and associated risks, free personnel for other missions, increase operational endurance, and enhance operational security and combat effectiveness.

Operational Contract Support

Operational contract support plays an ever increasing role in operations and is an integral part of the overall process of obtaining support across the range of military operations. Today, and for the foreseeable future, contract support will often be used to augment other support capabilities by providing an additional source for required supplies and services. These supplies and services include all classes of supply (Class VIII, subject to approval by medical personnel, and Class IX may be limited); labor; mortuary services (within specific parameters); laundry; showers; dining facility services; sanitation; transportation; and port operations (if not under the control of SDDC or Air Mobility Command [AMC]). Other contracted services may include billeting, maintenance and repair, printing and copier support, equipment leasing, and access to communication networks, temporary real property leasing, and limited minor construction. Currently, there are three broad types of contracted support: theater support, external support, and system support.

Financial Management Support

The integration of financial management capabilities with those of the TSC provides the basis for coordinated and synchronized financial management support to operations throughout the theater; ensuring consistency of financial management support and the most efficient use of all available financial management resources and services. Financial management capabilities are performed and coordinated by the financial management center (FMC) and the TSC assistant chief of staff, G-8. Financial management core competencies are procurement support; limited pay support; disbursing support; accounting support; banking and currency support; identify, acquire, distribute, and control funds; develop resource requirements; and track, analyze, and report budget execution. The paragraphs below provide a description of each core competency.

Human Resources Support

The human resources sustainment center (HRSC) plans, integrates, and coordinates human resources; casualty operations; reception, replacement, return to duty, rest and recuperation, and redeployment (R5) operations; and postal operations support of Army forces within the theater, theater of operations, AO, or JOA; coordinates and synchronizes human resources (HR) capabilities with those of the TSC SPO, and supports the ASCC G-1. The integration of HR capabilities with those of the TSC SPO provides the basis for coordinated and synchronized HR support to operations throughout the theater. When executed properly, integrated HR support is a combat multiplier.

theater automatic data processing service center (CTASC) then directs a sustainment brigade to fill the requirement based upon that brigade's ability to support the requirement or passes the requirement to the appropriate national inventory control point. This streamlined process permits the TSC to reach across TSC theater-wide resources and capabilities to satisfy Army requirements. Enabled by asset visibility, this approach not only reduces but also minimizes the level of stocks required to be stored throughout the theater.

This centralized materiel management is not exclusive. ESCs, if deployed, as well as sustainment brigades have unique roles in materiel management. Typically, ESC materiel management capabilities are focused on maintaining operational awareness of order status that enables effective distribution management within an AOR, theater of operations, or AO/JOA. In a similar manner, sustainment brigade materiel management capabilities are focused on the management of subordinate supply support activities (SSAs) that support the brigade's area support mission.

TSC materiel managers develop plans, policies, programs, and procedures involving supply activities; maintain liaison with supported and supporting units; and recommend allocation of resources and materiel management functions to support mission requirements.

F. Provide Army Special Operations Forces Support

Logistics support of ARSOF units is the responsibility of the Army except where otherwise provided for by support agreements or other directives.

The integration of ARSOF support cell capabilities with those of the TSC SPO facilitates synchronized and tailored support to specific ARSOF mission requirements and provides sufficient flexibility to respond to changing requirements. Additionally, the resulting coordination and synchronization between the ARSOF support cell and TSC provides the means for the TSC to leverage logistics resources and achieve greater operating efficiencies. This is particularly true in the case of demands for common user land transportation assets.

ARSOF units are supported in the same manner as conventional forces for common items and common-user support – on an area basis, through limited HN support, and contracting. Unique special operations items are resourced through special operations channels.

G. Provide Common-User Logistics (CUL) Support

3-110. GCCs must maximize limited distribution support systems, and leverage all logistical DOD and JIIM agencies; therefore, the option to designate a common-user logistic is optimal. There are two key considerations the GCCs may use to designate common-user responsibilities. These are the dominant user and the most capable force available. See JP 4-07 for additional information on common-user logistics (CUL) support.

Note. The TSC is usually the dominate user and most capable logistics mission command headquarters in theater.

When establishing CUL responsibilities within a subordinate joint force, the combatant commander must take into account existing CUL-related, DOD directed executive agencies as well as any existing CUL-related agreements. When CUL support is utilized, Service component commands retain overall responsibility for logistics support of their forces except when there are valid agreements or directives for the provision of CUL support outside of the normal Service component channels.

The TSC, as the ASCC's senior sustainment headquarters, plays a major role in optimizing resources and synchronizing materiel support to the joint force per geographic combatant commander's CUL guidance, approved executive agency responsibilities, inter-Service support agreements, and acquisition and cross-servicing agreements. When tasked to perform specified CUL responsibilities, the TSC must:

- Participate in the overall joint sustainment preparation of the operational environment effort in order to identify and gain access to key terrain, facilities, etc.
- Identify specific lead CUL responsibilities and CUL execution parameters.
- Review all service support requirements as they relate to determining CUL requirements.
- Coordinate CUL support in accordance with tasking assigned.
- Determine the source of support (military, civilian, HN, or other).
- Establish, maintain, and change priorities based on the operational situation and the combatant commander's guidance.
- Monitor critical classes of supply support capabilities for the purpose of mission tasking, economy of resources, and cross leveling of critical common-item resources in accordance with combatant commander directives for logistics.
- Coordinate requirements for agreements for inter-Service supply and support, local procurement and controls, and allocated indigenous facilities and logistics resources available within the theater.
- Coordinate transactions and implementing instructions for US and multinational support logistics exchange issues with the appropriate Service component, agency, and/or multinational points of contact.
- Prioritize the theater distribution and logistics effort by phase or operation.
- Manage intratheater movements.
- Allocate critical distribution and CUL resources in order to provide effective and efficient support.
- Establish a theater-wide capability to capture and maintain asset visibility of common-user materiel and services in accordance with the GCC's theater ITV and asset visibility data capture plan.
- Participate in functional boards or centers, if established, to centrally manage critical assets and more effectively react to unforeseen circumstances.

Managing the competing requirements of CUL and support to Army forces is a primary concern for TSC distribution managers. However, these managers must abide by the priorities established by the GCC and develop appropriate metrics to measure delivery objectives when executing CUL responsibilities. Above all, distribution managers must recognize that CUL offers greater economy and facilitates the rapid buildup of combat power by minimizing strategic lift requirements. However, the need to economize must be balanced with the requirement of timely delivery of supplies and materiel in support of GCC priorities.

H. Conduct Theater Closing Operations

Ref: ATP 4-94 (FM 4-94), Theater Sustainment Command (Jun '13), pp. 3-20 to 3-21.

Theater closing operations are complex events that require detailed planning and synchronized execution. Decisions made concerning withdrawal timetables, residual forces, and materiel to remain in the host country will influence the pace and nature of the redeployment. Theater closing and redeployment operations are conducted in accordance with the GCC redeployment operations plan (OPLAN) or GCC redeployment policy.

The ASCC redeployment OPLAN provides specific guidance to Army organizations preparing for redeployment; specifying the sequence for redeployment of units, individuals, and materiel. The plan also provides guidance on the support network, security requirements, and Army pre-position stocks turn-in procedures.

The TSC supports effective and efficient redeployment operations through its mission command of the intratheater distribution system by maintaining operational awareness of system capacity and leveraging joint capabilities. While many of the procedures used to rapidly build combat power – i.e. support RSOI, draw Army pre-position stocks, and operate the intratheater distribution system, apply to the redeployment process, two factors in particular complicate redeployment operations.

- First, the same elements that operate and manage the intratheater distribution system during deployment and sustainment operations perform similar roles during redeployment. When redeployment and deployment/sustaining operations occur simultaneously, the TSC may find it necessary to rebalance its forces or change the missions of subordinate organizations in order to effectively support simultaneous operations.
- Second, requirements vary widely depending on the scale and scope of redeployment operations, theater infrastructure, and other METT-TC considerations. For example, redeployment operations may range from personnel only to entire units and their equipment. Depending upon the political/military strategy, unit rotations may occur while decisive operations continue unabated or during operational transitions. Key TSC considerations include but are not limited to: size of the force redeploying/deploying; infrastructure requirements/limitations; security requirements; traffic circulation; staging areas; distribution system capacity; competing requirements for available resources; quantities of supplies and materiel to be redistributed; agricultural inspections; and establishing and maintaining accountability of retrograde cargo.

The TSC SPO provides staff oversight of TSC efforts in support of Army redeployment operations; coordinating and synchronizing unit movements to designated assembly areas in accordance with the ASCC redeployment OPLAN. Redeployment operations at the assembly areas are controlled and supervised by the TSC; supervising the actions necessary to prepare units for movement.

Key TSC planning considerations include the identification and allocation units, equipment, and supplies, to include HN and contractor support, required to support the redeployment operation. Coordination for medical support and other support functions such as communications, materiel handling equipment, and port of embarkation (POE) support is essential to mission success.

Close coordination with the ASCC, JDDOC and national/strategic partners facilitates the redeployment process. For example, coordination with the ASCC establishes the overall framework for the orderly sequencing of Army forces for redeployment. Issues such as responsibilities, reporting requirements, movement constraints, and resource allocation are addressed in a collaborative manner.

See pp. 1-26 to 1-27 for further discussion from ADP 4-0.

I. Logistics Over the Shore (LOTS)

Logistics over the shore (LOTS) or joint logistics over the shore (JLOTS) is the process of loading and unloading seaward supply ships without the benefit of deep-draft-capable fixed-port facilities. LOTS is a single service operation while JLOTS operations is normally a coordinated effort between Army and Navy personnel to receive, stage and push supplies, equipment and fuel inland to forward deployed forces. LOTS/JLOTS can be conducted in austere conditions ("bare beach") or at damaged port facilities. It may or may not necessarily follow amphibious assault operations. For the purpose of this publication, the term JLOTS will be used, as theater operations are inherently joint and multinational. The need for cooperation between Army, other Department of Defense components, and governmental, coalition, and host-nation organizations are critical when opening a theater in austere port/bare beach conditions.

In a joint environment the GCC will normally designate a JFLCC with an augmented JFC headquarters element for better mission command of theater operations. This normally falls on a service component commander to fulfill (for example, the ASCC). The GCC or JFLCC normally delegates mission command for JLOTS operations to a subordinate designated as the JLOTS commander. The JLOTS commander will be subordinate to the JFLCC while simultaneously coordinating logistical operations with the G4 and JDDOC elements. The JLOTS commander will have command authority to direct all aspects of the JLOTS operations, to include:

- Arrival of strategic maritime vessels within the JLOTS AOR
- Offshore offloading of seafaring vessels onto Army watercraft
- Consolidation of equipment and supplies to inland marshalling areas
- Movement of equipment and supplies to supported unit's AOR

The JLOTS commander will be assisted by an augmented JLOTS staff of personnel from the contributing service components and functional maritime units for subject matter expert input. This task may be assigned to the TSC to provide a JLOTS commander and the basis for a JLOTS staff. The TSC, as a functional higher-echelon headquarters, can greatly facilitate operational planning for and execution of theater JLOTS requirements. The TSC can provide a linear mission command structure to eliminate a multi-layered command structure by being responsible for ship to shore and shore to foxhole logistical movement. Mission command over specific JLOTS operations (with the possibility of multiple JLOTS operations being executed simultaneously) can be further delegated to an ESC for each operation with sustainment brigades for JLOTS port opening operations and inland movement of equipment and supplies. Further tactical mission command can be delegated to sustainment brigades with functional terminal battalions and companies attached for specific mission requirements.

The TSC will need to be augmented with joint and multi-national personnel to form a proper JLOTS staff, to include liaison teams and subject matter experts from:

- US TRANSCOM
- SDDC
- AMC
- DLA (for petroleum operations)
- Navy
- Marine Corps
- Air Force (weather data and air movement coordination inland)
- Coalition partners
- Host nation personnel

Units that can be placed under the mission command of a TSC to execute JLOTS operations may include:

- Terminal Battalion Headquarters
- Theater Terminal Opening Element (TTOE)
- Watercraft Units
- Movement Control Teams
- Seaport Operations Company
- Terminal Supervision Team
- Port Management Team
- Rapid Port Opening Element (in a direct support role)
- Causeway Element
- Units and personnel to handle communications, civil-military operations, human resources, Army health services

Each service component has the personnel and equipment needed to contribute to JLOTS operations within its area of core competency. The Army has transitioned in response to the need for a more expeditionary, flexible, joint-operations capable fighting force. The Army can provide but is not limited to:

- Logistics support vessels
- Lighterage
- Causeway Construction Units
- Ship offloading personnel
- Cargo transfer companies
- Surface movement teams and coordinators
- Medium to Heavy Transportation Assets
- Inland Petroleum Distribution Systems
- Divers for JLOTS site reconnaissance and examination

The Navy can provide but is not limited to:

- Divers for JLOTS site reconnaissance and examination
- Amphibious Construction Units (Seabees)
- Beach Group and Beachmaster ship-to-shore controllers
- Ship offloading personnel
- Tactical auxiliary cranes
- Fuel tanker vessels
- Lighterage
- Floating and elevated causeways and piers
- Beachhead force protection
- Offshore Petroleum Distribution System

Chap 8

Joint Logistics

Ref: JP 4-0, Joint Logistics (Feb '19), chap. I.

Sustainment

Sustainment—one of the seven joint functions (command and control [C2], information, intelligence, fires, movement and maneuver, protection, and sustainment)— is the provision of logistics and personnel services to maintain operations until mission accomplishment and redeployment of the force. Joint force commanders (JFCs) are called upon to maintain persistent military engagement in an uncertain, complex, and rapidly changing environment to advance and defend US values and interests, achieve objectives consistent with national strategy, and conclude operations on terms favorable to the US. Effective sustainment provides the JFC the means to enable freedom of action and endurance and to extend operational reach. Sustainment determines the depth to which the joint force can conduct decisive operations, allowing the JFC to seize, retain, and exploit the initiative. Joint logistics supports sustained readiness for joint forces.

I. Joint Logistics

The relative combat power that military forces can generate against a threat is constrained by their capability to plan for, gain access to, and deliver forces and materiel to points of application. Joint logistics is the coordinated use, synchronization, and often sharing of two or more combatant commands (CCMDs) or Military Departments' logistics resources to support the joint force. To meet the wide variety of global challenges, combatant commanders (CCDRs), subordinate commanders, and their staffs must develop a clear understanding of joint logistics, to include the relationship between logistic organizations, personnel, core functions, principles, imperatives, and the operational environment (OE). This publication provides logistics guidance essential to the operational capability and success of the joint force. It focuses on the integration of strategic, operational, and tactical support efforts while leveraging the global joint logistics enterprise (JLEnt) to affect the mobilization and movement of forces and materiel to sustain a JFC's concept of operations (CONOPS). Additionally, it provides guidance for joint logistics; describes core logistics functions essential to success; and offers a framework for CCDRs and subordinate commanders to integrate capabilities from national, multinational, Services, and combat support agencies (CSAs) to provide forces properly equipped and trained, when and where required. The identification of established coordination frameworks, agreements, treaties, theater distribution, and posture plans creates an efficient and effective logistics network to support the JFC's mission.

Joint logistics planning must account for the adversary's threat to logistics. It must also identify and reduce logistics and operational risks. The challenge for future joint logistics is to adequately support globally integrated operations given the combination of five ongoing trends:

- Increasing logistics requirements caused by global demand for US joint forces and operations.
- Constrained and degraded resources, both overall and within the logistics force structure.
- The growing complexity of logistics operations.
- The proliferation of advanced antiaccess/area denial capabilities by adversaries that would degrade logistics capabilities and capacities.
- The increase of cyberspace threats to joint and partner logistics networks and mission systems.

Logistics integrates strategic, operational, and tactical support efforts to project and sustain military power across the globe at a chosen time and place, and represents a comparative advantage that provides multiple options to leadership and multiple dilemmas to potential adversaries. A relevant and resilient JLEnt remains essential to the pursuit of national interests through assurance, deterrence, and responding to a full range of contingencies.

II. Joint Logistics Environment (JLE)

Military leaders conduct globally integrated logistics operations in a complicated, interconnected, transregional environment (see Figure I-1). These operations involve the total force, which consists of the Active Component and the Reserve Component and Department of Defense (DOD) civilians and contracted support. Additional capabilities in the area of responsibility (AOR) or joint operations area (JOA) could also include a variety of military forces, other governmental organizations, nongovernmental organizations (NGOs), and multinational forces (MNFs). The joint logistics environment is the sum of conditions and circumstances that affect logistics. The joint logistics environment exists at the strategic, operational, and tactical levels. Globalization, technology advancements, antiaccess/area denial, and flexible threats create a complex, ever-changing OE. The essential challenge is to support unified action by meeting increasingly demanding logistics requirements with constrained resources in a potentially contested environment. Globally integrated logistics is the capability to allocate and adjudicate joint logistics support on a global scale to maximize effectiveness and responsiveness, and to reconcile competing demands for limited logistics resources based on strategic priorities. Understanding the global environment is essential to plan, execute, synchronize, assess, and coordinate logistics operations.

Joint logistics takes place throughout the OE. Service components and CSAs provide the forces, materials, and capabilities while the JFC's staff focuses on integrating the capabilities with operations. Access to secure networks is necessary to sustain joint force readiness. Effective networks are used to find and access relevant information, facilitate collaboration, distribute data to forward deployed areas, increase performance and reliability, ensure the enterprise infrastructure for evolving DOD systems is resilient, and leverage partner nations' (PNs') capabilities.

See facing page for further discussion of the JLE operating framework.

Building Partnership Capacity (BPC)

Complicated supply lines, finite resources, the challenges of providing robust logistics in austere environments, and shared lines of communications (LOCs) require the ability to establish and foster nontraditional partnerships. For some operations, logistics forces may be employed in quantities disproportionate to their normal military roles and in nonstandard tasks. Further, logistics forces may precede other military

Joint Logistics Environment (JLE) Operating Framework

Ref: JP 4-0, Joint Logistics (Feb '19), fig. I-1, p. I-3.

Military leaders conduct globally integrated logistics operations in a complicated, interconnected, transregional environment (see Figure I-1). These operations involve the total force, which consists of the Active Component and the Reserve Component and Department of Defense (DOD) civilians and contracted support. Additional capabilities in the area of responsibility (AOR) or joint operations area (JOA) could also include a variety of military forces, other governmental organizations, nongovernmental organizations (NGOs), and multinational forces (MNFs). The joint logistics environment is the sum of conditions and circumstances that affect logistics. The joint logistics environment exists at the strategic, operational, and tactical levels.

Strategic Level Campaign Quality	Operational Level Coordinate, Integrate, and Synchronize	Tactical Level Effectiveness
• Industrial base capacity enables sustained operations • End-to-end processes drive efficiencies across Services, agencies, and industry • Effectiveness dependent upon optimizing processes against required outcomes	• Combatant commander integrates joint requirements with national systems • Must optimize component, agency, and other partner nation capabilities to meet requirements • Most significant impact for joint logistics and the joint force	• Outcome is measured • Operational readiness enables "freedom of action" • Desired outcomes should drive optimization–from strategic to tactical

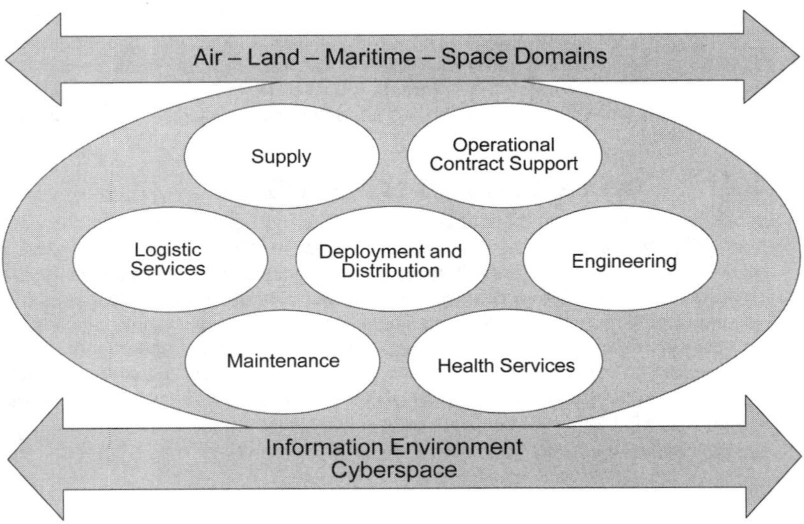

Ref: Figure I-1. Joint Logistics Environment Operating Framework.

forces or may be the only forces deployed. Logistics forces also may continue to support other military personnel and civilians after the departure of combat forces. BPC is important for sharing the costs and responsibilities, improving information flow, and establishing PN agreements. BPC includes coordination of resources with multinational partners, international organizations, and NGOs. BPC improves unity of effort within the entire JLEnt. BPC is an essential component of joint operations because the Services seldom have sufficient capability to support a joint force independently. BPC is an ongoing, long-term relationship development process that may not yield immediate results. The earlier the BPC efforts begin, the better the chance of success for securing partner logistics support when needed. By combining capabilities, commanders can provide maximum effectiveness and flexibility to the joint force focused on objectives that deliver sustained logistics support.

III. Joint Logistics Enterprise (JLEnt)

The purpose of the JLEnt is to protect and sustain military power across the globe at a time and place of our choosing and represents a US comparative advantage that provides multiple options to our nation's leadership and multiple dilemmas to potential adversaries. The JLEnt is a multi-tiered matrix of key global logistics providers cooperatively structured through an assortment of collaborative agreements, contracts, policy, legislation, or treaties utilized to provide the best possible support to the JFC or other supported organization. The key DOD organizations in the JLEnt include the Services, CCMDs, Defense Logistics Agency (DLA), Joint Staff J-3 [Operations Directorate], and Joint Staff J-4 [Logistics Directorate]. Other US Government departments and agencies, NGOs, and commercial partners also play a vital role in virtually all aspects of the JLEnt and function on a global scale, providing comprehensive, end-to¬end capabilities. The JLEnt may also include multinational partners and international organizations. Participants operate across the strategic, operational, and tactical levels— many are affiliated with either supported or supporting commands and operate under a variety of command relationships.

The JLEnt is interconnected among global logistics providers, supporting and supported organizations and units, and other entities. Knowing the roles, responsibilities, relationships, and authorities of JLEnt partners is essential to planning, executing, controlling, and assessing logistics operations. JLEnt partners must collaborate to ensure the coordinated employment and sharing of capabilities and resources. Global logistics providers manage end-to-end processes that provide capabilities to the supported CCDR to fulfill requirements. The transregional, multi-domain, and multifunctional nature of future threats, combined with budgetary pressures, require enterprise-wide tradeoffs; these tradeoffs can be mitigated through persistent responsiveness.

IV. Joint Logistics Imperatives

Joint logistics focuses on three imperatives to influence mission success: unity of effort, JLEnt visibility, and rapid and precise response. These imperatives define the desired attributes of a federation of systems, processes, and organizations that effectively adapt within a constantly changing OE to meet the emerging needs of the supported JFC. The joint logistics imperatives support operations, which is the primary purpose of logistics. Successfully meeting the needs of operational forces will build trust in the logistics process and between organizations during joint operations. These imperatives guide joint logisticians in the performance of the integrating functions needed for successful joint logistics operations.

See facing page for further discussion.

Joint Logistics Imperatives

Ref: JP 4-0, Joint Logistics (Feb '19, pp. I-5 to I-6.

Joint logistics focuses on three imperatives to influence mission success: unity of effort, JLEnt visibility, and rapid and precise response.

1. Unity of Effort

Unity of effort is the coordination and cooperation toward common objectives, even if the participants are not necessarily part of the same command or organization. Unity of effort is the product of successful unified action. For joint logisticians, unified action synchronizes and integrates logistics capabilities focused on the commander's intent. Unified action is critical to joint logistics objectives. To achieve unity of effort, joint logisticians must develop a clear understanding of how joint and multinational logistics (MNL) processes work, know the roles and responsibilities of the providers executing tasks in those processes, build agreement around common measures of performance, and ensure appropriate members of the JLEnt have visibility into the processes.

2. JLEnt Visibility

JLEnt visibility is access to logistics processes, resources, and requirements data to provide the information necessary to make effective decisions. JLEnt visibility is inclusive of the sub-components:

- **In-Transit Visibility (ITV)** is the ability to track the identity, status, and location of DOD units, non-unit cargo or supplies, passengers, patients, and personal property from origin to consignee or destination.
- **Defense Transportation System (DTS)** is that portion of the worldwide transportation infrastructure that supports DOD transportation needs in peace and war. It consists of three major sources of transportation resources and capabilities: military (organic), commercial (nonorganic), and host nation (HN). Resources include inland surface transportation (rail, road, and inland waterway), sea transportation (coastal and ocean), air transportation, and pipelines.

3. Rapid and Precise Response

Rapid and precise response is the ability of the core logistics functions, military and commercial, to meet the constantly changing needs of the joint force. The effectiveness of joint logistics can be measured by assessing the following attributes or key performance indicators.

- **Velocity** is at the core of responsiveness. Velocity does not mean everything moves at the same rate or fastest rate, but everything moves according to priority at the rate that produces a balance between efficiency and effectiveness to fully meet the CCDR's operational needs.
- **Reliability** is reflected in the dependability of the global providers and the development of a resilient distribution network able to deliver required support when promised. Reliability is characterized by a high degree of predictability or time-definite delivery of support. Time-definite delivery is the consistent delivery of requested logistics support at a time and destination specified by the requiring activity.
- **Efficiency** is related directly to the amount of resources required to achieve a specific objective. In the tactical and operational environments, inefficiency increases the logistics footprint, force protection requirements, and risk. At the strategic level, inefficiency increases the cost and risk for the operation.
- **Effectiveness** is the ability of the JLEnt to fully meet the CCDR's operational requirements within acceptable risk. Effectiveness is providing the right logistics solutions at the right time and place.

V. Joint Logistics Focus Areas

Ref: JP 4-0, Joint Logistics (Feb '19, p. I-7.

The joint logistics community must focus on the following five areas to influence mission success: warfighting readiness, competition below armed conflict, global integration, innovation, and the strengthening of alliance and partner networks. These areas discuss the desired attributes of a federation of systems, processes, and organizations that effectively adapt within a constantly changing OE to meet the emerging needs of the supported JFC. These focus areas will guide joint logisticians in the performance of the integrating functions needed for successful joint operations.

1. Warfighting Readiness

The Joint Staff will champion efforts to enhance and protect the JLEnt's capability, capacity, and comprehensive readiness to project/sustain military power globally at a chosen time and place. In particular, JFCs will assess and mitigate risk within operational planning activities to ensure the joint force is logistically positioned to support the range of military operations.

2. Competition Below Armed Conflict

Adversaries understand that strategic logistics is a comparative advantage of the US and will attempt to undermine its ability to project/sustain military power. These "supply chain wars" cut across all instruments of national power and include infiltration of traditional business systems. The joint force must understand this competition space, and our adversaries' capabilities/intentions in targeting the JLEnt, and pursue actions to protect mission assurance.

3. Global Integration

The JLEnt must be able to effectively allocate scarce resources to meet global priorities. Joint logisticians must have access to strategic logistics information and institute flexible processes that provide an accurate picture of the logistics environment to facilitate timely resource-informed decision making and enables operational success.

4. Innovation

A "data culture" improves the understanding of potential concepts like big data, artificial intelligence, machine learning, and modern computing power with regard to revolutionary improvements across the JLEnt. Adversaries will focus efforts on eroding the comparative competitive advantage in technology. Success in future conflicts may depend on the ability to expeditiously adopt and field new technologies that assure the continued ability to project and sustain power.

5. Strengthen Alliance and Partner Networks

Relationships with like-minded partners are essential to advancing US interests. It is critical to continue to advocate and support JLEnt efforts that increase joint force lethality, global agility, interoperability, and operational effectiveness through expanded access, visibility, and cooperation. Joint logisticians must understand the ability to project and sustain power is inextricably linked to the JLEnt, its array of partners and allies within DOD and broader US Government organizations, the industrial base, and aligned nations.

VI. Principles of Logistics

Ref: JP 4-0, Joint Logistics (Feb '19), pp. I-8 to I-9.

1. Responsiveness

Responsiveness is providing the right support when and where it is needed. Responsiveness is characterized by the reliability of support and the speed of response to the needs of the joint force. Clearly understood processes and well-developed decision support tools are key elements enabling responsiveness to emerging requirements. By monitoring the battle rhythm, the joint logistician can anticipate logistic issues.

2. Simplicity

Simplicity fosters efficiency in planning and execution, and allows for more effective control over logistic operations. Clarity of tasks, standardized and interoperable procedures, and clearly defined command relationships contribute to simplicity. Simplicity is a way to reduce the "fog of war" or the friction caused by combat. Clear objectives, relevant processes, and documented procedures assist unity of effort.

3. Flexibility

Flexibility is the ability to improvise and adapt logistic structures and procedures to changing situations, missions, and operational requirements. Flexibility is how well logistics responds in a dynamic environment. Where responsiveness is a commander's view of logistic support, flexibility is a logistician's view of being responsive. The logistician's ability to anticipate requirements in an operational environment allows for the development of viable options able to support operational needs.

4. Economy

Economy is the minimum amount of resources required to bring about or create a specific outcome. Economy is achieved when support is provided using the fewest resources within acceptable levels of risk. At the tactical and operational levels, economy is reflected in the number of personnel, units and equipment required to deliver support. Among the key elements of the logistic principle of economy is the identification and elimination of redundancy.

5. Attainability

Attainability is the assurance that the essential supplies and services available to execute operations will achieve mission success. Attainability is the point at which the CCDR or subordinate JFC judges that sufficient supplies, support, distribution capabilities, and LOC capacity exist to initiate operations at an acceptable level of risk. Some examples of minimal requirements are inventory on hand (days of supply), critical support and Service capabilities, theater distribution assets (surge capability), combat service support (CSS) sufficiency, and force reception throughput capabilities.

6. Sustainability

Sustainability is the ability to maintain the necessary level and duration of logistics support to achieve military objectives. Sustainability is a function of providing for and maintaining those levels of ready forces, materiel, and consumables necessary to support military action. Sustainability is focused on the long-term objectives and requirements of the supported forces. Sustainability provides the JFC with the means to enable freedom of action and extend operational reach.

7. Survivability

Survivability is the capacity of an organization to prevail in spite of adverse impacts or potential threats. To provide continuity of support, critical logistic infrastructure must be identified and plans developed for its protection. Survivability is directly affected by dispersion, design of operational logistic processes, and the allocation of forces to protect critical logistic infrastructure. Examples of critical logistic infrastructure include industrial centers, airfields, seaports, railheads, supply points, depots, LOCs, bridges, intersections, logistic centers, and military installations.

VII. Logistics Integration

Commanders and staffs apply basic principles, control resources, and manage capabilities to provide sustained joint logistics. Logisticians can use the principles of logistics as a guideline to assess how effective logistics are integrated into plans and execution. To achieve full integration, commanders and their logisticians coordinate, synchronize, plan, execute, and assess logistics support to joint forces during all phases of the operation.

Coordinating and Synchronizing

Effective coordination of joint logistics includes choosing organizational options to execute effective joint logistics operations.

See pp. 8-11 to 8-26 for further discussion.

Planning

Logistics planners at every level should set conditions for subordinate success. Timely, accurate, and responsive planning enables trade-offs, alternate courses of action (COAs), and, therefore, freedom of action for JFCs. Joint logistics planning links the mission and commander's intent to core logistics functions, procedures, and organizations. This establishes the JFC's ability to meet requirements in terms of forces, capabilities, movement, projection, sustainment, duration of operations, redeployment, and retrograde. Joint logistics operations overseas should be planned and conducted with appropriate consideration of their effect on the environment in accordance with applicable US and HN agreements, environmental laws, policies, and regulations. Joint logistics operations planned and conducted within the US and territories will be conducted in compliance with applicable federal, state, or local environmental laws and regulations. Early planning is essential to ensure all appropriate environmental reviews have been completed in accordance with Department of Defense Instruction (DODI) 4715.06, Environmental Compliance in the United States, and for installations outside the continental United States (CONUS), see DODI 4715.05, Environmental Compliance at Installations Outside the United States

See pp. 8-27 to 8-38 for further discussion.

Executing

Executing joint logistics involves the employment of capabilities and resources to support joint and multinational operations.

See pp. 8-39 to 8-42 for further discussion.

Assessing

Assessing joint logistics facilitates future success through plan refinement and adaptation. The joint logistician must be able to assess and respond to requirements by monitoring dynamic situations and providing accurate feedback to subordinates and decision makers.

Chap 8

I. Core Logistics Functions

Ref: JP 4-0, Joint Logistics (Feb '19), chap. II and pp. I-10 to I-12.

Core logistics functions provide a framework to facilitate integrated decision making, enable effective synchronization and allocation of resources, and optimize joint logistics processes. The challenges associated with support cut across all core logistics functions, especially when multiple joint task forces (JTFs) or multinational partners are involved.

Core Logistic Functions

Core Functions	Functional Capabilities
Deployment and Distribution	• Move the force • Sustain the force • Operate the joint deployment and distribution enterprise
Supply	• Manage supplies and equipment • Inventory management • Manage global supplier networks
Maintenance	• Depot maintenance operations • Field maintenance operations • Equipment reset
Logistics Services	• Food service • Water and ice service • Contingency base services • Hygiene services • Mortuary affairs
Operational Contract Support	• Contract support integration • Contracting support • Contractor management
Engineering	• General engineering • Combat engineering • Geospatial engineering
Joint Health Services	• Force health protection • Health service support

Ref: Figure II-1. Core Logistics Functions

The core logistics functions are: deployment and distribution, supply, maintenance, logistics services, operational contract support (OCS), engineering, and joint health services. The core logistics functions are considered during the employment of US military forces in coordinated action toward a common objective and provide global force projection and sustainment.

Core Logistics Functions

Ref: JP 4-0, Joint Logistics (Feb '19), pp. II-1 to II-13.

A. Deployment and Distribution

The global dispersion of the threats, coupled with the necessity to rapidly deploy, execute, and sustain operations worldwide, makes the deployment and distribution capability the cornerstone of joint logistics. These operational factors necessitate a shift from a supply-based system to a system that is primarily distribution-based with beginning-to-end synchronization to meet JFC requirements. Through sharing critical information, it is possible to create unity of effort among diverse distribution organizations to satisfy deployment, execution, and sustainment operations.

B. Supply

The joint logistician must understand the complexities of supply operations, the functions and processes that define them, and the organizations and personnel responsible for executing tasks to meet the JFC's requirements. The Services and DLA are primarily responsible for DOD supply chain operations and manage the supply processes to provide common commodities and services to joint forces. Planning for supply operations requires a collaborative environment to fully consider all major components of the JLEnt, to include the return and retrograde of equipment and supplies.

C. Maintenance

Maintenance supports system readiness for the JFC. The Services, as part of their Title 10, United States Code (USC), responsibilities, execute maintenance as a core logistics function. The Services employ a maintenance structure of depot- and field-level maintenance to improve the JFC's freedom of action and sustain the readiness and capabilities of assigned units.

D. Logistics Services

Logistics services comprise the support capabilities that collectively enable the US to rapidly provide global sustainment for our military forces. Logistics services include many scalable and disparate capabilities. Included in this area are food service, water and ice service, contingency base services, hygiene services, and mortuary affairs (MA).

E. Operational Contract Support (OCS)

OCS is a core logistics function and a critical component of total force readiness. DOD relies on contractors to perform many tasks. OCS provides the CCDR flexibility and options to employ commercially sourced logistics solutions from JLEnt partners such as BOS intra-theater transportation, logistics services, maintenance, storage, construction, security operations, and common-user commodities

F. Engineering

Engineer capabilities enable joint operations by facilitating freedom of action necessary for the JFC to meet mission objectives. Engineer operations integrate combat, general, and geospatial engineering to meet national and JFC requirements. Joint engineer operations facilitate the mobility and survivability of friendly forces; counter the mobility of enemy forces; provide infrastructure to position, project, protect, and sustain the joint force; contribute to a clear understanding of the physical environment; and provide support to civilian authorities and other nations.

G. Health Services

Joint health care services are conducted as part of an interrelated health system that shares medical services, capabilities, and specialists among the Service components and partners with multiple agencies and nations to implement a seamless unified health care effort in support of a joint force. Joint medical capabilities encompass both health service support (HSS) and force health protection (FHP) functions and are employed across the full range of military operations. These capabilities span the OA from prevention to point of injury/illness to definitive care.

Chap 8

II. Coordinating & Synchronizing Joint Logistics

Ref: JP 4-0, Joint Logistics (Feb '19), chap. III.

This section describes the authorities, organizations, and controls that synchronize logistics in support of the JFC. *JP 3-0, Joint Operations,* identifies C2 as a joint function. Command includes both the authority and responsibility for effectively using available resources and the art of motivating and directing people and organizations to accomplish missions. Control is inherent in command. However, logistics assets will rarely fall under one command, which makes control, coordination, collaboration, synchronization, and management of joint logistics more challenging. To control joint logistics, commanders direct forces and functions consistent with a commander's command authority. It involves organizing the joint logistics staff, operational-level logistics elements, CSAs, and their capabilities to assist in planning and executing joint logistics. Designating lead Service, assigning agency responsibilities, and developing procedures to execute the CCDR's directive authority for logistics (DAFL) will assist in planning, integrating, synchronizing, and executing joint logistics support operations. While logistics remains a Service responsibility, there are other logistics organizations, processes, and tasks to consider when developing a concept of logistics support (COLS) to optimize joint logistics objectives.

I. Logistics Authority

Directive Authority for Logistics (DAFL)

CCDRs exercise authoritative direction over logistics, in accordance with Title 10, USC, Section 164. DAFL cannot be delegated or transferred. However, the CCDR may delegate the responsibility for the planning, execution, and/or management of common support capabilities to a subordinate JFC or Service component commander to accomplish the subordinate JFC's or Service component commander's mission. For some commodities or support services common to two or more Services, the Secretary of Defense (SecDef) or the Deputy Secretary of Defense may designate one provider as the EA (see Appendix D, "Logistic-Related Executive Agents"). Other control measures to assist in developing common user logistics are joint tasks or inter-Service support agreements. However, the CCDR must formally delineate this delegated authority by function and scope to the subordinate JFC or Service component commander. The exercise of DAFL by a CCDR includes the authority to issue directives to subordinate commanders, including peacetime measures necessary for the execution of military operations in support of the following: execution of approved OPLANs; effectiveness and economy of operation; and prevention or elimination of unnecessary duplication of facilities and overlapping of functions among the Service component commands.

DAFL of a GCC applies to the entire AOR and affects all subordinate components, commands, and direct reporting units in the AOR. Some CCDR responsibilities include:

- Issuing directives to subordinate commanders, including peacetime measures necessary for the execution of military operations, in support of the following: execution of approved OPLANs, effectiveness and economy of operation, and prevention or elimination of unnecessary duplication of facilities and overlapping of functions among the Service component commands.

- Coordinating with USTRANSCOM to identify transportation-related requirements and initiatives (e.g., establishment of aerial port of debarkation [APOD]/ seaport of debarkation [SPOD], determining transportation routes and infrastructure to support).
- Coordinating with DLA to identify logistics requirements and initiatives (e.g., establishing storage locations, identifying pre-positioned material and equipment, determining fuel requirements, providing contingency contracting solutions).
- Establishing host-nation support (HNS) (e.g., acquisition and cross-servicing agreements [ACSAs]/mutual logistics support agreements, status-of-forces agreements, cost-sharing agreements).

Unless otherwise directed by SecDef, the Military Departments and Services continue to have responsibility for the logistics support of their forces assigned or attached to joint commands, subject to the following guidance:

- Under peacetime conditions, the scope of the logistics authority exercised by the CCDR will be consistent with the peacetime limitations imposed by legislation, DOD policy or regulations, budgetary considerations, local conditions, and other specific conditions prescribed by SecDef or the Chairman of the Joint Chiefs of Staff (CJCS). Where these factors preclude execution of a CCDR's directive by component commanders, the comments and recommendations of the CCDR, together with the comments of the component commander concerned, normally will be referred to the appropriate Military Department for consideration. If the matter is not resolved in a timely manner with the appropriate Military Department, it will be referred by the CCDR, through the CJCS, to SecDef.
- Under crisis, wartime conditions, or where critical situations make diversion of the normal logistics process necessary, the logistics authority of CCDRs enables them to use all facilities and supplies of all forces assigned to their commands for the accomplishment of their missions. The President or SecDef may extend this authority to attached forces when transferring those forces for a specific mission and should specify this authority in the establishing directive or order. Joint logistics doctrine and policy developed by the CJCS establishes wartime logistics support guidance to assist the CCDR in conducting successful joint operations.

A CCDR's DAFL does not:

- Discontinue Service responsibility for logistics support.
- Discourage coordination by consultation and agreement.
- Disrupt effective procedures or efficient use of facilities or organizations.
- Include the ability to provide contracting authority or make binding contracts for the US Government.

In exercising DAFL, CCDRs have an inherent obligation to ensure accountability of resources. This obligation is an acknowledgement of the Military Departments' Title 10, USC, responsibilities and recognizes that the Military Departments, with rare exceptions, do not resource their forces to support other DOD forces. In that regard, CCDRs will coordinate with appropriate Service components before exercising DAFL or delegating authority for subordinate commanders to exercise common support capabilities to one of their components. In keeping with the Title 10, USC, roles of the Military Departments, CCDRs should maintain an accounting of resources taken from one Service component and provided to another. This accounting can be used to reimburse the losing Service component in kind over time within the AOR when possible, or can be used to pass back a requirement to DOD for resource actions to rebalance Military Department resource accounts.

II. Joint Logistics Roles and Responsibilities

Clearly articulating responsibilities is the first step in fully synchronized and coordinated logistics support during joint operations.

A. Secretary of Defense (SecDef)

SecDef is the principal advisor to the President on defense matters and serves as the leader and chief executive officer of DOD. The offices of SecDef most concerned with logistics matters are the Under Secretary of Defense for Policy (USD[P]), Under Secretary of Defense for Acquisition and Sustainment (USD[A&S]) (formerly Under Secretary of Defense for Acquisition, Technology, and Logistics), and Assistant Secretary of Defense for Sustainment (ASD[S]) (formerly Assistant Secretary of Defense for Logistics and Materiel Readiness).

- **USD(P)** is SecDef's principal staff assistant (PSA) and advisor for all matters on the formulation of national security and defense policy and the integration and oversight of DOD policy and plans to achieve national security objectives.
- **USD(A&S)** is the PSA and advisor to SecDef and Deputy Secretary of Defense (DepSecDef) for all matters relating to logistics; installation management; military construction; procurement; environment, safety, and occupational health management; utilities and energy management; and nuclear, chemical, and biological defense programs.
- **ASD(S)** is the principal advisor to USD(A&S), SecDef, and DepSecDef on logistics and materiel readiness in DOD and is the principal logistics official within senior management. ASD(S) is the principal advisor to USD(A&S), SecDef, and DepSecDef for energy policy, plans, and programs, and advises the CJCS regarding the role of energy in the DOD planning process.

B. Chairman of the Joint Chiefs of Staff (CJCS)

The CJCS is the principal military adviser to the President and the National Security Staff (which consists of the National Security Council and the Homeland Security Council) and SecDef. The CJCS prepares joint logistics and mobility plans to support strategic and contingency plans and recommends the assignment of logistics and mobility responsibilities to the Armed Forces of the United States. The CJCS also advises SecDef on critical deficiencies in force capabilities (including manpower, logistics, intelligence, and mobility support).

C. Military Departments

The Military Departments exercise authority to conduct all affairs of their departments, including to recruit, organize, supply, equip, train, service, mobilize, demobilize, administer, and maintain forces; construct, outfit, and repair military equipment; adhere to environmental compliance; construct, maintain, and repair buildings, structures, and utilities; and acquire, manage, and dispose of real property or natural resources.

D. Services

In accordance with Title 10, USC, the Services are responsible for preparing for employment of Service forces. They recruit, supply, organize, train, equip, service, mobilize, demobilize, provide administrative support, and maintain ready forces. Services are the center of a collaborative network, and their logistics organizations form the foundation of the JLEnt. The Services are the primary force providers and executors of joint logistics, as well as the primary providers of logistics in support of their own Service organizations supporting the CCDR. They are responsible for operational logistics support systems, platforms, and their execution to support the force. They are responsible for maintaining systems' life-cycle readiness.

See pp. 8-20 to 8-21 for an overview of the Services' logistics execution.

E. Defense Logistics Agency (DLA)

As the nation's combat logistics support agency, DLA manages the global supply chain and in collaboration with JLEnt partners sustains the readiness and lethality of the Armed Forces of the United States. As a statutory CSA, DLA provides logistics advice, advocacy, and assistance to the Office of the Secretary of Defense, Joint Chiefs of Staff, the CCDRs, Military Departments, DOD components, and interagency partners.

F. The Joint Staff J-3

The Joint Staff J-3 is responsible for maintaining the global capability for rapid and decisive military force power projection. The Joint Staff J-3 is also responsible for leading the collaborative efforts of the joint planning and execution community to improve the joint deployment and redeployment processes, while maintaining the overall effectiveness of these processes so that all supported JFCs and supporting DOD components can execute military force power projection more effectively and efficiently. Additionally, the Joint Staff J-3 serves as the joint force coordinator and is responsible for coordinating the staffing of all force requirements among the joint force providers (JFPs), consolidating all execution and contingency sourcing recommendations, and performing the duties of a JFP for all conventional force requirements.

G. The Joint Staff J-4

The Joint Staff J-4 leads the DOD efforts in the JLEnt and coordinates policy and makes recommendations to improve the preparedness of the DOD global logistics force. Additionally, the Joint Staff J-4 advises the CJCS on the readiness assessments of the CCMDs and Services.

H. The Joint Staff J-5 [Directorate for Strategy, Plans, and Policy]

The Joint Staff J-5 collaborates with the Joint Staff J-4 to ensure contingency plans are resource-informed as the coordinating authority for global logistics.

I. Combatant Commands (CCMDs)

Unless otherwise directed by the President or SecDef, the CCDR exercises authority, direction, and control over the commands and forces assigned to that command through combatant command (command authority) (COCOM). CCDRs coordinate and approve the administration, support (including control of resources and equipment, internal organization, and training), and discipline necessary to carry out missions assigned to the command.

J. Executive Agent (EA)

A DOD EA is the head of a DOD component to whom SecDef or the DepSecDef has assigned specific responsibilities, functions, and authorities to provide defined levels of support for operational missions, or administrative or other designated activities that involve two or more of the DOD components. The DOD EA may delegate to a subordinate designee, within that official's component, the authority to act on that official's behalf for any or all of those DOD EA responsibilities, functions, and authorities assigned by SecDef or DepSecDef.

Refer to DODD 5101.1, DOD Executive Agent, and JP 4-0, Appendix D, "Logistics-Related Executive Agents," for details.

K. Combat Support Agencies (CSAs)

CSAs designated under Title 10, USC, Section 193, fulfill combat support (CS) or CSS functions for joint operating forces across the range of military operations and in support of CCDRs executing military operations. CSAs perform support functions or provide supporting operational capabilities, consistent with their establishing directives and pertinent DOD planning guidance.

For more information on CSAs, refer to DODD 3000.06, Combat Support Agencies.

L. US Transportation Command (USTRANSCOM)

USTRANSCOM is responsible for providing air, land, and sea transportation, terminal management, and aerial refueling to support the global deployment, employment, sustainment, and redeployment of US forces. USTRANSCOM serves as DOD's mobility JFP, DOD's single manager for defense transportation, and DOD's single manager for PM.

M. General Services Administration (GSA)

GSA provides logistics support for the functions and missions of the US Government, including DOD.

N. Defense Health Agency (DHA)

DHA is a CSA that enables the Army, Navy, and Air Force medical services to provide a medically ready force and ready medical force to CCMDs.

O. Lead Service

A Service or Service Component responsible for the programming and execution of common-user items, logistics functions, and/or service support. A CCDR may choose to assign specific CUL functions, to include both planning and execution to a lead Service. These assignments can be for single or multiple common logistics functions and may also be based on phases or OAs within the CCDR's AOR. In circumstances where one Service is the predominant provider of forces, or the owner of the preponderance of logistics capability, it may be prudent to designate that Service as the joint logistics lead for BOS-I. The CCDR may augment the lead Service logistics organization with capabilities from another component's logistics organizations, as appropriate. Key lead Service functions at operating areas typically include, but are not limited to, BOS-I, communications synchronization, and senior airfield authority (SAA) synchronization; budget programming; real property management; and provision (provide and fund) of common-user items or service support. The lead Service may consider a commercially contracted solution to meet the requirements in addition to, or in place of, organic support.

P. Base Operating Support-Integrator (BOS-I)

BOS-I is a sub-function of lead Service. The BOS-I is responsible for planning and synchronizing the efficient application of resources and contracting to facilitate unity of effort in the coordination of sustainment functions at designated contingency locations. When multiple Service components share a common base of operations, a GCC may designate a Service component or JTF as the BOS-I at each contingency location. The GCC, commensurate with special operations forces' (SOF's) capacity and capability, may assign SOF the synchronization of BOS functions in specific instances where SOF and their enablers are the only forces at a contingency location. The designated BOS-I is responsible for coordinating common user contract support, as well as the efficient use of other support resources, for all joint forces at the contingency location. Additional BOS-I responsibilities may include, but are not limited to: coordinating the issuance of war reserve materiel assets, collecting and prioritizing construction requirements, seeking infrastructure funding support, environmental management, emergency management, emergency services, force protection, and hazardous waste management. The BOS-I must closely coordinate with the SAA or single port or terminal manager. If no SAA or single port or terminal manager is assigned, the BOS-I is responsible for their functions.

III. Combatant Commander's Logistics Directorate

The logistics directorate of a joint staff (J-4) at the CCMD conducts logistics planning and execution in support of joint operations. They integrate, coordinate, and synchronize Service component and CSA logistics capabilities to support the joint force. The J-4 also advises the JFC on logistics support to optimize available resources. Although the organizational considerations outlined below could apply to a CCDR's J-4 staff, they will most frequently be applied to subordinate joint force J-4 organizations. The J-4 staff supports the operations directorate of a joint staff (J-3) in the planning and executing of requirements for the joint reception, staging, onward movement, and integration (JRSOI) process, as well as contingency base planning and sustainment. The J-4 coordinates, synchronizes, plans, and executes core logistics functions in joint and multinational environments.

- **Planning**. The J-4 provides logistics expertise as part of the joint planning process (JPP). In accordance with JP 5-0, Joint Planning, the J-4 establishes a logistics planning cell in coordination with the plans directorate of a joint staff to fulfill this responsibility. Planning occurs at every level of warfare in a networked, collaborative environment, which requires dialogue among senior leaders, concurrent and parallel plan development, and collaboration across multiple planning levels.
- **Execution**. The GCC's J-4 coordinates and synchronizes joint theater logistics. This includes communicating the logistics priorities of the GCC to the Services responsible for executing joint logistics operations. The J-4s organize their logistics staff functions to respond to anticipated or ongoing operations.

A. Joint Logistics Operations Center (JLOC)

The J-4 establishes a JLOC to monitor and control the execution of logistics in support of on-going operations. The JLOC is an integral part of the CCDR's operations element and provides joint logistics expertise to the J-3 operations cell. The JLOC is tailored to the operation and staffed primarily by the J-4 staff.

B. Joint Deployment Distribution Operations Center (JDDOC)

At time of need, a supported GCC can create a JDDOC and incorporate its capabilities into the staff functions. The GCC can place the JDDOC at any location required or under the operational control (OPCON) of other commanders. The JDDOC can reach back to the national partners to address and solve deployment and distribution issues for the GCC. The JDDOC develops deployment and distribution plans; integrates multinational and/or interagency deployment and distribution; and coordinates and synchronizes supply, transportation, and related distribution activities. The JDDOC synchronizes the strategic to operational movement of forces and sustainment into theater by providing advance notice to the GCC's air and surface theater movement C2 elements. In concert with the GCC's overall priorities, and on behalf of the GCC, the JDDOC coordinates common user and theater distribution operations above the tactical level. A joint movement center (JMC) may be established at a subordinate unified or JTF level to coordinate the employment of all means of transportation (including that provided by allies or HNs) to support the CONOPS. This coordination is accomplished through establishment of theater and JTF transportation policies within the assigned OA, consistent with relative urgency of need, port and terminal capabilities, transportation asset availability, and priorities set by a JFC. The JTF JMC will work closely with the JDDOC.

For more information, refer to JP 4-09, Distribution Operations.

C. Joint Logistic Boards, Centers, Offices, and Cells

Ref: JP 4-0, Joint Logistics (Feb '19), app. B.

The CCDR may also establish boards, centers, offices, and cells (e.g., subarea petroleum office [SAPO], joint facilities utilization board [JFUB], joint mortuary affairs office [JMAO], operational contract support integration cell [OCSIC]) to meet increased requirements and to coordinate the logistics effort.

Strategic-level Boards, Offices, and Centers

Strategic-level joint logistic boards, offices, and centers provide advice or allocation recommendations to the CJCS concerning prioritizations, allocations, policy modifications or procedural changes.

- Joint Logistics Board (JLB)
- Joint Materiel Priorities and Allocation Board (JMPAB)
- Joint Transportation Board (JTB)
- Joint Logistics Operations Center (JLOC)
- Deployment and Distribution Operations Center (DDOC)
- Defense Health Agency (DHA)
- Contingency Basing Executive Council (CBEC)
- Global Posture Executive Council (GPEC)
- Medical Logistics Division
- United States Transportation Command, Office of the Command Surgeon (TCSG)
- Armed Services Blood Program (ASBP)

Operational Joint Logistic Boards, Centers, and Cells

Operational-level joint logisticians must provide advice and recommendations to the supported CCDR concerning prioritizations, allocations, or procedural changes based upon the constantly changing operational environment.

- Joint Logistics Operations Center (JLOC)
- Joint Deployment and Distribution Operations Center (JDDOC)
- Combatant Commander Logistic Procurement Support Board (CLPSB)
- Joint Requirements Review Board (JRRB)
- Joint Contracting Support Board (JCSB)
- Joint Environmental Management Board (JEMB)
- Joint Facilities Utilization Board (JFUB)
- Logistics Coordination Board
- Joint Movement Center (JMC)
- Theater Patient Movement Requirements Center (TPMRC)
- Joint Patient Movement Requirements Center (JPMRC)
- Joint Blood Program Office (JBPO)
- Joint Petroleum Office (JPO)
- Sub-area Petroleum Office (SAPO)
- Joint Mortuary Affairs Office (JMAO)
- Explosive Hazards Coordination Cell (EHCC)
- Joint Munitions Office (JMO)
- Operational Contract Support Integration Cell (OCSIC)

IV. Logistics Execution Organizations

The fundamental role of joint logistics is to integrate and coordinate logistics capabilities from Service, agency, and other providers of logistics support and to facilitate execution of the Services' Title 10, USC, responsibilities while supporting the ever-changing needs of the JFC. Logistics may also be called upon to support the National Guard in Title 32, USC, status. It may also include special assignment airlift missions in addition to channel airlift, surface, and sealift movements. Joint logisticians should understand how each of the Services conducts logistics at the operational level.

See following pages (pp. 8-20 to 8-21) for discussion of the Services' logistics execution, and facing page for discussion of TRANSCOM, DLA, DCMA, and DSCA.

V. Logistics Control Options

The CCDR's logistics authority enables use of all logistics capabilities of the forces assigned as necessary for the accomplishment of the mission. The President or SecDef may extend this authority to attached forces when transferring those forces for a specific mission and should specify this authority in the establishing directive or order. The CCDR may elect to control logistics through the J-4 staff tailored and augmented as discussed previously in "Combatant Commander's Logistics Directorate." The CCDR may also decide to control joint logistics by designating a subordinate logistics organization. In these instances, the CCDR will delineate the authorities and command relationships that will be used by the subordinate commander to control logistics. In both cases, the CCDR exercises effective control of joint force logistics by fusing procedures and processes to provide visibility and control over the logistics environment and integrating joint logistics planning with operations planning. Control of joint logistics is enhanced by how effectively the logistician combines the capabilities of the global providers and the Services' logistics elements with the JFC's requirements in a way that achieves unity of effort.

A. Staff Control

The J-4 staff may be used to support a wide range of operations, including campaigns; complex or long-duration major operations; or complex operations involving multiagency, international organizations, NGOs, or MNFs, if properly augmented. For example, the staff may be sized and tasked to provide increased movement control or material management capabilities; it could be augmented with a robust OCS planning and integration capability; the J-4 could receive augmented capability to coordinate multinational support operations or execute JOA-wide infrastructure repair/restoration missions. J-4 staff augmentation can come from a combination of military, civilian emergency workforce, and contractor personnel. When exercising this option, the CCDR will specify the control authorities delegated to the J-4 over the components logistics elements. Taskings to Service component logistics elements in this case must come from formal tasking orders issued through the CCDR's J-3. The logistics taskings, which could come in the form of a fragmentary order (FRAGORD), formalizes the authorities given the J-4 by the JFC and enables the rapid response to operational logistics requirements.

B. Organizational Control

As another alternative for controlling the major operations outlined above, the CCDR may elect to assign responsibility to establish a joint command for logistics to a subordinate Service component. The senior logistics headquarters of the designated Service component will normally serve as the basis for this command, an organization joint by mission (e.g., campaigns, major operations, humanitarian missions), but not by design. When exercising this option, the CCDR retains DAFL and must specify the control and tasking authorities being bestowed upon the subordinate joint

Logistics Execution Organizations

Ref: JP 4-0, Joint Logistics (Feb '19), p. III-10 to III-11.

The fundamental role of joint logistics is to integrate and coordinate logistic capabilities from Service, agency and other providers of logistic support, and to facilitate execution of the Services' Title 10, USC responsibilities while supporting the ever changing needs of the JFC.

US Transportation Command (USTRANSCOM)

Serves as the global distribution synchronizer, as outlined in the Unified Command Plan, on behalf of and in coordination with the JDDE COI to establish processes to plan, apportion, allocate, route, schedule, validate priorities, track movements, and redirect forces and supplies per the supported commander's intent. This coordination and synchronization will not infringe upon either the supported CCDRs, or Services Title 10, USC, designated responsibilities, but serves to facilitate a unity of effort throughout the JDDE to support the CCDR or subordinate JFC. The supported GCC is responsible to plan, identify requirements, set priorities, and redirect forces and sustainment as needed to support operations within the respective AOR. USTRANSCOM, as the Mobility Joint Force Provider, exercises responsibility for planning, resourcing, and operating a worldwide defense transportation system in support of distribution operations, to include reviewing taskings and analyzing supported CCDR's requirements for transportation feasibility, and advising on changes required to produce a sustainable force deployment. During the deployment, sustainment, and redeployment phases of a joint operation, CCDRs coordinate their movement requirements and required delivery dates with USTRANSCOM, and supported GCCs are responsible for deployment and distribution operations executed with assigned/attached force in their respective AORs.

Defense Logistics Agency (DLA)

DLA manages, integrates, and synchronizes suppliers and supply chains to support the Armed Forces of the US, allies, and multinational partners. The ASD(L&MR), under the USD(AT&L), exercises authority, direction, and control over DLA. As a statutory CSA, DLA provides logistics advice and assistance to the Office of the Secretary of Defense, the CJCS, Joint Chiefs of Staff, the CCDRs, Military Departments, DOD components, and interagency partners, as appropriate. Additionally, DLA operates as part of the JLEnt in providing humanitarian assistance. DLA manages nine diverse supply chains and serves as the DOD EA for four supply chains: subsistence, construction and barrier materiel, bulk petroleum, and medical materiel. DLA also directs a network of distribution depots located throughout the US, Europe, Pacific, and South West Asia.

Defense Contract Management Agency (DCMA)

DCMA is the CSA responsible for ensuring major DOD acquisition programs (systems, supplies, and services) are delivered on time, within projected cost or price, and meet performance requirements. DCMA's major role and responsibility in contingency operations is to provide contingency contract administration services for external and theater support contracts. DCMA is also responsible for selected weapons system support contracts with place of performance in the operational area and theater support contracts when contract administration services are delegated by the procuring contracting officers.

Defense Security Cooperation Agency (DSCA)

DSCA arranges DOD funded and space available transportation for NGOs for delivery of humanitarian goods to countries in need; coordinates foreign disaster relief missions; and, in concert with DLA, procures, manages, and arranges for delivery of humanitarian daily rations and other humanitarian materiel in support of US policy objectives.

Service Logistics Execution

Ref: JP 4-0, Joint Logistics (Feb '19), p. III-10 to III-14.

Army

The overarching theater-level headquarters is the theater Army/Army Service component command (ASCC), which provides support to Army forces and other Services as directed. It is important for the ASCC and theater special operations command (TSOC) J-4 to enhance conventional forces and SOF synchronization of sustainment. The theater sustainment command (TSC) is the logistics C2 element assigned to the ASCC and is the senior Army logistics headquarters within a theater of operations.

The TSC is responsible for executing port opening, theater opening, theater surface distribution, and sustainment functions in support of Army forces and provides lead Service and EA support for designated CUL to other government departments and agencies, MNFs, and NGOs as directed. The TSC is also responsible for establishing and synchronizing the intratheater segment of the surface distribution system in coordination with the JDDOC with the strategic-to-theater segment of the global distribution network.

The TSC establishes C2 of operational-level logistics in a specified area of operations by employing one or more expeditionary sustainment commands (ESCs), which provide a rapidly deployable, regionally focused, forward-based C2 capability until a TSC can assume that function. When the Army is the predominant land force operating within an OA, the TSC or ESC, at the discretion of the JFC, has the capability to become a joint logistics headquarters providing logistics support to all joint forces within the OA. This is contingent upon the other Services, DOD agencies, and CCMDs providing the appropriate augmentation of personnel and capabilities to support this joint mission. Though the TSC can be sourced from any component of the Army, the preponderance of the Army's logistic capability is in the Reserve Component, either Army Reserve or Army National Guard.

Marine Corps

The Marine expeditionary force (MEF) is the principle warfighting organization in the Marine Corps, capable of conducting and sustaining expeditionary operations in any geographic environment. The Marine logistics group is responsible for providing tactical logistics above the organic capability of supported units to all elements of the MEF. It is a permanently organized command structured with functional and multi-functional units which are organized to support a MEF possessing one Marine division and one Marine aircraft wing. Integration with strategic- and operational-level logistics support is coordinated through the Marine Corps component commander.

Navy

For numbered fleets, the senior logistician is the assistant chief of staff for logistics. The assistant chief of staff for logistics is normally the logistics readiness center (LRC) director. Coordination and unity of effort between the LRC and logistics supporting staffs and commands providing logistics resources and support is key to effectively controlling and executing logistics support.

The logistics forces of each numbered fleet are organized into standing task forces, and the commanders of these task forces are the principal logistics agents for the fleet commander. The logistics task force commander is responsible to the fleet commander for management of logistics support forces for maritime sustainment of Navy, United States Coast Guard (USCG), and Marine Corps units. The logistics task force commander has tactical control of Military Sealift Command Combat Logistics Force ships, plans resupply for all classes of supply, and plans and manages theater ship repairs in military and commercial yards outside the CONUS.

Fleet operational forces are normally organized into task forces under the command of a task force commander. The task force commander exercises control of logistics through a fleet logistics coordinator, task force logistics coordinator, or task group logistics coordinator and coordinates the replenishment of forces at sea.

Air Force

The air expeditionary task force (AETF) is the organizational structure for deployed US Air Force forces. AETF presents a scalable, tailorable organization with three elements: a single commander, embodied in the commander, Air Force forces (COMAFFOR); appropriate C2 mechanisms; and tailored and fully supported forces. The Air Force forces staff is the vehicle through which the COMAFFOR fulfills operational and administrative responsibilities for assigned and attached forces, and is responsible for long-range planning that occurs outside the air tasking cycle. The PSAs to the COMAFFOR for JOA-wide integration of agile CS capabilities and processes are the director of manpower, personnel, and services (A1); the director of logistics, engineering, and force protection (A4); and the surgeon general (SG).

A1 is responsible for the functions of billeting; MA assistance; and food service, to include bottled water to support planned meals. Responsibility for planning daily consumable water outside of planned meals resides with A4, civil engineering. Contracting is the responsible agent to procure bottled water (when the requirements have been established) from approved sources that are coordinated with bioenvironmental engineers and public health. A4 controls logistics planning; distribution; material management; fuels; maintenance; and munitions; civil engineering; fire emergency services; explosive ordnance disposal; chemical, biological, radiological, and nuclear (CBRN) defense and response elements of emergency management; and force protection. The SG advises on FHP and HSS.

USCG

USCG maritime patrol and deployable specialized forces are capable of supporting joint military operations worldwide. In order to accomplish the many missions, deployable units and assets consist of high-endurance cutters, patrol boats, buoy tenders, aircraft, port security units, maritime safety and security teams, maritime security response teams, tactical law enforcement teams, and the National Strike Force. Logistics support for the USCG is provided by the Deputy Commandant for Mission Support and its subordinate elements. When USCG forces operate as part of a JTF, they may draw upon the logistics support infrastructure established by/for the JTF. These general support functions normally include, but are not limited to, the following: berthing, subsistence, ammunition, fuel, and accessibility to the naval supply systems. The Navy logistics task force commander coordinates the replenishment, intratheater organic airlift, towing, salvage, ship maintenance, and material control, as well as commodity management for the task force group.

Special Operations (SOF)

Commander, United States Special Operations Command (CDRUSSOCOM), exercises COCOM over all SOF and the TSOCs unless otherwise directed by SecDef. SOF are dependent on Service and joint logistics support as the primary means of support. As directed, GCCs exercise OPCON over assigned TSOCs and SOF.

When a GCC establishes and employs subordinate JTFs and task forces, the GCC or commander, theater special operations command (CDRTSOC), may establish and employ a special operations joint task force (SOJTF), joint force special operations component, special operations command-forward, or joint special operations air component to control SOF assets and accommodate special operations requirements. Accordingly, the GCC establishes command relationships between SOF commanders and other JTF/task force commanders. CDRUSSOCOM can establish and employ a SOJTF or a joint special operations task force as a JFC in coordination with GCCs for special operations in their AOR.

Refer to JP 3-05, Special Operations.

command for logistics, as well as the command relationships it will have with the Service components. This command would control logistics taskings as directed by the CCDR and must not infringe on the authorities and responsibilities as specified in "Joint Logistics Roles and Responsibilities." The CCDR, through the global force management process, would request augmentation with joint, agency, and other Service capabilities to effectively integrate and control logistics requirements, processes, and systems and with forces made available.

C. CUL Control

Planners should consider areas where CUL organizational options are best suited. CCMD and subordinate logistics planners must keep in mind that while CUL support can be very efficient, it may not always be the most effective method of support. By its very nature, CUL support will normally take place outside routine support channels, which may lead to reduced responsiveness if not properly planned, coordinated, and executed. CCDRs, along with their subordinate commanders, must review, coordinate, and direct CUL requirements with DLA; functional CCDRs, which include their supporting contracting activities; and Service component commanders to provide an integrated joint logistics system from the strategic to tactical levels. All parties must ensure that the advantages and disadvantages of each CUL-related COA are properly considered, to include the extent of reliance on commercially sourced, contracted support. However, the GCC has overall responsibility for deciding the amount and type of CUL support for a particular joint operation. The CCDR's decision to use DAFL to direct CUL support within a subordinate joint force must be deliberate and coordinated to ensure proper CUL execution.

Cross-Leveling CUL Assets

It must be clearly understood that only the CCDR has the authority to direct the cross-leveling of supplies within a joint force. Cross-leveling of a supply for one Service component will be only for common items, should be accomplished in a very prudent and deliberate manner, and consider reimbursement between Services. CUL suitability for commodities is displayed in Figure III-1, as well as other potential CUL areas that should be considered in reducing redundancy, risks, and costs.

Organizational Control Options

Based on the operational situation, the CCDRs can modify or mix two major control options: single-Service logistic support or lead Service/agency support.

- **Single-Service Logistic Support.** . In this organizational option, each Service retains primary responsibility for providing support to their subordinate organizations. CUL would be limited to existing support relationships between Services as identified in inter-Service support agreements. If delegated by the CCDR, the J-4 may coordinate limited CUL support to other Services or agencies in certain situations. This method would most likely be used in major operations where the operational situation allows for, and calls for, the deployment of the requisite Service component logistics assets in a timely manner and where logistics effectiveness is paramount.
- **Lead Service or Agency CUL Support.** The CCDR may designate a lead Service or DOD agency to provide selected CUL support to one or more Service components, governmental organizations, and/or NGOs in a joint or multinational operation. This CUL option is normally based on the dominant user and/or most capable Service concepts and may or may not involve OPCON or tactical control of one Service component logistics units to the lead Service.

D. Control Option Selection Considerations

Ref: JP 4-0, Joint Logistics (Feb '19), p. III-15.

After determining what commodities and functions will be joint, the CCDR must decide how to control those logistics operations. The selection of a control option should benefit from a careful analysis to include the following considerations. These considerations are not designed to stand alone. They should be considered comprehensively to properly inform the commander's decision.

Mission

The mission is the foremost consideration from the commander when selecting the option that will be used to control joint logistics. Mission analysis helps identify the complexity and scale of the joint logistics requirements the command will face during execution. Generally, the more complex operations have greater need for an organizational control option.

The Most Capable Service Component

This consideration aligns with the most prevalent Service capabilities in the OA. It is one of the most important considerations to analyze because no Service component's logistics organization or supporting contracting activity is staffed or equipped to plan and execute joint logistics or joint contracting support. To some degree, the most capable Service component organization will have to be augmented to provide common-user support responsibilities. Without adequate Service component logistics C2 capability available, the staff control option would be the most appropriate.

The Geographic and Physical Infrastructure in the Operational Area

This consideration is related to the most capable Service component consideration. The geographic and physical infrastructure in the OA usually dictates the nature of the LOCs needed to support the joint force and the need for contingency basing. The LOCs will influence the distribution system, to include the location of distribution points and the challenges brought on by the ITV technology need to support the operation. Additionally, the condition of the LOCs may force CUL, common-user land transport, and intratheater plans. The GCC should coordinate with USTRANSCOM, DOD agencies, and other stakeholders when analyzing the geography and physical infrastructure in the OA and when selecting the control option.

Geographic Combatant Commander (GCC) Option Selection and Design

Figure III-2 (following page) details a logical sequence that can be used by GCCs when evaluating, selecting, and designing the option they will use to control joint logistics. For more amplifying information detailing the joint logistics factors and enablers with regard to the staff and organization control

See following pages (pp. 8-24 to 8-25) for an overview and further discussion.

Geographic Combatant Commander (GCC) Option Selection and Design

Ref: JP 4-0, Joint Logistics (Feb '19), p. III-19 to III-21 and app. D.

JP 4-0, appendix E provides amplifying information detailing the joint logistics factors and enablers with regard to the staff and organization control options.

GCCs require visibility over the JLEnt to meet the command priorities. The factors below should be considered when the GCC is establishing the logistics control required by the JFC. These factors are not absolute nor all inclusive; but they do reflect the best practices observed in the field. These factors are applicable regardless of the control option selected by the GCC.

Centralized Joint Logistics Planning

This factor implies a capability to match joint logistics planning with the planning done during the execution of a mission.

Maintenance of Situational Awareness

This factor represents more than using radio signals and internet-based application data to track cargo movement (ITV). It involves elements such as the design and use of logistics situation reports and the building of ground truth in logistics input to the JFC's COP.

Adjudication of Conflicting Priorities

This factor is to have processes in place to identify conflicts when following the commander's priorities. For example, a reliable logistics input to the JFC's COP may provide the means to identify conflicts, and a fusion cell may provide the capability to adjudicate.

Timely Identification of Factors and Shortfalls

To meet this factor a process that links the logistics portion of the battle rhythm with the planning windows must exist.

Clear Understanding of Component Capabilities

This factor involves the building of databases that reflect current Service component and support agencies logistics capabilities. Fulfilling this factor may require liaison and physical presence of logisticians representing all appropriate Service components within the selected joint logistics control option.

Ability to Synchronize Components Capabilities

This factor matches the best capability, regardless of Service component, to the joint logistics need.

Integrated Logistics Processes

This factor is founded on the notion that the joint logistic staff comprehends the Service components logistic processes and uses this understanding to build the visibility required by the JFC to control joint logistics.

Integrated Distribution

This factor deals with the establishment of the JDDOC and its integration within the joint theater logistics construct. It maximizes the capabilities of the JDDOC to fill the seams between strategic and operational level deployment and distribution tasks. The JDDOC also strives to maximize and synchronize the use of common user land transportation and intratheater lift.

Cross Component Supply

This factor involves the establishment of CUL responsibilities and the processes required to achieve their execution.

GCC Option Selection and Design

Joint Logistics Factors →	Enabled by
Centralized joint logistics planning	FC, LCB, CLPSB
Maintenance of situational awareness	FC, LCB, GCSS-J, logistics COP
Adjudication of conflicting priorities	GCSS-J, logistics COP, JDDOC, JRRB
Timely identification of requirement/shortfall	FC, LCB, LSM, CLPSB, battle rhythm
Clear understanding of component capability	FC, LCB, LSM, battle rhythm
Ability to synchronize component capabilities	FC, LCB, GCSS-J, logistics COP, LSM, battle rhythm, contract support
Integrated logistics processes	Logistics SOP
Integrated distribution	JDDOC, LSM, battle rhythm
Coordinated component supply	LSM, battle rhythm
Cross component asset visibility	GCSS-J, logistics COP
Improved capability to direct the process	FC, battle rhythm, CCDR directives

Staff

- Base–existing joint staff
- Authorized to coordinate and deconflict joint logistics issues
- Empowered to recommend in coordination with the J-3 preparation of fragmentary orders to direct logistic actions
- Augmented as required by the mission

Option Selection Considerations*

- Mission
- Most capable service component
- Geographic and physical infrastructure of the operational area

*Options are staff or organizational

Organizational

- May establish joint command for logistics using existing Service led logistics organization
- Augmented as required to provide joint perspective
- Operates with clear delineation of command relationships and authorities

Legend

CCDR	combatant commander	J-3	operations directorate of a joint staff
CLPSB	combatant commander logistics procurement support board	JDDOC	joint deployment and distribution operations center
COP	common operational picture	JRRB	joint requirements review board
FC	fusion cell	LCB	logistics coordination board
GCSS-J	Global Combat Support System-Joint	LSM	logistics synchronization matrix
		SOP	standing operating procedure

Ref: Figure III-2. Geographic Combatant Commander Option Selection and Design.

Cross Component Visibility

This factor refers to the ability for the Service components to see and understand assets available from other components.

Improved Capability to Direct the Process

This factor proposes the establishment of a decision-making process to direct logistics actions. These actions usually are directed in the form of further guidance to enhance the planning or assessment processes, or the publication of a FRAGORD to direct an action.

Designation of Contracting Construct

It is imperative that a detailed analysis of the OCS aspects of the OE be prepared to help shape COA development and determine the possible intended and unintended outcomes of OCS.

VI. Technology

The rapid advance of technology, if leveraged effectively, can enable the JFC to effectively control logistics within the OA. Technology, in the form of information systems, decision support tools, and communications capabilities, can improve visibility of logistics processes, resources, and requirements and provide the information necessary to make effective decisions.

VII. Interorganizational Cooperation

Interorganizational cooperation that results in operational arrangements regarding joint logistics are bound together by a web of relationships among global providers. These relationships are critical to joint logistics success because logistics capabilities, resources, and processes are vested in a myriad of organizations, which interact across multiple physical domains, the information environment, and span the range of military operations.

Multinational Operations

In today's OE, logisticians will likely be working with multinational partners. While the US maintains the capability to act unilaterally, it is likely that the requirement, and the desire, to operate with multinational partners will continue to increase. MNL is a challenge. However, leveraging MNL capabilities increases the CCDR's freedom of action. Additionally, many multinational challenges can be resolved or mitigated by having a thorough understanding of the capabilities and procedures of our multinational partners before operations begin. Integrating and synchronizing logistics in a multinational environment requires multinational information sharing, developing interoperable logistics concepts and doctrine, as well as clearly identifying and integrating the appropriate logistics processes, organizations, and C2 options. Careful consideration should be given to the broad range of MNL support structures.

Interorganizational and NGOs

Integration and coordination among military forces, NGOs, and international organizations are different from the coordination requirements of a purely military operation. These differences present significant challenges to coordination. First, NGO and international organization culture is different from that of the military. Their operating procedures will undoubtedly differ from one organization to another and with DOD. However, their similar needs (e.g., distribution, materials handling equipment, shelter, water, and power) in a contingency environment will add another requirement for resources that must be addressed early in any operation. Ultimately, some NGOs and international organizations may even have policies not in consonance with those of DOD. In the absence of a formal command structure, the joint logistician will need to collaborate and elicit cooperation to accomplish the mission. NGOs and international organizations possess unique skills and capabilities that can assist in providing the joint warfighter more robust logistics.

Logistics Support of US Government Organizations and Agencies

Logistics must be integrated at lowest echelons and are complicated by the creation of more support relationships of greater variety across Service lines and at lower levels. Likewise, multinational operations and support to interagency partners can complicate logistics by introducing a wider variety of potential partners. This complication is both the challenge and the solution, as it demands working with partners with a variety of requirements while also providing access to external resources and expertise. Partner logistics capabilities vary, as do their specific materiel resources, procedures, and information systems.

III. Planning Joint Logistics

Ref: JP 4-0, Joint Logistics (Feb '19), chap. IV.

Joint logistics planning provides the process and the means to integrate, synchronize, and prioritize joint logistics capabilities toward achieving the supported commander's operational objectives during all phases of plan development. This section is applicable to combatant command campaign plans (CCPs), subordinate campaign plans, campaign support plans, and contingency plans tasked in Chairman of the Joint Chiefs of Staff Instruction (CJCSI) 3110.01, (U) 2015 Joint Strategic Campaign Plan (JSCP) (commonly referred to as the JSCP), or as directed by the CCDR. This section also addresses planning considerations, input and output products used by joint logisticians to create OPLANs and operation orders (OPORDs) that enable transition from peacetime activities to execution of orders. Focus is on the JPP in development of the theater logistics overview (TLO) as a segment of the CCP.

Joint logistics planning is conducted under the construct of joint planning and the JPP addressed in JP 5-0, Joint Planning. Joint planning consists of planning activities associated with joint military operations by CCDRs and their subordinate commanders in response to contingencies and crises. It transforms national strategic objectives into activities by development of operational products that include planning for the mobilization, deployment, employment, sustainment, redeployment, and demobilization of joint forces and supporting contractors. Joint planning occurs at multiple strategic national and operation levels using process, procedures, tactics, techniques, and facilitating information technology tools/applications/systems aligned to the Joint Operation Planning and Execution System (JOPES) and the Adaptive Planning and Execution (APEX) enterprise.

The theater logistics overview (TLO) segment of the CCP articulates the overarching logistic architecture of the GCC's AOR. It is the start point of subsequent JPP logistics planning for regional OPLAN development and other contingencies.

I. Planning Functions

Joint planning encompasses a number of elements, including four planning functions: strategic guidance, concept development, plan development, and plan assessment. Depending upon the type of planning and time available, these functions can be sequential or concurrent. Joint planning features detailed planning guidance and frequent dialogue between senior leaders and commanders to promote a common understanding of planning assumptions, considerations, risks, COA, implementing actions, and other key factors. Plans may be rapidly modified throughout their development and execution. This process involves expeditious plan reviews and feedback, which can occur at any time, from SecDef and the CJCS. The intent is to give SecDef and the CCDR a mechanism for adapting plans rapidly as the situation dictates.

Integrated planning coordinates resources, timelines, decision points, and authorities across CCMD functional areas and AORs to attain strategic end states. Integrated planning produces a shared understanding of the OE, required decisions, resource prioritization, and risk across the CCMDs. JFCs and component commanders need to involve all associated commands and agencies within DOD in their plans and planning efforts. Moreover, planning efforts must be coordinated with other US Government department and agency stakeholders in the execution of the plan to assure unity of effort across the whole-of-government. The integrated planning process is the way the joint force will address complex challenges that span multiple CCMD

AORs and functional responsibilities. Integrated planning also synchronizes resources and integrates timelines, decision points, and authorities across multiple GCCs to achieve GEF-directed campaign objectives and attain contingency end states.

Supported CCDR

The supported CCDRs lead integrated logistics planning for their problem sets, inclusive of all associated plans related to the logistics problem both intertheater and intratheater. As such, supported CCDRs have coordinating authority for logistics planning. They lead the logistics planning process with all supporting CCMDs to develop a common understanding of logistics requirements, synchronize logistics planning activities, identify problem set logistics resource requirements, and provide logistics supportability analyses (quantitative and qualitative), as well as risk and supportability assessments associated with the plans. The supported commander designates and prioritizes objectives, timing, and duration of the supporting action. The supported commander ensures supporting commanders understand the operational approach and the support requirements of the plan. If required, SecDef will adjudicate competing demands for resources when there are simultaneous requirements amongst multiple supported CCDRs.

Supporting Commander

Supporting commanders will ensure their logistics planning is sufficiently integrated and synchronized across the problem set. They assist the supported CCMDs' efforts to develop a unified view of the logistics environment and synchronize resources, timelines, logistics C2, decision points, and authorities. The supporting commander determines the forces, tactics, methods, procedures, and communications to be employed in providing support. The supporting commander advises and coordinates with the supported commander on matters concerning the employment and limitations (e.g., logistics) of required support, assists in planning for the integration of support into the supported commander's effort, and ensures support requirements are appropriately communicated throughout the supporting commander's organization.

A. Strategic Guidance

The primary end product of the strategic guidance function and an in-progress review (IPR) is an approved CCDR's mission statement for contingency planning and a commander's assessment (operational report-3 pinnacle command assessment) or commander's estimate for crisis planning.

B. Concept Development

During concept development, if an IPR is required, the CCDR outlines COAs and makes recommendations to higher authority for approval and further development. Products from concept development include an approved mission statement, preliminary COAs, and prepared staff estimates. The CCDR recommends a COA for SecDef approval in the commander's estimate. The SecDef's approved COA from a concept development IPR is the basis for CONOPS.

C. Plan Development

This function is used to develop a feasible plan or order that is ready to transition into execution. This function fully integrates mobilization, deployment, employment, sustainment, conflict termination, redeployment, and demobilization activities through all phases of the plan. When the CCDR believes the plan is sufficiently developed, the CCDR briefs the final plan to SecDef (or a designated representative) for approval. Plan development solidifies the CONOPS and the OPLAN, concept plan (CONPLAN), or OPORD and required supporting documents are prepared.

Logistics Planning Integration (Strategic Guidance, Plans, & Operations)

Ref: JP 4-0, Joint Logistics (Feb '19), fig. IV-1, p. IV-5.

Using the JPP framework for planning, Figure IV-1 reflects the cascading relationship from strategic guidance and tasking to planning and developing OPORDs with a focus on CCP and associated key logistics area products. These key logistics area products, TLO, logistics estimate, and COLS support the CCP and provide the basis for plan and OPORD development. These products are key to the GCC's conduct of missions throughout the AOR. Figures IV-2 and IV-3 (following pages) reflect the joint logistics planning process combined with elements of the joint planning activities, functions, and products depicted in Figure IV-1 (below).

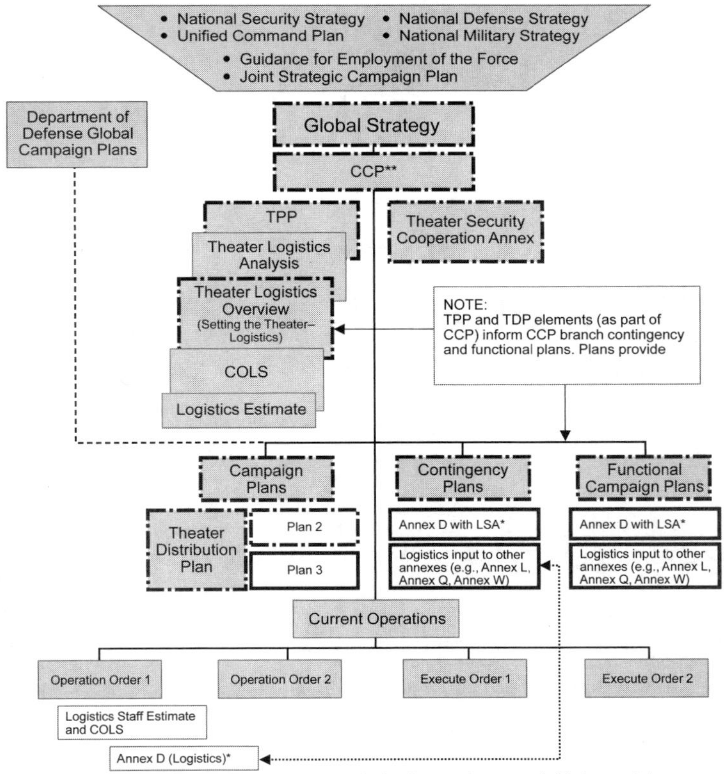

* Principal annex with logistics/sustainment segments also in others, such as annex L (Environmental Considerations), annex Q (Medical Services), annex W (Operational Contract Support), etc.

**CCP replaces the term TCP [theater campaign plan] per Chairman of the Joint Chiefs of Staff Instruction 3110.01, *(U) Joint Strategic Campaign Plan (JSCP).*

Legend

CCP	combatant command campaign plan
COLS	concept of logistics support
GEF	Guidance for Employment of the Force
JSCP	Joint Strategic Campaign Plan
LSA	logistics supportability analysis
TDP	theater distribution plan
TPP	theater posture plan

II. Joint Logistics Planning Considerations

Ref: JP 4-0, Joint Logistics (Feb '19), pp. IV-3 to IV-6 (figs. IV-2 and IV-3).

Figures IV-2 and IV-3 reflect the joint logistics planning process combined with elements of the joint planning activities, functions, and products depicted in Figure IV-1 (pervious page).

Steps I - III (Joint Logistics Planning)

Strategic Guidance/Concept Development/Plan Development

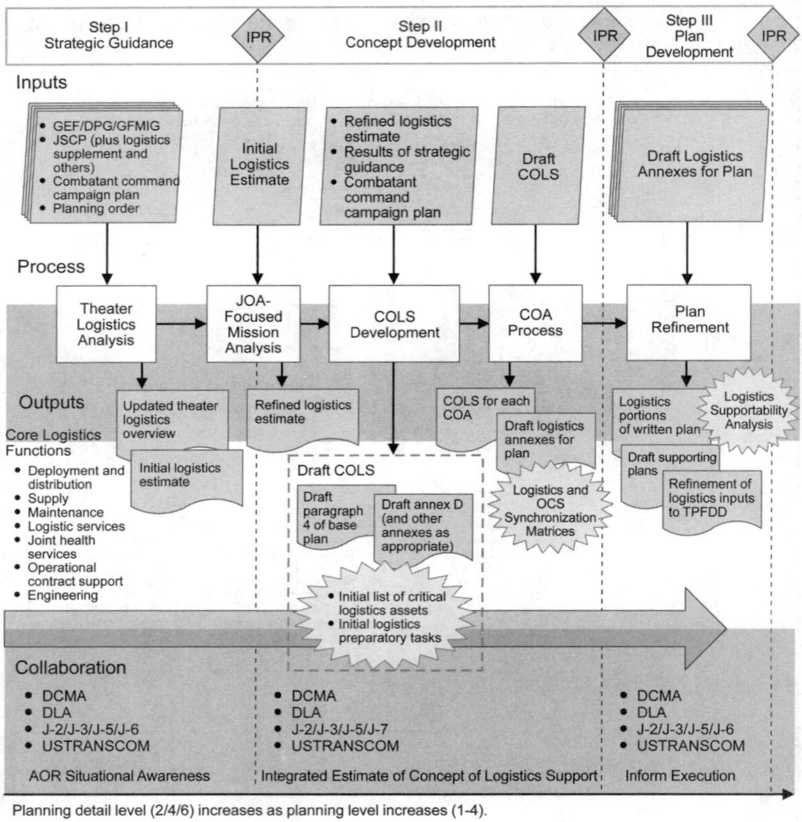

Legend

AOR	area of responsibility	J-3	operations directorate of a joint staff
COA	course of action	J-5	plans directorate of a joint staff
COLS	concept of logistics support	J-6	communications systems directorate of a joint staff
DCMA	Defense Contract Management Agency	J-7	operational plans and interoperability directorate of a joint staff
DLA	Defense Logistics Agency		
DPG	Defense Planning Guidance	JOA	joint operations area
GEF	Guidance for Employment of the Force	JSCP	Joint Strategic Campaign Plan
GFMIG	Global Force Management Implementation Guidance	OCS	operational contract support
		TPFDD	time-phased force and deployment data
IPR	in-progress review	USTRANSCOM	United States Transportation Command
J-2	intelligence directorate of a joint staff		

Ref: JP 4-0 (Feb '19), fig. IV-2. Joint Logistics Considerations, Steps I-III.

Joint Logistics

A means of anticipating future requirements is through the theater logistics analysis (TLA) process supporting TLO development and codification, logistics estimate, and logistics planning process Anticipating requirements is essential to ensuring responsiveness and determining adequacy of support. The purpose of the logistics planning process is to ensure the logistics facts, assumptions, information, and considerations are properly analyzed and effectively synthesized within an integrated plan that supports the CONOPS.

Steps III - IV (Joint Logistics Planning)

Plan Development/Plan Assessment

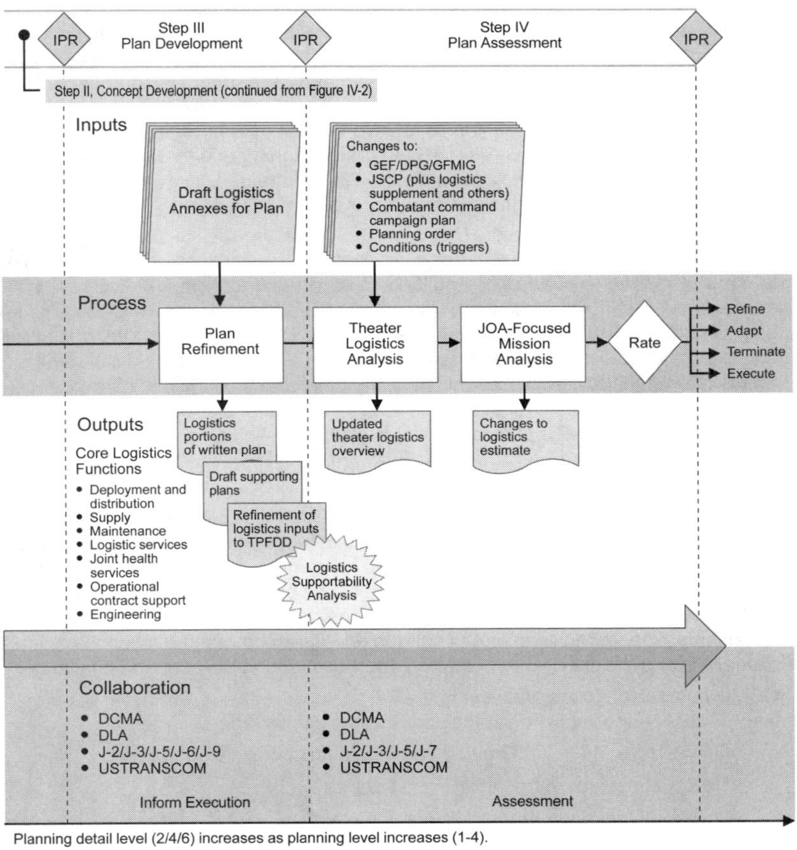

Planning detail level (2/4/6) increases as planning level increases (1-4).

Legend

DCMA	Defense Contract Management Agency	J-5	plans directorate of a joint staff
DLA	Defense Logistics Agency	J-6	communications systems directorate of a joint staff
DPG	Defense Planning Guidance	J-7	operational plans and interoperability directorate of a joint staff
GEF	Guidance for Employment of the Force		
GFMIG	Global Force Management Implementation Guidance	J-9	civil-military operations directorate of a joint staff
		JOA	joint operations area
IPR	in-progress review	JSCP	Joint Strategic Campaign Plan
J-2	intelligence directorate of a joint staff	TPFDD	time-phased force and deployment data
J-3	operations directorate of a joint staff	USTRANSCOM	United States Transportation Command

Ref: JP 4-0 (Feb '19), fig. IV-3. Joint Logistics Considerations, Step III-IV.

D. Plan Assessment

The joint planning and execution community continually reviews and evaluates the plan; determines one of four possible outcomes: refines, adapts, terminates, or executes; and then acts accordingly. Commanders and the joint planning and execution community continue to evaluate the situation for any changes that would require changes in the plan. The CCDR will brief SecDef during routine plan update IPRs of modifications and updates to the plan based on the CCDR's assessment of the situation, changes in resources or guidance, and the plan's ability to achieve the objectives and attain the end states.

E. Concept Development

This planning function includes the following JPP steps: COA development, COA analysis and wargaming, COA comparison, and COA approval. The staff, in coordination with supporting commands, Services, and agencies, develops, analyzes, and compares valid COAs and prepares staff estimates. The output is an approved COA. Critical elements include a common understanding of the situation, interagency coordination requirements, multinational involvement (if applicable), and capability requirements. Logistics planners must integrate planning efforts with operation planners, as deployment, redeployment, distribution, contracted support requirements, and sustainment requirements are an integral part of COA development.

The logistician must also identify requirements, critical logistics assets (CLAs), and services needed. A CLA is a logistics asset that is essential to completing key tasks that ensure mission accomplishment; if nonoperational or absent, it would have a seriously debilitating effect on the ability of a CCMD to execute their mission. The logistician must be aware of force structure planning, TPFDD development, and existing contracts and task orders, as well as the limitations of the OCS and JRSOI requirements. The logistician also uses this planning data during concept of support development to meet sustainment requirements from theater entry and operations to redeployment and reset. Logistics planners address all the core joint logistics functions.

During COA refinement, phasing of joint operations is done to ensure joint capabilities are available in the proper sequence to meet the operational requirements. Events drive phase changes, not time. Phasing helps the planning community visualize the entire operation to define requirements in terms of forces, resources, time, space, and purpose. The CCDR determines the number and nature of the phases during the operational design. Transitions between phases are designed to be distinct shifts in joint force focus and may be accompanied by changes in command relationships. Phase transition often changes priorities, command relationships, force allocation, or even the design of the OA, thereby creating new support challenges.

CCMD campaigns focus on shaping the OE to support the CCDR's overall objectives. Questions to ask when setting the theater can include:

- Do we have the right C2 and communications systems?
- Do we have access to critical infrastructure?
- Do we have a good theater distribution plan?
- Have we coordinated at the higher levels with the strategic partners (e.g., DLA and Army Materiel Command)?
- Have we properly positioned logistics assets at the tactical level?

Military engagement, security cooperation, and deterrence activities occur during shaping operations, seeking to improve cooperation with allies and other partners. These activities complement broad diplomacy and economic development in support of a friendly government's own security activities and setting the theater for major combat operations. Military engagement, security cooperation, and deterrence activities may also occur with new emerging governments and those previously considered as non-friendly to US national interests.

SOF are normally highly engaged during CCMD operations. Joint logistics planners must be aware of SOF requirements during day-to-day operations. SOF logistics support includes the sustainment and replenishment of all classes of supply, maintenance, transportation, joint health services, facilities, BOS, and services. Logistics support of SOF units is the responsibility of the parent Service, except where otherwise provided for by support agreements and/or directives. This may include Service support, joint in-theater support, nonstandard support, special operations-peculiar support.

Shaping offers logisticians the opportunity to expand knowledge of and access to additional capabilities in anticipation of future events. If it becomes apparent that an event will occur, the logistician can begin preliminary actions, such as pre-positioning (PREPO) of materiel, preparing organic or commercial JLEnt partners to surge capabilities, coordinating award of contracts (e.g., external and theater), and readying the assets to move on short notice. Shaping is a critical period to identify potential risks in terms of access, capabilities, and capacities so alternatives and mitigating measures can be developed. Planners must identify and assess critical infrastructure and installation needs and compare the results to current and programmed military construction requirements and authorities. Analysis of required logistics support for deployment and sustainment of flexible deterrent options may occur during this phase. This analysis must carefully balance joint logistics capabilities currently assigned, projected early joint deployers, and changes to OCS requirements.

F. Plan Development

During the plan development function, the CCDR's staff creates a detailed OPLAN, OPORD, or CONPLAN, with required annexes. The supported CCDR, subordinate commanders, supporting commanders, CSAs, and staff conduct a number of different planning activities, to include force planning, support planning, deployment planning, redeployment or unit rotation planning, shortfall identification, feasibility analysis, refinement, documentation, plan review and approval, and supporting plan development. Planning activities culminate in training and wargaming exercises to provide feedback on the planned concept of support. The joint logistics concept of support specifies how capabilities will be delivered over time, identifies who is responsible for delivering a capability, and defines the critical logistical tasks necessary to achieve objectives during all phases of the operation. Annex W (Operational Contract Support) is closely tied to the COLS since contracted support may fill critical operational and logistics capability gaps. The COLS encompasses joint capabilities of all force capabilities, to include multinational, HN, interagency partners, international organizations, NGOs, DOD OCS, plus Active Component and Reserve Component forces.

G. Plan Assessment

The supported commander extends and refines planning, while supporting and subordinate commanders and CSAs complete their support plans. Branch plans and other options may be developed. The CCDR and staff continually evaluate the situation for changes which trigger plan refinement, adaptation, termination, or execution. Additional means of assessing joint logistics planning are LSAs completed as appendix 4 (Logistics Supportability Analysis), annex D (Logistics) during plan development, Service component analysis, joint combat capability assessments-plans assessment, Global Logistics Readiness Dashboard, and Defense Readiness Reporting System assessments.

Guidance for development of an LSA is available in CJCSI 3110.03, (U) Logistics Supplement (LOGSUP) to the Joint Strategic Capabilities Plan (JSCP), and Chairman of the Joint Chiefs of Staff Manual (CJCSM) 3130.03, Adaptive Planning and Execution (APEX) Planning Formats and Guidance, provides the LSA format.

III. Joint Planning Process (JPP)

Joint planning is the overarching process that guides CCDRs in developing plans for the employment of military power within the context of national strategic objectives and national military strategy to shape events, meet contingencies, and respond to unforeseen crises. Logisticians provide key inputs, analysis, and assessments throughout the process. Logistics input is derived from mission analysis; COA development, analysis, and selection; and plan development, to include preparation and submission of LSA. Previously completed TLA, TLO (setting the theater-logistics), and COLS prepared for the CCP provide a foundational basis for complementary sections for tasked contingency plans. This foundation can also assist with transition to OPORD preparation for crisis execution under a plan and/or no-plan scenario.

IV. Planning Levels

JP 5-0, Joint Planning, identifies four levels of planning detail and establishes a minimum level of effort for each.

See facing page (fig. IV-4) and following pages (pp. 8-36 to 8-37) for discussion of key logistics planning process outputs supporting and/or included in CCP development and execution planning.

A. Level 1 Planning Detail—Commander's Estimate

This level of planning involves the least amount of detail and focuses on producing multiple COAs to address a contingency. The product for this level can be a COA briefing, command directive, a commander's estimate, or a memorandum with a required force list.

B. Level 2 Planning Detail—Base Plan (BPLAN)

A BPLAN describes the CONOPS, major forces, concepts of support, and anticipated timelines for completing the mission. It does not normally contain annexes. Unless the CCDR opts to produce an annex D or the JSCP requires an annex D, there will be a paragraph 4 (Administrative and Logistics) only within the BPLAN summary. A BPLAN may contain alternatives, including flexible deterrent options, to provide flexibility in addressing a contingency as it develops or aid in developing the situation. Command logisticians should develop a Logistics Estimate (paragraph 4).

C. Level 3 Planning Detail—CONPLAN

This level is an abbreviated OPLAN with selected annexes and a CCDR's estimate of the plan's feasibility with respect to forces, logistics, and transportation. It will produce, if applicable, a COLS to include a "gross¬transportation-feasible" TPFDD, thus, the further delineation of a 3T plan (i.e., a CONPLAN or TPFDD). The COLS for CONPLANs or 3T plans will mirror the level of detail contained in the supported annex D. Appendix 4 to annex D provides the LSA for the plan. Level 3T plans (level three plan with TPFDD) and above require an annex W.

D. Level 4 Planning Detail—OPLAN

This plan requires a full description of the CONOPS, a complete set of annexes, and a TPFDD. Figure IV-4 (facing page) depicts logistics planning products by level of plan. Within the JPP, key logistics outputs are OPORD TLO, logistics estimate supporting development of the commander's estimate, and COLS. The COLS further supports annex D plans and OPORDs. Appendix 4 to annex D provides the LSA for the plan. In terms of operations execution, logistics supportability is addressed and status update reported in the JFC's situation report (SITREP) per CJCSM 3150.05, Joint Reporting System Situation Monitoring Manual. Logistics input to the SITREP provides shared situational awareness and visibility within and across echelons of command to address the core logistics functions, force, and sustainment tracking; JRSOI supporting declaration of force closure for operational employment; and other conditions that increase, or materially detract from, the adaptability and readiness of forces.

Likely Expected Logistics Outputs to JPP

Ref: JP 4-0, Joint Logistics (Feb '19), pp. IV-13 to IV-14.

Figure IV-4, below, depicts logistics planning products by level of plan. Within JPP, key logistics outputs are OPORD TLO; logistics estimate supporting development of the commander's estimate and COLS. The COLS further supports annex D for deliberate plans and OPORDs. Appendix 4 to annex D for deliberate plans provides the LSA for the plan. In terms of operations execution, logistics supportability is addressed and status update reported in the JFC's situation report (SITREP) per CJCSM 3150.5, Joint Reporting System Situation Monitoring Manual.

Likely Expected Logistics Outputs

Plan Level	Strategic Guidance	Concept Development	Plan Approval	Plan Review
Level 1 "Commander's Estimate"	TLO, ILE, and RLE (briefing)			Δs to TLO and RLE (briefing)
Level 2 Base Plan	TLO, ILE, and RLE (briefing)	Paragraph 4 (written and briefing)		Δs to TLO and RLE (briefing)
Level 3 Base Plan with Select Annexes	TLO, ILE, and RLE (briefing)	Paragraph 4, Annex D, logistics enablers, preparation tasks, COLS, and LSM (written and briefing) Annex Q		Δs to TLO and RLE (briefing)
Level 3 with TPFDD		Transportation feasible TPFDD, Annex W, CSSM		
Level 4 Base Plan with Annexes and Detailed TPFDD	TLO, ILE, and RLE (briefing)	Paragraph 4, Annex D, Annex W, logistics enablers, preparation tasks, COLS, LSM, and CSSM (written and briefing) Annex Q	Logistics portions of plan, draft supporting plans, logistics inputs to TPFDD, and LSA (written and briefing)	Δs to TLO, RLE, COLS, and LSA; status of supporting plans (briefing)

Legend

Δ	change
COLS	concept of logistic support
CSSM	contracted support synchronization matrix
ILE	initial logistics estimate
LSA	logistics supportability analysis
LSM	logistics synchronization matrix
RLE	refined logistics estimate
TLO	theater logistics overview
TPFDD	time-phased force and deployment data

Ref: JP 4-0 (Feb '19), fig. IV-4. Likely Expected Logistics Outputs.

Logistics input to the SITREP assists in providing shared situational awareness and visibility within and across echelons of command to address the core logistics functions, force and sustainment tracking, JRSOI supporting declaration of force closure for operational employment, and other conditions that increase, or materially detract from, the adaptability and readiness of forces.

V. Key Logistics Planning Process Outputs

Ref: JP 4-0, Joint Logistics (Feb '19), pp. IV-13 to IV-16.

A. Theater Logistics Analysis (TLA)

The TLA is a supporting process facilitating development of the TLO through examination, assessment, and codification of an understanding of current conditions of the OE. Analysis determines infrastructure, logistics assets/resources, and environmental factors in the OE that will optimize or adversely impact means for supporting and sustaining operations within the theater. To facilitate developing the TLA, logistics planners leverage all interactions with PN logistics professional counterparts (e.g., during multinational exercises logistics planning and execution) to capture insights into their capabilities, processes, and policies by writing and distributing detailed after action reports (AARs). To effectively share best practices and lessons learned from logistics operations across DOD, observations, insights, and AARs should be entered into the Joint Lessons Learned Information System (JLLIS). Entry into JLLIS facilitates awareness of issues and may provide solutions to logistics planning issues. The TLA provides a detailed country-by-country analysis of key infrastructure by location or installation (main operating base/forward operating site/cooperative security location); footprint projections (including contingency locations); HN agreements; and available contracted support capabilities, existing contracts, and task orders to logistically support the theater during peacetime through contingency operations. Work completed supports TLO development as a segment of the CCP and development of directed plans and OPORDs. Information and data collected and codified during the TLA process are the basis for analysis which assists in identifying, resolving, and/or mitigating risk associated with theater shaping operations. Additionally, the TLA provides the framework for conceptual planning, which involves understanding the OE and the problem, determining the operation's end state, and visualizing an operational approach. Using the TLA, the operational approach is initially addressed in a logistics estimate and transitions to culminate in the TLO. Detailed planning works out the scheduling, coordination, or technical problems involved with moving, sustaining, and synchronizing the actions of force as a whole to achieve objectives. Effective planning requires the integration of both the conceptual and detailed components of planning. The TLA assists in improving the JFC's situational awareness and understanding of theater logistics support capabilities and readiness to support/execute theater operations.

B. Theater Logistics Overview (TLO)

The TLO is a segment of the iterative planning process which addresses identification, understanding, and framing the theater's mission at the campaign level, not for a specific operation. The TLO uses TLA information to inform decisions about the approaches to be used for sourcing and distribution of logistics support for theater operations. Having captured influencing elements in the TLA as a frame, the JFC's logistics staff elements develop and codify an overarching approach to theater operations in the TLO. The TLO then serves as an important link between conceptual planning and the detailed planning tasked in the GEF/JSCP. Additionally, the TLO helps the JFC and operations and logistics staff segments measure the overall effectiveness of employing forces, force sustainability, and logistics capability readiness to ensure that the operational approach remains feasible and acceptable. As such, the TLO is key to help identify and address capability gaps, risk mitigations, and residual risk. If risk cannot be resolved or mitigated to an acceptable level then the operational concept may be reframed. Reframing involves revisiting earlier COAs, conclusions, and decisions that underpin the current operational CONOPS. Reframing can lead to a modification of the current CONOPS or result in preparation of a branch plan or entirely new plan. In developing the TLO, logistics planners, in coordination with intelligence and operations staff segments, identify opportunities/initiatives by anticipating events. This allows them to identify decision points to operate inside the threat's decision cycle or to react promptly to deteriorating situation advancing beyond shaping operations. Time to complete the TLA and resulting TLO assists in optimizing available planning time for associated detailed plans. Based on their understanding and learning gained during TLO development, the JFC and senior logistics staff representative issue logistics planning guidance to support and enable the operational approach expressed in the CONOPS and to guide more detailed planning.

Refer to JP 4-0, Appendix A, for an example of a TLO format.

C. Logistics Estimate

Logistics estimate supports the commander's estimate, COLS, OPORD development, and execution. Execution planning may involve abbreviated and compressed timelines from situational awareness/initiating event and reporting to potential JFC planning guidance or CJCS planning order to OPORD and execution. The TLA and TLO provide a foundation for rapid review and response development. Due to accelerated timelines, availability, and incorporation of TLA information and TLO segments, preparation of the logistics estimate may be compressed supporting the commander's estimate and initial work for COA development, analysis, and selection. Updating the TLA/TLO baseline, the logistics estimate supporting the commander's estimate informs the COLS and OCS concept prepared for OPORD annex D and annex W development and iterative planning during operations execution. The logistics estimate is an analysis of how CSS factors can affect mission accomplishment. It contains the logistics staff's comparison of requirements and capabilities, conclusions, and recommendations about the feasibility of supporting a specified COA. This estimate includes how the core logistics functions affect various COAs. Preparation of the logistics estimate provides a coordinated and formalized means for the staff to identify and consider logistics shaping in support of the operational CONOPS. Planners should evaluate the feasibility of OPLANs in light of strategic lift capabilities and limitations. The logistics effort and development of the logistics estimate refined as COLS for OPORD annex D must be integrated into the JPP and OPORD development upfront. Using the TLA/TLO baseline, logistics staff segments will be able to identify if specific operational actions to augment or expand theater logistics capabilities to support the operational CONOPS must be taken. The previously developed TLA/TLO assists the logistics planners in providing logistics characteristics of the AOR and area of operations/area of interest for the specified operations. The TLA/TLO aids planners in identification of logistics infrastructure of the OE (what exists in the OA that may be put to use).

D. Concept of Logistic Support (COLS)

In support of the CCDR and preparation of plans/OPORDs, the logistics staff elements prepare a logistics estimate which is further refined and developed into a COLS. The COLS provides a foundational basis in preparation of annex D for assigned contingency plans and/or OPORD development tasks. The COLS establishes priorities of support across all phases of operations to support the JFC's CONOPS. Logistics staff elements' active participation within and across JPP activities at all echelons facilitates CONOPS and associated COLS development. A COLS addresses the sustainment of forces, to include identification and status of contingency basing. Through exercising DAFL, the CCDR may assign a component commander with the responsibility for conducting various theater logistics functions, as well as base support at designated theater locations. Logistics functions may include management of afloat assets; identification and status of theater sustainment elements, to include identification and/or forecast of required augmentation; priority of sustainment by class of supply with guidance on days of supply to be maintained (minimum and maximum); movement priorities for airlift and sealift aligned to JFC's CONOPS; guidance for employment of sea-air interfaces to facilitate JRSOI; controlling CUL; JFC's declaration of force closure; actions by phase; logistics assets required; and designation of contracting construct (e.g., lead Service for contracting [LSC], joint theater support contracting command [JTSCC]).

For more information on the COLS, refer to CJCSI 3110.03, (U) Logistics Supplement (LOGSUP) for the 2015 Joint Strategic Capabilities Plan (JSCP).

VI. Transition to Execution

Planning does not cease with development, submission, and approval of a plan or OPORD. Planning is iterative and continues throughout as actions and assessments evolve in a dynamic manner across command echelons from the strategic national to operational to tactical levels. Strategic guidance for plans, as well as plan segments and resulting OPORDs, is refined as situational awareness and understanding evolves. Through assessment, guidance and/or plans may be reframed. Assessment is a determination of the progress toward accomplishing a task, creating an effect, or achieving an objective. Assessment is a continuous activity to support the operation process and associated planning and execution activities. During planning, assessment focuses on understanding current conditions of an OE and assumptions to address mission, enemy, terrain and weather, troops and support available, time available, and civil considerations. During preparation, assessment focuses on determining force readiness to execute the operation and verifying the assumptions on which the plan is based. During execution, assessment focuses on evaluating progress of the operation. Based on their assessments, commanders at various echelons direct adjustment to the plan/OPORD ensuring the plan/OPORD/operation stays focused on mission.

VII. Sustainment Distribution Planning and Management Process

USTRANSCOM's sustainment distribution planning and management process supports its JDDC role by enhancing the JDDE ability to ensure an agile, scalable, and resilient distribution network. Sustainment distribution planning and management provides the JDDE with a suite of five capabilities: distribution lane validation, distribution workload forecast/demand planning, advanced air route planning, strategic surface route plan, and sustainment distribution plans. Sustainment distribution plans are codified in USTRANSCOM's Campaign Plan for Global Distribution 9033.

For more information on distribution planning, refer to JP 4-09, Distribution Operations.

IV. Executing Joint Logistics

Ref: JP 4-0, Joint Logistics (Feb '19), chap. V.

The term "executing joint logistics" is used to describe actions and operations conducted by joint logistics forces in support of the JFC mission. Force reception, theater distribution, and MA are examples of joint logistics operations. Since joint logistics operations span the strategic, operational, and tactical levels, the transition from planning to execution is critical.

I. Essential Elements for Joint Logistics Execution

A. Organizing for Execution

The CCMD J-4 monitors, assesses, plans, synchronizes, and directs logistics operations throughout the theater. This transition may occur through the directed expansion of the JLOC and/or the CCDR's JDDOC. The CCDR's or JFC's staff is augmented (either physically or virtually) with representatives from Service components, USTRANSCOM, other supporting CCDRs, CSAs, and other national partners or agencies outside the command's staff. For example, each GCC has established a JDDOC to synchronize and optimize the flow of arriving forces and materiel between the intertheater and intratheater transportation. As the operating tempo increases during a contingency or crisis, additional joint logisticians and selected subject matter experts (e.g., maintenance, ordnance, supply) can augment JDDOCs and use established networks and command relationships instead of creating new staffs with inherent startup delays and inefficiencies. This expanded organization must be organized and situated to ensure increased coordination and synchronization of requirements in the deployment and distribution process. This organization must have clear roles and responsibilities between the various elements and clearly understood relationships between the logistics elements and the CCMD staff.

B. Expeditionary Capabilities

The joint logistician should understand the expeditionary theater opening capability options available to the commander. Expeditionary theater opening capabilities provide GCCs critical initial actions for rapid insertion/expansion of force capabilities into an OA that directly affects the JFC's ability to expand and adjust force flow to allow flexible, agile response to asymmetric and dynamic operational requirements. Expeditionary theater opening capabilities support the first critical OA entry missions with the eventual transition of theater port of debarkation (POD) operations to a JFC-designated Service component and establish conditions to facilitate the arrival of larger Service theater distribution and sustainment forces where/when appropriate.

C. Technology and Communications

Logisticians use a variety of automated tools to assist in planning and execution. Effective execution of logistics plans requires a robust data communications architecture. Planning should anticipate communications in degraded environments at all levels and phases of operations and include considerations for alternate routing, redundant systems, use of other systems, protocols, and message standards. These degradations may be imposed by the threat, the environment, by the JFC as part of operational security, or a combination of all of them. Sustained impaired/ inadequate information exchange capability must be anticipated and incorporated into risk management considerations during logistics operations planning and execution.

D. Achieving Situational Awareness

A role of the joint logistician is to provide situational awareness of the current logistics posture to support the JFC in making decisions and disseminating and executing directives. Maintaining situational awareness requires visibility of the status and location of resources. This includes status of existing contracts and task orders over the current and future requirements of the force and over the joint and component processes that deliver support to the joint force. In order to provide this visibility, timely, and accurate data and information are required for all equipment, sustaining supplies, repair parts, munitions, fuel and etc., moving into, within, exiting, or being stored in the GCC's AOR. This kind of visibility is the key to continuously monitoring progress and is enabled by operational inputs which serve to inform joint logisticians about the current situation. Service reports, operational summaries, logistics SITREPs, and HN reports all serve to expand the joint logistician's awareness of the JOA.

E. Battle Rhythm

The JFC will establish a battle rhythm for the operation along with mechanisms to establish and maintain visibility for all functional areas, to include logistics. The joint logistician must develop a supporting battle rhythm for the sustainment staff that supports the JFC's battle rhythm and is designed to provide proactive logistics options. Synchronizing logistics reporting with operational updates, ensuring that the operational planning cycle is part of the logistics battle rhythm, and minimizing shift changes at critical points in the battle rhythm will enable more effective execution. Additionally, tying the component logistics elements to the JFC's battle rhythm will provide more accurate and timely situational awareness and promote better integrated support to the joint force.

F. Joint Logistics Boards, Offices, Centers, Cells, and Groups

The joint logistician will often use boards, centers, or other organizations to assist the J-4 staff in executing joint logistics operations, by prioritizing and/or allocating resources, controlling functions, or prioritizing requirements.

See p. 8-17 for an overview and further discussion.

G. Execution Synchronization

A synchronization matrix or decision support tool/template can establish common reference points to help assess the progress of an operation. Joint logisticians may use a matrix to display progress against actual execution and recommend adjustments as needed. A logistics synchronization matrix is built around the concept of the operation and normally contains the phasing of the operation over time along the horizontal axis. The vertical axis normally contains the functions the joint logistician integrates into a concept of support. The body of the matrix contains the critical tasks, arrayed in time and linked to responsible elements for execution. This decision support tool enables logisticians to graphically display the logistics concept of support, see potential gaps, develop options to mitigate those gaps, and respond to a changing OE.

H. Commander's Critical Information Requirements (CCIRs)

CCIRs are elements of friendly and enemy information the commander identifies as critical to timely decision making. Joint logisticians update the critical information requirements related to logistics. Joint logisticians will most often use friendly forces information requirements to guide decision making. Those requirements are often a direct reflection of resources (force availability, unit readiness, or materiel availability).

II. Joint Logistics Execution

Ref: JP 4-0, Joint Logistics (Feb '19), pp. IV-1 to IV-3.

JFCs adapt to evolving mission requirements and operate effectively across a range of military operations. These operations differ in complexity and duration. The joint logistician must be aware of the characteristics and focus of these operations and tailor logistics support appropriately. This range of military operations extends from shaping activities to major operations and campaigns. US and multinational partners collaborate to expand mutual support and leverage capabilities to quickly respond to future contingencies.

Military Engagement, Security Cooperation, and Deterrence

The GEF directs development of CCPs focused on current operations, military engagement, security cooperation, deterrence, and other shaping or preventative activities. Specific issues that can be addressed in the CCMD campaigns include securing interagency approvals; addressing PN and regional sensitivities, changing politics, and overall stability; determining optimal presence and posture; BPC; and developing formal agreements/permissions between the US and PNs. Developing mutually supportive relationships to enhance coordination is an important enabler for joint logistics operations. ACSAs are bilateral international agreements that allow for the provision of cooperative logistics support under the authority granted in Title 10, USC, Sections 2341-2350. They are governed by DODD 2010.9, Acquisition and Cross-Servicing Agreements, and implemented by CJCSI 2120.01, Acquisition and Cross-Servicing Agreements.

Crisis Response and Limited Contingency Operations

US military history indicates crisis response and limited contingency operations are typically single, small-scale, limited-duration operations. Many of these operations involve a combination of military forces, the private sector, and capabilities in close cooperation with other US Government departments and agencies, international organizations, and NGOs. Logisticians must understand multinational, private-sector, and interagency logistics capabilities and coordinate mutual support, integrating them into the joint operation when appropriate. Many crisis response missions, such as foreign humanitarian assistance and disaster relief operations, require time-sensitive sourcing of critical commodities and capabilities, and rapid delivery to the point of need. In these operations, joint logistics is often the main effort, often operating in support of the Department of State.

Major Operations or Campaigns

Major operations or campaigns typically involve the deployment, sustainment, redeployment, and retrograde of large combat forces. Joint logistics can be executed by an appointed lead Service or agency for CUL. Joint logisticians develop support plans for the duration of the operation, as well as the return of personnel and equipment to CONUS or other locations. These plans often leverage contractor support to augment Service logistics capabilities. The primary challenges for logisticians during these types of operations are identifying the requirements, ensuring logistics issues are considered among competing priorities and adjusting to the situation to ensure sustained readiness and synchronized timelines as the operation transitions across phases. Logistics plans must account for and have the flexibility to mitigate the impact of CBRN-contaminated APODs and SPODs on force flow. This includes identifying locations for transload and exchange zone operations. A critical planning requirement during any operation is to plan for the transition to the final phase, where logisticians will have competing requirements to support stability activities, provide basic services while conducting contract closeout and changes to the contractor management plan, support foreign humanitarian assistance, and assist with reconstruction efforts all while conducting movement of redeploying forces and equipment. The retrograde of contaminated materiel will require special handling to control contamination and protect the force and mission resources.

III. Joint Logistics Assessment

Assessment is an integral part of planning for and execution of any operation, fulfilling the necessary requirement for analyzing changes in the OE and determining progress of an operation. Assessment activity involves the entire staff and other sources such as higher and subordinate headquarters, interagency and multinational partners, and other stakeholders. Logisticians not only feed assessment data to the commander to determine progress towards objectives but also assess the adequacy of logistics support, making adjustments to the logistics plan as required. Logisticians collect information from both the end-user and service providers to adjust and improve logistics support. Assessment is a continuous process throughout a campaign or operation that measures the overall effectiveness of employing joint force logistics capabilities. It involves monitoring and evaluating the current situation and progress of logistics support toward mission completion and requires input from not just the logistician but the end-user and JLEnt membership as well.

IV. Terminating Joint Operations

Terminating joint operations is an aspect of the CCDR's strategy that links to achievement of national strategic objectives. The supported CCDR can develop and propose specified conditions approved by the President or SecDef that must be met before a joint operation can be concluded. These termination criteria help define the desired military end state, which normally represents a period in time or set of conditions beyond which the President does not require the military instrument of national power as the primary means to achieve remaining national objectives.

A. Concluding Joint Logistics Operations

Joint logistics operations are always ongoing, but it is possible that some aspects of logistics operations could be completed before the operation has concluded. For example, force reception operations could be completed when forces have been placed under the control of the commander for integration and employment, and no other forces are flowing into the JOA. Joint logisticians monitor transitional activities and ensure resources are fully utilized or redeployed. Withdrawal and redeployment from an operation are challenging and require a synchronized and holistic effort by joint logisticians. Maintenance support planning should address the process for determining equipment disposition and the requirements for preparing equipment for shipment. In addition, maintenance support planning should ensure that equipment is available for movement when required while minimizing the impact on readiness. In accordance with DOD policies, logisticians plan for the disposition of materiel, such as retrograde and demilitarization, scrap removal, and disposal of hazardous waste, and, when required, clearance decontamination of supplies and equipment.

B. Theater Closure

When it has been determined that joint operations should be terminated, joint logistics operations focus on tasks that include redeploying personnel and materiel from the JOA to a new OA or home station/demobilization station, departure of contractor personnel, disposal of equipment, transitioning materiel and facilities to HN, foreign military sales, or disposal of materiel. Joint logistics operations also play a major role in closing ports to military operations and terminating operational contracts and agreements. Plans should be developed to monitor or assist the retrograde of contractor equipment and personnel. DOD must receive back any government-furnished property loaned to contractors as part of their mission. Operational contracts and agreements are not considered closed out until the force has confirmation of receipt of all goods and services and full payment has been made. Contracting and payment officials should not redeploy until all contracts and agreements are closed out.

Chap 9 Deployment & Redeployment

Ref: ATP 3-35 (w/Chg 2), Army Deployment and Redeployment (Mar '15), chap. 1.

Force projection is the ability to project the military instrument of national power from the United States or another theater, in response to requirements for military operations. (JP 3-0) It is a demonstrated ability to alert, mobilize, rapidly deploy, and operate effectively anywhere in the world. The Army, as a key member of the joint team, must be ready for global force projection with an appropriate mix of combat forces together with support and sustainment units. Moreover, the world situation demands that the Army project its power at an unprecedented pace. The flexible and rapid deployment of Army forces with sufficient depth and strength to sustain multiple, simultaneous operations enables Army units to seize, retain, and exploit the initiative to gain and maintain a position of relative advantage in unified land operations to create conditions for favorable conflict resolution.

See following pages (pp. 9-2 to 9-3) for further discussion.

Deployment Phases

A Deployment Planning

B Predeployment Activities

C Movement
- Fort-to-Port
- Port-to-Port

D Reception, Staging, Onward Movement, and Integration (RSOI)

Ref: ATP 3-35, Army Deployment and Redeployment (Mar '15), chap. 1.

Deployment is composed of activities required to prepare and move forces, supplies, and equipment to a theater. This involves the force as it task organizes, tailors itself for movement based on the mission, concept of operations, available lift, and other resources.

The employment concept is the starting point for deployment planning. Proper planning establishes what, where, and when forces are needed and sets the stage for a successful deployment. Consequently, how the geographic combatant commander (GCC) intends to employ forces is the basis for orchestrating the deployment structure. All deployment possibilities must be examined as they dramatically influence employment planning. Deployment directly impacts the timing and amount of combat power that can be delivered in order to achieve the GCC's desired effects.

I. Force Projection

Ref: ATP 3-35 (w/Chg 2)), Army Deployment and Redeployment (Mar '15), p. 1.

Force projection is the ability to project the military instrument of national power from the United States or another theater, in response to requirements for military operations. (JP 3-0) It is a demonstrated ability to alert, mobilize, rapidly deploy, and operate effectively anywhere in the world. The Army, as a key member of the joint team, must be ready for global force projection with an appropriate mix of combat forces together with support and sustainment units. Moreover, the world situation demands that the Army project its power at an unprecedented pace. The flexible and rapid deployment of Army forces with sufficient depth and strength to sustain multiple, simultaneous operations enables Army units to seize, retain, and exploit the initiative to gain and maintain a position of relative advantage in unified land operations to create conditions for favorable conflict resolution.

Force Projection Processes

Force projection encompasses a range of processes including mobilization, deployment, employment, sustainment, and redeployment. These processes have overlapping timelines, are continuous, and can repeat throughout an operation. Force projection operations are inherently joint and require detailed planning and synchronization. Decisions made early in the process directly impact the success of the campaign.

1. Mobilization

Mobilization is the process of assembling and organizing resources to support national objectives in time of war and other emergencies. Mobilization includes bringing all or part of the industrial base and the Armed Forces of the United States to the necessary state of readiness to meet the requirements of the contingency.

2. Deployment

Deployment is the movement of forces to an operational area in response to an order.

3. Employment

Employment prescribes how to apply force and/or forces to attain specified national strategic objectives. Employment concepts are developed by the combatant commands (COCOM) and their component commands during the planning process. Employment encompasses a wide array of operations—including but not limited to—entry operations, decisive operations, and post-conflict operations.

4. Sustainment

Sustainment is the provision of personnel, logistics, and other support necessary to maintain and prolong operations or combat until successful accomplishment or revision of the mission or national objective.

5. Redeployment

Redeployment is the transfer of forces and materiel to home or demobilization stations for reintegration and/or out-processing. This manual is the proponent for this definition which differs from the joint definition.

Each force projection activity influences the other. Deployment and employment cannot be planned successfully without the others. The operational speed and tempo reflect the ability of the deployment pipeline to deliver combat power where and when the joint force commander requires it. A disruption in the deployment will inevitably affect employment.

Global Force Management (GFM)

Ref: JP 3-35, Deployment and Redeployment Operations (Jan '18), p. I-2 to I-3.

GFM allows SecDef to strategically manage the employment of the force among CCDRs. It aligns force assignment, allocation, and apportionment methodologies in support of the Defense Strategy Review, joint force availability requirements, and joint force assessments.

Assignment

By statute, the Secretaries of the Military Departments assign forces to CCDRs as directed by SecDef to perform missions assigned to those commands. SecDef promulgates this direction via the assignment tables communicated in the Global Force Management Implementation Guidance (GFMIG) in even years and Forces For Unified Commands Memorandum ("Forces For") in odd years. The Secretaries of the Military Departments then identify the actual units to satisfy SecDef's direction in the assignment tables. Those forces henceforth become "assigned" to that combatant command (CCMD) and the CCDR exercises combatant command (command authority) (COCOM) over them.

Allocation

SecDef has the authority to allocate forces between CCDRs. The allocation process temporarily adjusts the distribution of forces among the CCDRs to meet force requirements in support of current operations and campaign plans to mitigate near-term military and strategic risk. SecDef publishes decisions to allocate forces in the CJCS annual deployment order (DEPORD) (i.e., the Global Force Management Allocation Plan [GFMAP] and its associated annexes), as well as periodic modifications to the base plan. CCDRs usually exercise operational control (OPCON) over allocated forces; however, SecDef will specify the command relationship the gaining CCDR will exercise and the losing CCDR will relinquish.

Apportionment

Forces are apportioned quarterly by the CJCS to provide an estimate of the Military Departments' capacity to generate capabilities that can reasonably be expected to be made available along general timelines. This estimate informs and shapes CCDR resource-informed planning but does not identify the actual forces that may be allocated for use if a plan transitions to execution. Apportionment is dependent on the number of operational forces, the readiness and availability of the forces, and the number of forces employed globally and serves as the starting point for planning.

Further discussion of the GFM processes can be found in the GFMIG; CJCSM 3130.06, (U) Global Force Management Allocation Policies and Procedures; CJCSM 3122.02, Joint Operation Planning and Execution System (JOPES) Volume III (Time-Phased Force and Deployment Data Development and Deployment Execution); CJCS Guide 3130, Adaptive Planning and Execution Overview and Policy Framework; and Joint Publication (JP) 5-0, Joint Planning.

Refer to Joint/Interagency SMARTbook 1: Joint Strategic & Operational Planning (Planning for Planners), 2nd Ed. (JIA1-2). At 408-pgs, topics include planning fundamentals (strategic organization, campaigning, sequencing actions), planning functions (contingency planning, planning in a crisis, APEX), global force management (GFM, force ID & sourcing, force planning), joint intel prep of the operational environment (JIPOE), intel prep of the information environment (IPIE), joint planning process (JPP), plan or order development (plan review & assessment), and APEX execution functions.

II. Deployment Phases

Ref: ATP 3-35 (w/Chg 2)), Army Deployment and Redeployment (Mar '15), pp. 1-2 to 1-4.

The Joint deployment process is divided into four phases -- deployment planning; predeployment activities; movement; and JRSOI. The terminology used to describe the Army deployment phases is in synch with the Joint process. The Joint process includes a planning phase at the outset whereas the Army considers planning to be woven through all the phases. Moreover, the movement phase in the Army process is discussed in two segments – fort to port and port to port. The Army relies on U.S. Transportation Command (USTRANSCOM) to provide the strategic lift to and from the port of embarkation (POE).

For the Army, deployments consist of four distinct but interrelated phases which are addressed in the following subparagraphs. A successful deployment requires implementation of each phase with seamless transitions and interactions among all of them. The phases are not always sequential and could overlap or occur simultaneously.

A. Deployment Planning

The initial activity in planning a development plan using the military decision making process. The objective is to synchronize deployment activities to facilitate execution. The steps used in planning and preparation during predeployment activities include: analyze the mission, structure forces, refine deployment data, prepare the force, and schedule movement. Successful deployment planning requires knowledge of the unit's deployment responsibilities, an understanding of the total deployment process, and an intellectual appreciation of the link between deployment and employment.

1. **Analyze the Mission.** The mission is examined and courses of action (COA) are developed bearing in mind that the employment considerations are paramount. The primary purpose of a deployment is to provide the right force at the right place and at the right time.
2. **Structure Forces.** The COAs outline the ways (employment) and the means (forces) to accomplish the mission. Initially, capabilities are identified; however, as the COAs are further defined, the requirements are being translated into type units.
3. **Refine Deployment Data.** As forces are identified, the development of the time-phased force and deployment data (TPFDD) is begun. The supported combatant commander defines the intent for deployment which may be very specific and direct the sequence of units or just identify a general deployment timeline. In any case, the intent should clearly express how the deployment postures the force for employment.
4. **Prepare the Force.** Force packages are developed, ensuring the right capabilities are in the proper combinations to meet the intentions of the supported combatant commander.
5. **Schedule the Movement.** The supporting combatant commands must clearly and completely define their mobility requirements and milestones based on the concept of operations. The right sequencing of forces will provide the commander with the capabilities required to achieve the desired objectives. Once the strategic lift schedule is put in motion, it is difficult to change without losing the identified transportation asset and its position in the lift schedule.

B. Predeployment Activities

An expeditionary Army requires that units are prepared for potential deployments consistent with ARFORGEN model. During predeployment units constantly plan for various contingencies and hone their deployment skills. When units train and exercise their predeployment activities, they become second nature and are accomplished efficiently. Not only should units be trained, personnel must be nearly 100 percent compliant with respect to Soldier readiness processing (SRP), encompassing those administrative, medical, and dental checks required to prepare a Soldier for deployment. This level of readiness and training requires school-trained, dedicated mobility officers, UMOs, hazardous cargo certifiers, and load planners. Their requirements are documented in organizational equipment lists (OEL), UDLs, and loaded into the Transportation Coordinators Automated Information for Movements System II (TC-AIMS II). In addition, units must acquire movement expertise, knowledgeable deployment support teams, joint deployment process improvement tools, and an understanding of the Joint Operation Planning and Execution System (JOPES) to enable seamless deployment operations.

See pp. 9-7 to 9-20 for further discussion.

C. Movement

1. Fort-to-Port

The receipt of the Air Mobility Command (AMC) air tasking order and Surface Deployment and Distribution Command (SDDC) port call message initiates POE operations and specifies the dates on which units must arrive at the POE. At the installation staging areas unit movement data is verified and equipment is inspected and configured for movement. It is then typically moved to the POEs by convoy or commercial surface transport. The installation coordinates and/or provides support to assist the deploying force by using non-deploying units, installation resources, or contracted support. Support may include load teams, materiel handling equipment (MHE), maintenance teams, arrival/departure airfield control groups (A/DACGs), and deployment support teams. Other support should be identified during deployment exercises and then written into installation deployment support plans. The Mission Support Element is a TDA-augmentation capability used by the mission commander to develop and maintain the deployment support plan.

2. Port-to-Port

USTRANSCOM operates the Defense Transportation Service (DTS) and provides common user strategic transportation to support the CCDR for deployment. The port to port phase begins with strategic lift departures from POEs and ends with lift assets arrival in the designated theater PODs. Fundamental to the success of the port-to-port movement is the ability of the CCDR to synchronize the arrival of airlift and sealift force packages so that vessels can be brought to a berth or offloaded in-stream with minimal delay. This ensures that cargo can be received and cleared from the port in a timely manner.

See pp. 9-21 to 9-28 for further discussion.

D. Reception, Staging, Onward Movement, and Integration

RSOI is the process that delivers combat power to the Joint Force Commander (JFC) in the operational theater. The very nature of seizing the initiative demands expeditious processing of personnel and equipment throughout the deployment pipeline. Consequently, facilities must be available on or near the PODs for personnel reception and equipment staging and preparation (to include refueling). One of the essential requirements at the Aerial Port of Debarkation (APOD) is adequate parking and operational areas to sustain the required number of aircraft to meet the throughput requirements. RSOI support, whether provided by theater support contracts, external support contracts (primarily the Army Logistics Civil Augmentation Program), or regionally available commercial host nation support, and/or military assets, must be sufficient to immediately support the arrival of deploying units. Effective RSOI matches personnel with their equipment, minimizes staging and sustainment requirements while transiting the PODs, and begins onward movement as quickly as possible. A plan to accomplish integration and maintain combat readiness must be understood, trained, and ready to implement upon arrival.

See pp. 9-29 to 9-38 for further discussion.

III. Deployment Principles

Ref: ATP 3-35 (w/Chg 2)), Army Deployment and Redeployment (Mar '15), p. 1-2.

Four principles apply to the broad range of activities encompassing deployment:

1. Precision

Precision applies to every activity and piece of data. Its effect is far-reaching, and the payoff is speed. For example, precise unit deployment lists (UDL) ensures that correct lift assets are assigned against the requirement. Precision includes accurate weights, dimensions, and quantities. This degree of precision eases loading requirements and improves departure speed and safety. Precision allows units to meet the GCC's timeline and supports the concept of employment.

2. Synchronization

Deployment activities must be synchronized to successfully close the force. Effective synchronization of scarce lift assets and other resources maximizes their use. Synchronization normally requires explicit coordination among the deploying units and staffs,supporting units and staffs, a variety of civilian agencies, and other Services. Realistic exercises and demanding training are paramount to successful synchronization.

3. Knowledge

One of the more critical pieces in deployment planning is the knowledge upon which decisions are made. There is a short period of time during which the deploying commander must make crucial decisions on employment. These decisions set the tone for the remainder of the deployment. Many decisions are very hard to change and have significant adverse impacts if changed; others are irrevocable.

4. Speed

Speed is more than a miles per hour metric. The proper focus is on the velocity of the entire force projection process, from planning to force closure. Critical elements of speed associated with force projection include agile (state-of-the-art) ports, submission of accurate information,safe and efficient loading, trained unit movement personnel at all levels, timely arrival of throughput enablers, maintaining unit integrity, delivering capability rather than entire units, and force tracking information

Chap 9 I. Predeployment Operations

Ref: ATP 3-35 (w/Chg 2)), Army Deployment and Redeployment (Mar '15), chap. 2.

The Army's effort to be more responsive begins at home station. Predeployment activities are actions taken to prepare forces for deployment and are not limited to the deploying unit, but include supporting units and the installation staff. Planning, document preparation, equipment readiness, and training are the foremost predeployment unit activities. This chapter discusses predeployment activities of the deploying unit as well as those in a support role.

I. Planning

GCCs use assigned forces to perform missions in their areas of responsibility. If additional forces are required to support an operation or contingency plan, the GCC requests forces via the Global Force Management allocation process. The Global Force Management allocation process begins with the supported GCC requesting the forces necessary to support the mission. The Joint Staff validates the submitted force requirements and assigns the appropriate joint force provider, Service force provider, or supporting combatant commander to develop sourcing recommendations. The Joint Staff, along with the Service Headquarters through the assigned Service components, and the joint force providers develop sourcing recommendations. The Joint Staff consolidates all recommendations and then staffs the recommendations with all Services, GCCs and Department of Defense (DOD) agencies. The sourcing recommendations are then presented to the Secretary of Defense. Following Secretary of Defense approval of the sourcing recommendations, the Chairman of the Joint Chiefs of Staff (CJCS) orders deployments via the published Global Force Management Allocation Plan and attached Annexes. The Global Force Management Allocation Plan directs force providers to provide forces to meet the GCC force and Joint Individual Augmentation requirements. Ordered force providers, Service Secretaries, GCCs with assigned forces, and Directors of DOD agencies implement the orders in the Global Force Management Allocation Plan Annexes through deployment orders. This process may cover a period of several months or be compressed to days or even hours for crisis action planning.

A. Deployment Planning

Contingency planning is typically used in those cases where the deployment and employment of forces is in response to anticipated operations. It is designed to produce a detailed operational plan for a potential event and relies heavily on a number of assumptions ranging from the threat to anticipated host nation support. Conversely, crisis action planning is accomplished in response to a time-sensitive, imminent threat that may result in an actual military operation. The plan is based on circumstances existing at the time planning occurs. In either contingency or crisis planning, prescribed procedures are followed to formulate and implement a response. Deployment planning is a key element of both contingency and crisis action planning and aims at delivering the right force, at the right place, and at the right time.

Planning for deployment is based on mission requirements and time. During deployment operations, supported combatant commanders are responsible for building and validating movement requirements, determining predeployment standards, and balancing, regulating, and effectively managing the transportation flow. Supporting combatant commands and agencies source requirements not available to the geographic combatant commander and are responsible for verifying supporting unit movement data, regulating the support deployment flow, and coordinating during deployment operations. Each of these activities is an element of effective deployment planning. To facilitate these processes, the joint planning and execution community uses a common framework of directives, guidance, and decision support tools within JOPES.

See p. 9-20 for further discussion of JOPES and TPFDD development.

B. Movement Planning

To meet contingency support requirements, units develop movement plans and SOPs.

Unit Movement Plans

Movement plans define responsibilities, functions, and details for each part of a unit deployment from mobilization station or installation to reception in theater. There may be more than one plan required depending on the number of contingencies/operations plans (OPLANs) the unit must prepare to support. Movement plans are written in a five-paragraph OPLAN format.

Containerization must be addressed during deployment planning. The key to successfully using containerization operations to maximize shipping options is to identify units with high percentages of equipment compatible with containers.

The USTRANSCOM's component commands schedule lift against the unit line number (ULN) to meet the earliest arrival date (EAD)-latest arrival date (LAD) window. AMC publishes airflow schedules to call forward personnel and equipment to the APOE. These schedules are in GCCS. The call forward schedules are movement directives that specify when units must have their equipment at the POE to meet the available-to-load dates (ALD). Based on these schedules, deploying units and their respective commands backward plan movements to the POE to meet the ALD. Movement directives (if published) provide windows by mode for cargo arrival at the POE.

See following pages (9-10 to 9-11) for discussion on developing a deployment movement plan.

Unit Movement SOP

The unit movement SOP defines the day-to-day as well as alert functions. The SOP defines the duties of subordinate units/sections that will bring the unit to a higher state of readiness. These duties can be written in separate annexes that can be easily separated and issued to leaders for execution. Functions addressed in the SOP could include unit property disposition, supply draw, equipment maintenance, vehicle and container loading, security, marshalling procedures, purchasing authorities, unit briefings, in-transit visibility (ITV), and other applicable deployment activities.

Movement Binders

Units maintain movement binders containing the unit movement plan; unit movement SOP; appointment orders; training certificates; recall rosters; a current OEL; copies of load cards and container packing lists; prepared copies of transportation requests; convoy movement requests and special handling permits; and blocking, bracing, packing, crating, tie-down (BBPCT) requirements. The binder also serves as a continuity bridge from one UMO to the next.

Route and Location Reconnaissance and Rehearsal

Reconnaissance of the route to pre-designated POEs and of the POEs themselves should be an ongoing activity. It may be accomplished through passive means such as map surveillance or, optimally, through site visits. Walking the terrain at the power projection platform and designated port facilities allows commanders to understand space limitation, see choke points, survey facilities, understand the simultaneous nature of the operation, and visualize the deployment operation. Terrain walks can be useful as a unit level activity, but are more beneficial when they involve all participating and supporting units. Rehearsals validate deployment plans and permit commanders and unit movement officers to see possibilities and limitations. The physics of the operation can become plainly evident.

Unit Movement Dates

Ref: ATP 3-35 (w/Chg 2), Army Deployment and Redeployment (Mar '15), p. H-1.

Movement plans define responsibilities, functions, and details for each segment of a unit deployment from origin to reception in theater. There may be more than one movement plan required depending on the number of contingencies/OPLANS the unit must plan to support. Movement plans are written in five paragraph OPLAN format.

GCCS is the system used to manage deployments. Deployment related information is contained in the GCCS database and is accessible through ad hoc queries or via JOPES, a GCCS application. Units deploying under a JOPES OPLAN must increment their movements consistent with OPLAN TPFDD requirements, as delineated by ULNs. A ULN is a seven-character alphanumeric code that defines a unique increment of a unit. For example, it may be used to identify an advance party, the main body, and the equipment sealift and airlift requirements for any specific segment of the move. The Army command or other designated deploying unit higher headquarters assigns ULNs to units. It is essential that deploying units use the correct ULN for equipment to be scheduled for movement at the right time by the correct mode. This is the key to the JOPES database validation process. An incorrect ULN could overstate or understate airlift requirements and delay movements.

Unit line numbers (ULNs) available on JOPES OPLAN reports divide the unit by transportation mode, ports of embarkation or debarkation, and dates. Dates correspond to the established **C-day (day on which a deployment operation begins or is scheduled to begin)** for the designated plan TPFDD.

The unit movement is phased by the following dates relative to C-day:

1. Ready-to-Load Date (RLD)

The RLD is the date in the TFPDD when the unit will be prepared to depart its origin.

2. Available-to-Load Date (ALD)

The ALD is the TPFDD specified date when the unit will be ready to load on an aircraft or ship at the POE.

3. Earliest Arrival Date (EAD)

The EAD is the date specified by the supported combatant commander that is the earliest that a unit, a resupply shipment or replacement personnel can be accepted at a POD during a deployment. It is used with the latest arrival date to define a delivery window for transportation planning.

4. Latest Arrival Date (LAD)

The LAD at the APOD/SPOD is the date specified by the supported combatant commander that is the latest date when a unit, sustainment, or replacement personnel can be accepted at a POD and support the concept of operations. It is used with the earliest arrival date to define a delivery window for transportation planning.

5. Required Delivery Date (RDD)

The RDD is the date when a unit must arrive at its destination and complete offloading to properly support the concept of operation.

At the predeployment conference the unit requests a ULN for each element and/or separate movement. The Army Command or ASCC assigns ULNs to units. It is essential that deploying units use the correct ULN for equipment scheduled for movement at the right time by the correct mode. Using the correct ULN is key to the JOPES database validation process. An incorrect ULN could overstate or understate strategic and inter-theater lift requirements and delay passenger and cargo movements.

Developing a Unit Movement Plan

Ref: ATP 3-35 (w/Chg 2), Army Deployment and Redeployment (Mar '15), pp. H-2 to H-3. Refer to ATP 3-35, app. I, for a sample unit movement plan.

The following paragraphs describe a recommended step-by-step process for developing deployment movement plans:

1. Identify what needs to be moved

Based upon mission, enemy, terrain and weather, troops and support available, time available, and civil considerations (METT-TC) and command guidance, deployment planning must reflect personnel, equipment, supplies, and how the unit will accomplish the move. For planning purposes, units plan to deploy with assigned personnel and on-hand equipment. Upon execution, the plan may need to be modified if additional personnel are assigned or equipment cross-leveled to bring the unit to the required readiness level. Units should plan to move their basic load of supplies to sustain their operations upon arrival in the theater. The quantities to be deployed are normally defined in OPLANs, unit or Army Command SOPs, or ASCC instructions. The UMO must have a detailed listing of each piece of equipment to be deployed based on the automated unit equipment list (AUEL). All outsize, oversize, overweight, or hazardous equipment/cargo must be identified, as it will require special considerations.

2. Identify equipment to accompany troops (TAT)

- **Yellow to accompany troops (TAT)** equipment must accompany troops and be accessible enroute. Examples include Class I basic load items, individual carry-on baggage, and weapons. For personnel traveling via commercial air, this is generally only the baggage that will fit under the seat.
- **Red TAT items** must be available at the destination before or upon unit arrival. This equipment may be sensitive cargo that requires special security or handling at the POE or POD. Red TAT must be unitized/palletized and reported on the AUEL/deployment equipment list (DEL). Not to accompany troops (NTAT) equipment is normally shipped by surface and does not accompany the troops. It consists of all other equipment required by the unit to perform its mission.

3. Identify what needs to move by air

This could include personnel, advance parties, baggage, and some equipment. The balance of equipment normally moves by sea. For deployments supporting combatant commander OPLANs/operation orders (OPORD), the TPFDD will designate the strategic movement mode.

4. Identify hazardous (also sensitive and classified) cargo for packaging

Identify hazardous (also sensitive and classified) cargo for packaging, labeling, segregation, and placarding for movement. Codes of Federal Regulation (CFR) 49 Transportation provides guidance on the packaging, labeling, placarding, and movement of hazardous materials.

5. Identify bulk cargo that needs to be moved and develop packing lists

All consolidated cargo (boxed, crated, etc.) loaded in vehicles, containers, and on 463L pallets must display a separate packing list, DD Form 1750 (Packing List) that shows all contents. Packing lists are not required for items that do not need identification, such as empty vehicles, nested cans, or bundled shovels, however, they must be listed on the load diagram if loaded in a truck or container.

6. Develop vehicle load plans for unit equipment

Equipment that cannot be loaded on organic vehicles should be planned for movement by other means (container, commercial rail or highway, or military assets). Vehicle load plans are recorded on DD Form 1750 (Packing List) for organic vehicles and trailers carrying secondary loads. FORSCOM units may use FORSCOM Form 285-R (Vehicle Load Card) for preparing vehicle load plans. The installation transportation officer (ITO) is the installation point of contact (POC) for obtaining commercial transportation to move equipment to POE that is beyond the unit's organic capability. Unit cargo (vehicles and equipment) is prepared for shipment according to the mode of transportation. Depending on the strategic lift full reduction may or may not be required. Reduction details are normally in the SDDC port call message or the operations order for sealift. For deployment by air, reduction is determined by type of aircraft. Transportation Engineering Agency (TEA) Pamphlet 55-24 provides guidance for preparing vehicles for airlift. Vehicle modifications (that is, shelters, bumper modifications, and so forth) made by the unit which change the vehicle configuration/dimensions/weight normally must be approved by the unit's Army Command/ASCC and ultimately by TEA. Vehicle modifications must be reflected on the AUEL and DEL. Information on dimensions, weights, and cubes for all Army equipment is in CD-ROM and World Wide Web versions of TB 55-46-1. The hard copy version only contains major end items.

7. Identify Blocking, Bracing, Packing, Crating & Tie-Down (BBPCT) requirements

All crates, containers, boxes, barrels, and loose equipment on a vehicle must be blocked, braced, and tied-down to prevent shifting during transit. The POC for blocking and bracing requirements is normally the UMC.

Chapter 6 of FORSCOM/ARNG Regulation 55-1 describes policy for obtaining and stocking BBPCT materials and related railcar loading equipment for deploying units. Additional tie down guidance is in TEA Pamphlet 55-19 and TEA Pamphlet 55-20.

8. Translate what needs to be moved into transportation terms

Personnel and equipment data are translated into transportation terminology as unit movement data (UMD) and recorded on the OEL. Upon deployment execution, units use TC-AIMS II to update the OEL and create the DEL. The UMC provides assistance to deploying units for OEL updates and DEL development.

9. Determine how the personnel and equipment will move to the POEs

Determine how the personnel and equipment will move to the POEs. In continental United States (CONUS) wheeled vehicles and tracked vehicles move via commercial rail, truck, or barge. Unit personnel usually move to the POE by military or commercial buses. Army rotary wing aircraft normally self-deploy to the POE, where they will be disassembled for shipment.

10. Prepare the unit deployment plan

The administrative, logistical and coordination requirements for the plan must be determined. Items such as enroute medical, messing, and maintenance for movement to POEs must be coordinated and documented.

11. Maintain the plan

Update the OEL as changes occur in the OPLAN, equipment, commander's intent, and mission execution. The OEL is used to produce the unit's equipment manifest and MSLs and errors can result in the equipment being lost while in transit.

Unit Movement Officer (UMO)

Ref: ATP 3-35 (w/Chg 2), Army Deployment and Redeployment (Mar '15), app. D.

The UMO is the commander's appointed representative and attends to the details of getting the unit ready for movement. This appendix outlines the responsibilities of the UMO and provides the references for them. Commanders appoint, in writing, an officer or NCO (E6 or above) as a UMO and an alternate (E5 or above). The designated persons will attend an approved school.

The UMO must focus on thorough planning, coordination, training, and execution of unit deployment. Specific responsibilities of the UMO include—

- Preparing and maintaining documentation for unit movement in TC-AIMS II. This includes maintaining the unit movement data, from which the OEL is generated, and creating and processing the UDL.
- Preparing the unit movement plan
- Planning convoy movements
- Requesting commercial and military transportation
- Coordinating with higher headquarters and support activities for unit movements
- Coordinating logistical support for the move
- Coordinating with the A/DACG and CRE at the APOE and APOD
- Coordinating with SDDC representatives at the SPOE and SPOD
- Transporting of the units' organic equipment and cargo
- Establishing and training unit loading teams
- Obtaining 463L pallets, containers; and BBPCT materials
- Ensuring all cargo is properly labeled with MSLs and RF tags when directed
- Ensuring unit personnel are authorized to certify HAZMAT
- Ensuring packing lists are prepared for containers
- Maintaining movement binders or continuity books that include appointment orders, training certificates, recall rosters, OEL, transportation requests, and BBPCT requirements

The UMO will maintain a deployment binder for reference and continuity. The following is a list of recommended contents—

- Unit movement SOP
- Appointment orders and training certificates for UMOs, load teams, and HAZMAT certifiers
- Recall rosters and instructions
- Coordination requirements for plan execution and a list of supporting agencies and POCs
- Major equipment shortage list
- Supply list by supply support activity, coordination requirements, and prepared requisitions
- List of BBPCT on hand and due out
- OEL
- Copies of packing lists
- Prepared copies of transportation requests, convoy movement requests, and special hauling permits

- Strip maps for convoy routes
- Advance party composition and instructions
- Sample forms required for personnel support during deployment
- Transportation requirements
- Rear detachment and family support group operations plans
- Unit-generated checklists or SOPs to assist in deployment planning and execution
- Off-duty hours contact telephone numbers for deployment support organizations

Reference List

The following table identifies the primary references for the functions performed by the UMO.

FUNCTION	***REFERENCE***
Duties of UMO	DTR Part III, AR 525-93
Unit movement data	DTR Part III
Unit movement plans	TEA Pamphlet (PAM) 700-2
Transportability of unit equipment	TEA Pam 700-5; TB 55-46-1
Motor transport operations	ATP 4-11; TEA Pam 700-2; TEA Pam 55-20
Convoy plan	ATP 4-11
Rail operations	ATP 4-14; Technical Manual (TM) 55-2220-058-14; TEA Pam 55-19; TEA Pam 700-2
Port operations	ATP 4-13
Airlift operations	DTR; TM 38-250; TEA Pam 55-24; TEA Pam 700-2
Sealift operations	DTR; TEA Pam 700-2; TEA Pam 700-4
Vehicle load plans	TEA Pam 55-20
Aircraft load plans	TEA Pam 55-24
Tiedown of equipment in aircraft	TEA Pam 55-24
Containerization	DTR Part IV; ATP 4-12; TEA Pam 55-23; TEA Pam 700-2
HAZMAT	Title 49 code of federal regulation (CFR); TM 38-250
AITand RF tags	DTR; AR 700-80

Ref: ATP 3-35, Army Deployment and Reployment (Mar '15), fig. D-1, p. D-2.

II. In-Transit Visibility (ITV)

In-transit visibility (ITV) is the capability to employ information technology resources to track the identity, status, and location of DOD units, nonunit cargo (excluding bulk petroleum, oils, and lubricants), passengers, patients, and personal property from origin to consignee or destination across the range of military operations. (JP 4-01) Commanders, whose employment planning depends on the closure of forces, base their decisions on accurate and reliable ITV.

Deploying units contribute to the operational picture and the employment planning of commanders by properly marking and tagging their equipment for input into the DOD automated identification technology (AIT) system and the DTS. Prior to departing their deployment stations, units must properly prepare vehicles and equipment for shipment to include securing basic issue items, weighing, and affixing the necessary markings, labels, and in-transit visibility devices to their equipment, containers, and 463L pallets. This includes Military Shipping Labels (MSL), pallet IDs, and radio frequency identification (RFID) Tags. ITV is achieved as the progress of the container or vehicle is recorded as it passes through the network of interrogators. ITV preparation begins during predeployment and continues through the load out of vehicles and equipment. Ensuring AIT storage devices are accurate, properly attached, and readable facilitates ITV throughout the transportation pipeline. AIT readers and interrogators report the movement to automated information systems (AIS) allowing deployment managers to track and control the flow of equipment.

Refer to ATP 3-35, appendix K for further information on AIT.

III. Training

Individual and unit deployment training are essential in developing the skills required to rapidly project combat power.

A. Deployment Training

Units with deployment missions are required to have an appropriate number of personnel trained to perform special deployment duties. These duties include unit movement officer, unit loading teams, hazardous cargo certifying officials, and air load planners. Army Commands and ASCCs have specific policies for appointing and training personnel to assume these positions, and this is a synopsis of the more important ones.

See facing page for further discussion.

B. Collective Training

Companies and battalions train to meet unit and individual training requirements for deployment operations. Deployments can occur at any time leaving the deploying unit with little or no time to correct training deficiencies. The objective of collective deployment training is to implant the knowledge, skills, attitudes, and abilities so it becomes a reflex activity executed with precision. Units must identify deployment as a mission essential task, annotate it on their mission essential task list (METL) and gain and maintain proficiency. Many Army training programs offer the opportunity to include deployment training in major training events.

Units with deployment missions execute their command deployment discipline program and participate in periodic deployment readiness exercises. These events are designed to evaluate unit deployment readiness and exercise unit or command movement plans for overseas deployment. Exercises may involve the unit moving to POEs and loading unit equipment on strategic sealift or airlift assets. Major commands, installation, and brigade level commands normally have SOPs and deployment regulations and policies establishing subordinate unit required activities in an hourly deployment sequence. These documents guide unit activities during these events. Furthermore, synchronizing an annual IDSP exercise or evaluation (*refer to ATP 3-35, app. F*) with unit deployment training and or exercise can maximize readiness of both the unit and installation.

Deployment Training

Ref: ATP 3-35 (w/Chg 2), Army Deployment and Redeployment (Mar '15), pp. 2-4 to 2-5.

Units with deployment missions are required to have an appropriate number of personnel trained to perform special deployment duties. These duties include unit movement officer, unit loading teams, hazardous cargo certifying officials, and air load planners. Some commands and installations maintain a local capability to provide deployment training to ensure the supported units have ready access to the required training.

1. Unit Movement Officer (UMO)

The commander is responsible for all aspects of deployment preparation, training, and execution and appoints the UMO as his designated representative. The UMO must know the unit's mission and the commander's intent for the appropriate coordination, planning, and execution to take place.

FM 3-35, Appendix D provides more detailed information on the UMO.

2. Loading Teams

Units must have personnel trained in vehicle preparation and aircraft and rail loading/unloading techniques. The type and quantity of equipment to be loaded and the time available for loading determines the composition of the team. Training is arranged through the installation unit movement coordinator (UMC) and once completed the load teams are put on unit orders. Load team composition is tailored to the type and quantity of equipment being loaded and time available for loading:

- **For rail movements**, a well trained team of five operators, using prefabricated tiedown devices, can complete loading and lashing of equipment on a chain equipped flatcar. Units are normally provided 72 hours for loading once the cars are spotted.
- **For air movement**, a six person team can provide efficient loading and tie down of equipment. United States Air Force Mobility Command offers the Equipment Preparation Course to train unit load teams to prepare, load and tie down unit equipment on military aircraft.

3. Hazardous Cargo Certification

At least one individual will be on orders and trained to certify hazardous cargo at each unit level. The hazardous cargo certifying official is responsible for ensuring the shipment is properly prepared, packaged, and marked. The certifying official is also responsible for personally inspecting the item being certified and signing the hazardous material (HAZMAT) documentation. Hazardous cargo certifiers must be trained at a Department of Defense (DOD) approved school within the past 24 months and receive refresher training every two years. Upon training completion, they are authorized to certify documentation for commercial and military truck, rail, sea and air. A common mistake occurs when the HAZMAT certifier is sent with the advance party leaving no one to accomplish the HAZMAT inspections during departure operations.

4. Air Load Planning

Air load planners are appointed and trained to prepare, check, and sign unit aircraft load plans and in the planning and execution of airlift operations. The Air Mobility Command offers an Airlift Planners Course to those units aligned under the AMC Affiliation Program.

IV. Hazardous, Classified, and Protected Sensitive Cargo (Special Cargo)

Ref: ATP 3-35 (w/Chg 2), Army Deployment and Redeployment (Mar '15), app. J.

1. Hazardous Material (HAZMAT)

Packaging, shipping, handling, and inspecting of HAZMAT is mandated by US and international laws. These laws also apply to the use of intermodal containers and container equipment. This appendix provides an overview of doctrinal guidance and tactics, techniques, and procedures that are common to Department of Defense (DOD) and other US government agencies and organizations. This appendix also applies to the selection of standard American National Standards Institute/International Standards Organization (ANSI/ISO) commercial- or military-owned intermodal containers that meet the standards for shipment of Class I explosives and other HAZMAT.

Refer to MIL-HDBK 138 for compliance with container standards criteria.

HAZMAT must be properly prepared and documented IAW DOD Regulation 4500.9-R, Volume II and III; TM 38-250; and other service or command regulations. Documentation must include the total HAZMAT quantity and a certification statement stating that the HAZMAT is properly classified, described, packaged, marked, and labeled. Only specially trained individuals have authority to certify HAZMAT for transportation. Contact the Installation Transportation Officer (ITO) or Movement Control Team (MCT) for assistance in determining what certification requirements apply to each HAZMAT item being prepared for shipment. The deploying unit must ensure that:

- All ammunition and explosives are secured properly in containers and vehicles. SDDC issues authorization for ammunition to be in the port and aboard vessels.
- Provisions for DOT exemptions which may be used for shipment are followed. (For example, vehicle fuel tanks will be no more than three-quarters full when shipping under DOT Exemption 7280. Otherwise, fuel tanks must be only one-quarter full when shipping aboard a commercial vessel that is carrying civilians in addition to military cargo.)
- Fire extinguishers in racks designed expressly for them will not be removed from motor vehicles.
- Oxygen and acetylene tanks are labeled and marked with the UIC and shipment unit number and removed from the vehicle and placed on a separate pallet.
- Fuel tanks of trailer mounted equipment containing combustion engines are only 50 percent full.
- Five-gallon fuel cans, field cans, water heaters, gasoline lanterns, portable generators, blow torches, and similar equipment (in which combustibles other than diesel fuel are stored) are completely drained and cleaned before shipment. In a declared national emergency, 5-gallon cans can contain fuel.
- Battery boxes and covers are serviceable and positioned so as not to touch the terminals and to prevent arcing.
- Batteries of non-self-propelled equipment are disconnected and terminal ends protected from arcing and corrosion.
- When mode or other regulatory guidance requires, bulk fuel carriers are drained and purged and the proper placards affixed to them. A purge certificate should be prepared and kept available.
- Fueled vehicles shipped in closed freight containers have their battery cables disconnected and secured. Moreover labels should be affixed to the access doors of the container warning of a potential explosion when the doors are opened.

Refer to ATP 3-35, Table J-1, for references related to preparing and documenting hazardous materials.

2. Ammunition

Ammunition shipments are usually scheduled through military ammunition ports. Designated military ammunition ports serve the strategic purpose of routinely handling shipments of ammunition. To meet deployment requirements, ammunition may be moved through a commercial port. If the unit is deployed through a commercial seaport and must carry basic load ammunition with them, the SDDC manager for the port must first be notified of the intent to ship ammunition. The unit submits the following data through the ITO:

- DOD Ammunition Code.
- DOT proper shipping name.
- Total quantity and Number of packages.
- Total net explosive weight in pounds.
- Weight of each package in pounds.
- Cube of each package.
- United Nations identification number.
- Classification code consisting of hazard class and division number followed by compatibility group letter.
- Shipment configuration (to allow processing of DOD safety waivers and CG permits).

3. Classified Material

Classified material is cargo that requires protection in the interest of national security. The nature of classified cargo requires that shippers and transporters handle it in a way that it be identified, accounted for, secured, segregated, or handled in a special way to safeguard it. Detailed instructions are included in DTR 4500.9R. Do not identify classified cargo on the outside of the shipping containers.

When transporting classified material, enclose it in two sealed containers, such as boxes or heavy wrappings. Detailed instructions for packing classified material are contained in AR 380-5. Among its implementing instructions are the following excerpts from Chapter 8, AR 380-5: Classified information will be transmitted and transported only as specified in this Chapter 8, AR 380-5. Communications security information will be transmitted in accordance with AR 380-40. Special Access Programs material will be transmitted and transported in accordance with Appendix I of AR 380-5, AR 380-381 and applicable special access program procedure guides. Commands will establish local procedures to meet the minimum requirements to minimize risk of compromise while permitting use of the most effective transmission or transportation means.

4. Sensitive Cargo

Sensitive material is cargo that could threaten public safety if compromised. Sensitive cargo must be properly secured and identified to port personnel so sufficient security can be provided. Do not identify security cargo on the outside of the shipping containers. Detailed instructions are included in DTR 4500.9R. J-12. For sensitive cargo, units must adhere to the following:

- Remove crew-served weapons from vehicles. Place them in containers that are sealed and secured with an approved device.
- Ensure packaging material is strong and durable enough to provide security protection while in transit.
- Secure containers, vehicles, or compartments with an appropriate locking device as directed by the installation security officer. Also, place a serial-numbered seal on the door. Enter the serial number on the shipment unit packing list.
- Identify sensitive items in the commodity code on the unit's OEL and UDL.
- Eliminate indications of sensitive items from outside of the container, vehicle, or compartment that it contains sensitive items. Identify this fact on the unit's OEL/UDL.
- Provide guards or escorts when shipping sensitive material by rail.

V. Installation Deployment Support

Ref: ATP 3-35 (w/Chg 2), Army Deployment and Redeployment (Mar '15), app. F.

Heretofore deploying forces were supported by other military units as they prepared to move from the installation to the POE. The garrison staff, with civilians and contractors as the majority of their work force, has now assumed this role because military units are no longer available. The purpose of this appendix is to identify the functions performed by the garrison staff to support deploying and redeploying forces.

Installation Support

Installations are an integral part of the deployed force from home station to the foxhole. Operational deployments and rotational assignments across the globe mean installation capabilities will transcend more traditional expeditionary support requirements associated with mobilizing, deploying, and sustaining the force. More than a jump point for projecting forces, installations serve a fundamental role in minimizing their footprint through robust connectivity and capacity to fully support reach back operations.

Installation facilities must readily adapt to changing mission support needs, spiraling technology, and rapid equipment fielding. Installation connectivity must also support en route mission planning and situational awareness. Education and family support will use the same Installation support connectivity to sustain the morale and emotional needs of our Soldiers and their families.

NOTIONAL Installation Deployment Processing

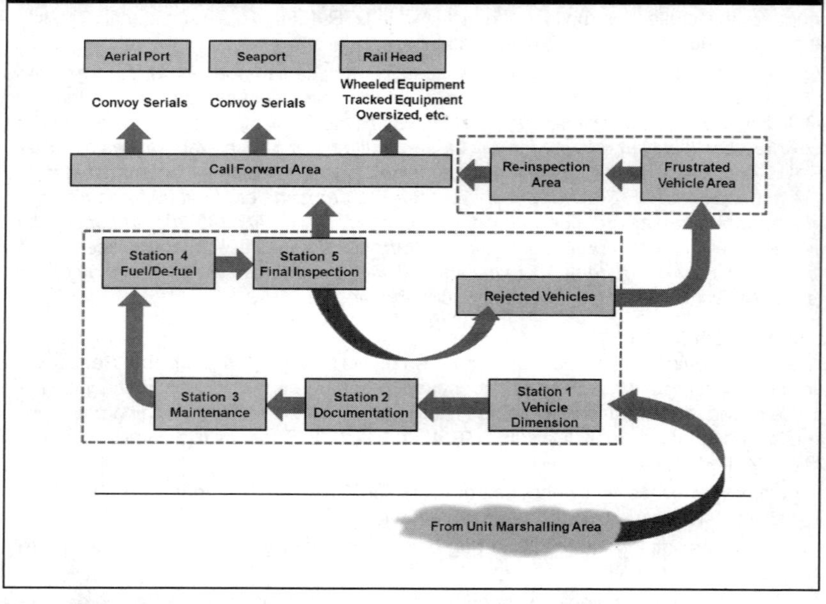

Ref: ATP 3-35. Army Deployment and Replovment. fig. F-1. p. F-2.

A typical installation deployment processing flow diagram (above) provides an indication of the magnitude of support involved. The installation deployment-processing site is a centralized location where deploying units process and assemble their equipment for movement to the POE. The site has several different labels, such as alert holding area, installation support area, and deployment readiness reaction fields, but the functionality is the same.

Redeployment Support

The significance of returning units to their home station in a timely manner has risen dramatically to meet the Army Force Generation and Reset requirements. All systems and procedures used during deployment operations apply to redeployment operations. Installations designated as demobilization sites complete processing of units and individuals being separated or released from active duty and returning to reserve status. It should be noted that when forces are relocating to their home station their focus is on reestablishing force readiness for subsequent contingency mission requirements.

Installations provide support at the POD to receive inbound unit personnel, equipment and cargo and to provide onward movement to the next destination. Installations will coordinate with the supporting commander(s) to—

- Prepare returning units for onward movement to home station or mobilization station.
- Receive and ship outbound unit equipment and cargo.
- Provide readiness processing for returning Soldiers.
- Sustain returning forces and individuals until they arrive at final destination.
- Assist in obtaining transportation of all modes for movement of unit equipment.

The installation deployment support plan should include measures to ensure the quality of its services as it prepares to receive redeploying units. Effective redeployment support at the installation level means transportation is coordinated, Unit equipment is clean and in proper shipping configuration, HAZMAT identified and documentation completed to include unit movement documentation. The ultimate challenge of installations is to maintain unit integrity to ensure the successful return of a unit's equipment to aerial and sea ports of debarkations and ultimately its final destination.

Installation Deployment Support Plan (IDSP) General Guidance and Instructions

The following installation deployment support plan guidance, instructions for the exercise program and a sample template follows:

1. Senior commanders are responsible to ensure installations provide the support necessary for units to deploy off their installation within the timeframe required to meet Combatant Commander requirements. This will require integration of installation support staff from all organizations with deployment support responsibilities.

2. The IDSP, along with published local procedures, provides guidance to installation organizations that support deploying and redeploying units. The Installation Directorate of Plans, Training, Mobilization and Security is responsible for integrating across installation organizations to create and exercise the IDSP.

3. For standardization, the IDSP should follow the five-paragraph operations order per FM 6-0, Commander and Staff Organization and Operations, with the modified annexes listed in table F-1 on page F-8 (ATP 3-35). The five required paragraphs are: situation, mission, execution, sustainment, and command and signal.

4. Installations differ in geographical location, infrastructure and types of deployable units. The IDSP should include installation specific capabilities, assumptions, limitations, and other relevant factors to describe policies, processes and procedures used to receive and deploy units. Installations with Reserve Component mobilization missions should use Annex F in table F-1 on page F-8 (ATP 3-35) to integrate installation deployment and mobilization support plans.

Deployment Planning (JOPES & TPFDD)

Ref: ATP 3-35 (w/Chg 2), Army Deployment and Redeployment (Mar '15), p. 2-2.

Joint Operational Planning & Execution System (JOPES)

JOPES is the integrated, joint command and control system used to support military operational planning, execution, and monitoring activities. JOPES incorporates policies, procedures, personnel and systems, and underlying Global Command and Control System (GCCS) information technology support to provide senior-level decision-makers and their staffs with enhanced capability to plan and conduct joint operations. JOPES provides the mechanism to submit movement requirements to lift providers in the form of a TPFDD. The TPFDD is both a force and a transportation requirements document.

Time-Phased Force and Deployment Data (TPFDD)

TPFDD is the JOPES data-based portion of the operational plan; it contains time-phased force data, non-unit related cargo and personnel data, and movement data for the operational plan including:

- In-place units
- Units to be deployed to operational plan with a priority indicating the desired sequence for their arrival at the POD
- Routing of forces deployed
- Movement data associated with deploying forces
- Estimates of non-unit related cargo and personnel movements to be conducted concurrently with the deployment of forces
- Estimates of transportation requirements, which are fulfilled by common user lift resources, as well as those requirements that can be fulfilled by assigned or attached transportation resources

The geographic combatant commander begins preparation of COAs based on the tasking received from the Chairman, Joint Chiefs of Staff (CJCS). Combatant commanders receive tasking through the Joint Strategic Capabilities Plan for contingency planning or guidance through a CJCS warning order during crisis action planning. These directives establish command relationships, identify the task or mission, and provide any planning constraints. In addition, these directives will either identify forces and strategic mobility resources and establish tentative timing for execution, or request the supported combatant commander develop these factors.

Force requirements are initially identified in the planning process during mission analysis and course of action (COA) development. Force composition is derived from the troops apportioned for deployment and employment planning and the supported combatant commander's need for a particular unit capability to accomplish the mission.

COAs outline the scheme of employment and force requirements to accomplish the assigned mission. The Services monitor development of COAs and begin planning for support forces, sustainment, and mobilization, if required. As force requirements are identified, TPFDD development commences for each COA.

II. Movement

Ref: ATP 3-35 (w/Chg 2), Army Deployment and Redeployment (Mar '15), chap. 3.

I. Movement to the Port of Embarkation (POE)

Movement to the POE normally begins with receipt of an execute order from the Army Command; the order, along with any additional guidance is forwarded to the appropriate subordinate commands, deploying units, and installations. SDDC issues a port call message that identifies the date the unit must have their equipment at the seaport of embarkation (SPOE) to meet the ALD. The port call message or the operations order for sealift will also normally include details for vehicle reduction and preparation. AMC enters the aerial port of embarkation (APOE) and airflow scheduling information into Global Transportation Network (GTN). Scheduling information is also available in the JOPES. Based on port call messages and air schedules, the organization backward plans movements to the POEs. Deploying unit equipment normally moves from unit marshalling areas to a central staging area on the installation for further processing. The name, organization and responsibilities for these installation level staging or marshalling areas may differ; however, the functions performed to prepare units for movement are essentially the same.

Each installation has an associated strategic aerial port and/or seaport, but must be prepared to deploy from other ports as necessary. The proximity of the port facilities to the installation determines the type of movement and the numbers and types of assets required to complete the movement to the port. In some cases, the distance to APOE and SPOE is short, allowing units to maximize the use of organic equipment and convoys. In other cases, the distance to the APOE or SPOE is longer; in which case, units may have to rely heavily on commercial road and/or rail transport to complete the move to the port.

Convoy Operations

In some cases, units convoy their vehicles and equipment to the POE. A convoy is a group of vehicles organized for the purpose of control and orderly movement with or without escort protection that moves over the same route at the same time and under one commander. (JP 1-02) Vehicles in a convoy are organized into groups to facilitate command, control, and security and normally move at the same rate.

Refer to ATP 4-01.45, TCO Multi-Service Tactics, Techniques, and Procedures for Tactical Convoy Operations, for specific guidance on planning and conducting convoy operations.

Per the Defense Transportation Regulation (DTR) 4500.9-R, Part III, Appendix F, State Adjutants General establish the State Movement Control Center (SMCC) and appoint a Defense Movement Coordinator (DMC), who is located at the SMCC. The SMCC and the DMC receive and approve convoy movements on public highways. The DMC ensures convoy movements conform to federal, state, and local laws. Units with convoy requirements will submit a Department of Defense form (DD Form) 1265 (Request for Convoy Clearance) through the installation UMC. The unit will also submit a DD Form 1266 (Request for Special Hauling Permit) if required. The UMC will either fax or email the forms to the SMCC in their state (where the convoy is to start). The SMCC enters the request for a convoy into TC-AIMS II, which deconflicts the request on the national server. TC-AIMS II provides an approved convoy movement order which the SMCC will electronically provide to the unit.

Procedures for processing convoy clearances (including the number of vehicles that constitute a convoy) are established by theater policy, standardization agreement, or the host nation support agreement

Rail Operations

Responsibility for planning and executing rail movements of vehicles and equipment is split between the deploying units and the installation transportation office (ITO). The deploying unit—

- Determines its movement requirements and submits them to the ITO
- Prepares their equipment for rail loading
- Load railcars and chock, block and tie down equipment under the technical supervision of the ITO, who is ultimately responsible for approving all rail loads

The ITO is responsible to--

- Obtain rail cars based on deploying unit requirements
- Validate railcar requirements based on unit rail load plans
- Maximize the use of the available rail assets
- Serve as the official liaison with SDDC and the railway agent and inspect all railcars for serviceability before units begin loading.

The movement control team (MCT) performs the ITO functions in OCONUS locations and obtains the rail cars, validates railcar requirements, serves as the liaison with the railway agent, and inspects the railcars before the units begin loading.

Movement of Passengers

Once the equipment and material are moved to the POE the movement of troops is addressed. Most troops move long distances by air and are configured as advance party, main body, and trail party. In addition, some troops move with the equipment to provide security, property accountability, and assist in reception activities.

The unit prepares personnel rosters for each chalk or plane load. The UMO requests busses and trucks to move the personnel and their baggage from the unit area to the A/DACG. As the personnel arrive at the A/DACG manifests are prepared by the A/DACG and the personnel information verified by checking the CAC (common access card). The baggage is palletized and moved to the ready line. The Soldiers remain in the sterile area until the chalk is called forward to load and the AF moves the chalk to the aircraft for loading.

II. Movement to the Port of Debarkation (POD)

The combination of strategic airlift, sealift, and prepositioned equipment, referred to as the Strategic Mobility Triad, provides the capability to respond to contingencies. Each element of strategic lift has its own unique advantages and disadvantages. In general, airlift transports light, high priority forces and passengers required to rapidly form units with prepositioned equipment and supplies. Airlift is fast and fairly flexible but has limited capacity; it is also expensive and depends on airfield availability. On the other hand, sealift is slower (compared to airlift) and has limited flexibility; however, it is cheaper and has much greater capacity. Sealift is also dependent on port availability or assets for in-stream discharge. Afloat prepositioning of unit equipment reduces the reaction time required to move the force packages to the JOA. Forward stationing of Army watercraft reduces the sail time required to link up with arriving vessels and begin intra-theater water transportation operations. The disadvantages associated with prepositioning are the high costs associated with the periodic offload of vessels and the maintenance of equipment; also there is a risk that the forward deployed assets may not be close to the contingency.

Strategic Mobility Triad

Ref: ATP 3-35, Army Deployment and Redeployment (Mar '15), fig. 3-3, p. 3-7.

Strategic Airlift

During the early stages of a deployment strategic airlift is the primary means of moving forces and remains so until the sea line of communication is established. Strategic airlift assets are provided by Air Mobility Command and include both military aircraft and commercial aircraft activated as required under the Civil Reserve Air Fleet program.

Refer to JP 3-17 for more information about air mobility.

Strategic Sealift

Strategic sealift normally moves the majority of unit equipment identified for deployment. Strategic sealift assets are provided by Military Sealift Command and include both military and commercial vessels. Additional capacity can be mobilized under the Voluntary Intermodal Sealift Agreement which is a partnership between government and industry to provide commercial sealift and intermodal shipping services.

Refer to JP 4-01.2 for more information on sealift support.

Commercial Liner Service

In many cases unit equipment is moved by commercial liner service operating over scheduled routes on a regular basis. The carrier often picks up the equipment, moves it to the POE, and loads it aboard the ship, reducing the requirement for DOD transportation assets. In that most liner service vessels are container ships the equipment being moved must be containerized or loaded on flat racks. In-transit visibility of unit equipment moved by liner services is an issue that is being addressed by the Joint Deployment and Distribution Enterprise.

Army Prepositioned Stocks (APS)

If sited appropriately, Army prepositioned stocks reduce the amount of strategic lift required to support a rapid buildup of forces to demonstrate US resolve. Forward stationing of Army watercraft in a combatant command AOR reduces the sailing times required to link up with afloat Army prepositioned stocks (APS). Army watercraft are also co-located with land-based APS to reduce movement times if these stocks need to be shifted in a theater.

Refer to FM 3-35.1 for more information on Army prepositioned operations.

III. Seaport of Embarkation (SPOE)

Ref: ATP 3-35 (w/Chg 2), Army Deployment and Redeployment (Mar '15), pp. 3-2 to 3-4. *See p. 7-28 for discussion of SPOE responsibilities.*

There are essential activities that occur at the SPOE during deployment operations as units prepare for shipment by strategic sealift. The tasks are performed by a number of DOD and Army units and ad hoc organizations.

Notional Seaport of Embarkation

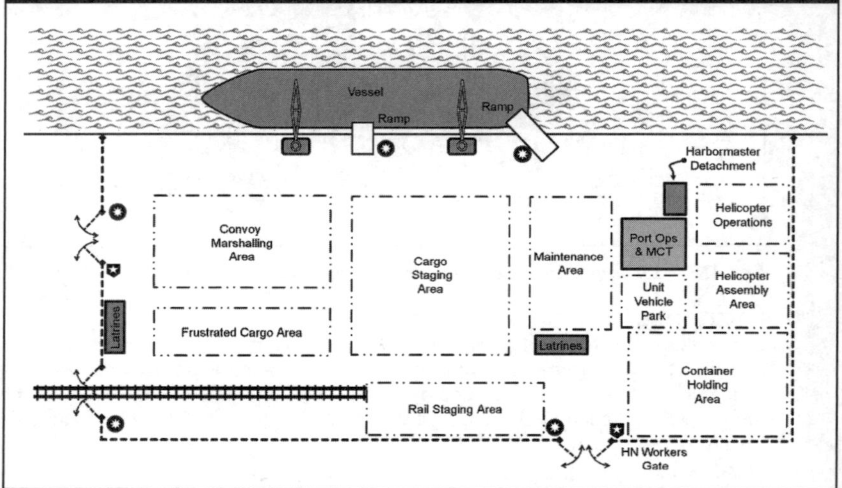

Ref: ATP3-35, Army Deployment and Redeployment (Mar '15), fig. 3-1, p. 3-3.

1. Marshalling Area

For movement to SPOEs deploying units and equipment may use an en route marshalling area. These areas are ideally located near the port staging area and in the immediate vicinity of rail and truck discharge sites. The SPOE marshalling area is the final en route location for preparation of unit equipment for overseas movement prior to the equipment entering the port staging area. Establishment of a marshalling area reduces congestion within the terminal area and provides space for sorting vehicles for vessel loading. The layout of a marshalling area is not fixed but is contingent on available space and needs of the unit. Equipment arriving in the marshalling area is normally segregated in accordance with the vessel stow plan.

2. Staging Area

The equipment is moved from the marshalling area to the staging area based on the vessel work plans and as directed by the port commander. The SDDC port commander assumes custody of the cargo at this point. Activities within the area include equipment inspection for serviceability, packing lists and load card, accuracy of dimensions and weights, properly secured secondary loads, and documentation of any cargo requiring special handling. Military shipment labels affixed to equipment will be scanned using bar code readers. The data will then be loaded into the Global Air Transportation Execution System (GATES). GATES has a module to produce ship manifests and serve as the basis for status reports. Additionally, GATES feeds data to IGC.

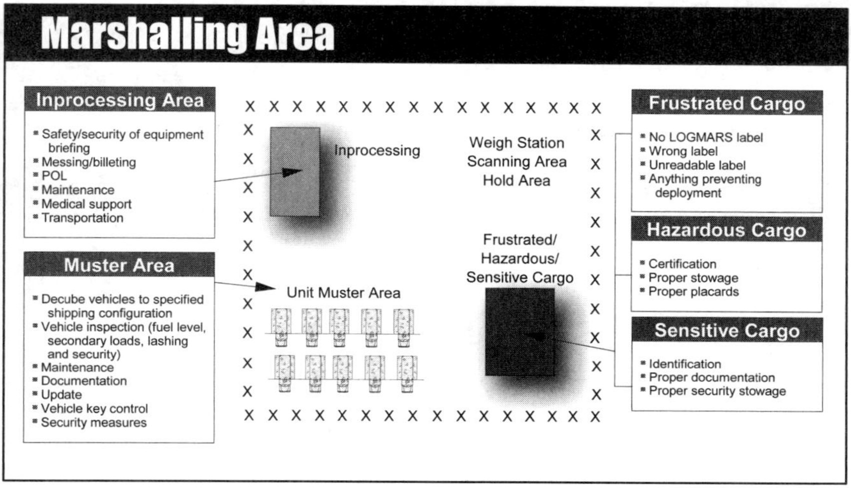

Ref: Adapted from FM 4-01.011, Unit Movement Operations, fig. 4-1, p. 4-4.

The **port support activity (PSA)** is a flexible support organization designed to assist SDDC with the loading of equipment at seaports. SDDC provides PSA capability through Stevedore and Related Terminal Services contracts. SDDC will coordinate with IMCOM, FORSCOM, and supporting Army Service Component Command (ASCC) for requirement outside of SDDC capabilities. SDDC also has the capability to assist deploying units with documentation, ITV, and vehicle inspection.

When processed, equipment may be segregated into different lots within the staging area by type, size, and any other special considerations such as hazardous materials, sensitive and classified items, and containerized equipment. From the staging area, vehicles are called forward to load the ship based on the stow plan and call forward schedules.

Additional information on seaport roles and responsibilities can be found in ATP 4-13 Army Expeditionary Intermodal Operations.

3. Supercargo

Supercargoes are unit personnel designated on orders to accompany, secure, and maintain unit cargo on board ships. They perform liaison during cargo reception at the SPOE, vessel loading and discharge operations, and seaport of debarkation (SPOD) port clearance operations. The supercargoes are attached to the port operator and remain with the port manager at the SPOD until the offload is complete and they are released back to their units.

Deploying unit commanders recommend the composition of supercargoes based on several factors including the amount and types of equipment loaded aboard the ship and the number of units with equipment on the ship. Military Sealift Command (MSC) determines the actual number of supercargo personnel permitted onboard, based on the berthing capacity of the ship.

IV. Aerial Port of Embarkation (APOE)

Ref: ATP 3-35 (w/Chg 2), Army Deployment and Redeployment (Mar '15), pp. 3-4 to 3-6.

The APOE is the transition point for Army units deploying by air. There are four distinct areas—marshalling area, alert holding area, call forward area, and loading ramp area associated with an APOE as shown below.

Notional Aerial Port of Embarkation

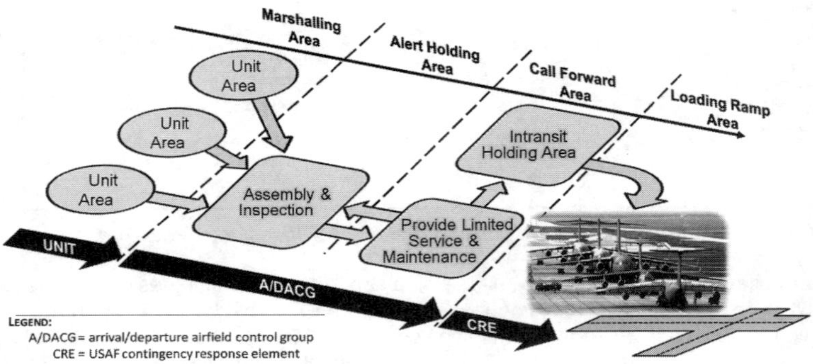

Ref: ATP 3-35, Army Deployment and Reployment (Mar '15), fig. 3-2, p. 3-5.

Preparation for air movement begins with receipt of the mission directive or order and continues through the planning phase until execution. A series of local joint conferences is required during the planning phase for close coordination and to ensure a clear understanding of responsibilities. As a minimum, a joint planning conference will be held as soon as possible after receipt of the air movement order or directive. A final coordination conference will be held immediately before the move. Participating elements should be represented at these conferences by key personnel who can make decisions for their organization. These conferences do not rule out the need for continuous coordination throughout the planning cycle. The GCC or representative will conduct a final joint coordination meeting with the representative of the deploying unit, the A/DACG, and the CRE. The CRE is a deployed Air Mobility Command organization that provides on-site management of airfield operations. At this meeting, the deploying unit, A/DACG, and CRE will present planning status and identify any problems. Air movement requirements will be based upon unit movement data (UMD) and provided to FORSCOM or ASCC in accordance with published command guidance or as otherwise specified by the movement order or tasking directive. FORSCOM or ASCC consolidates and provides these refined lift requirements to the supported combatant commander and USTRANSCOM. USTRANSCOM uses the unit movement requirements to generate airlift schedules that are loaded in JOPES which feeds to IGC and COMPASS.

See following page (p. 9-28) for further discussion of the Contingency Response Element (CRE) and Arrival/Departure Airfield Control Group (A/DACG).

Normally there will be Army and USAF personnel working together to process, inspect, document, and load the personnel and equipment. At some locations, the Army must carry out all of the APOE functions and in those cases the Army personnel are trained and certified to fulfill the USAF roles.

The duties of the departure airfield control group (DACG) and the CRE are discussed in ATP 3-35, Appendix G.

1. Marshalling Area

The primary purpose of the marshalling area is to provide a location near the APOE to assemble personnel, supplies, and equipment and make final preparations for air shipment. Unit marshalling areas are used to receive convoys and process vehicles before they are staged for loading. Marshalling areas is the responsibility of the deploying commander, assisted by the ITO, supporting units, or other designated organizations; operations are based on local policy and standing operating procedures.

The deploying unit:

- Establishes liaison with the DACG
- Coordinates a joint planning conference with the DACG and CRG to discuss aircraft allowable cabin load, pallet restrictions, aircraft configuration, equipment preparation requirements, airflow schedule, and any other issues impacting deploying unit preparation and processing
- Prepares vehicles and equipment
- Ensures adequate shoring material is available
- Prepares personnel and cargo manifests
- Assembles personnel, supplies, and equipment into aircraft loads
- Ensures planeload commanders are appointed and briefed
- Provides escorts for sensitive items
- Builds 463L pallets.

2. Alert Holding Area

The alert holding area is the equipment, vehicle, and passenger control area. It is normally located in the vicinity of the departure airfield and is used to assemble, inspect, hold, and service aircraft loads. Control of loads is transferred from the individual unit to the A/DACG at this point.

The deploying unit:

- Ensures the aircraft loads arrive on time
- Provides manifests to the DACG
- Corrects load discrepancies identified during pre-inspection
- Ensures vehicle drivers remain with the vehicles until released
- Passes control of aircraft loads to DACG

3. Call Forward Area

The call forward area is the joint responsibility of the CRE and the DACG and is the location for the joint inspection of deploying unit equipment and cargo. The DACG, the deploying unit, and the CRE conduct the inspection. They complete a DD Form 2133 (Joint Airlift Inspection Record) to indicate to the loadmaster that it has completed the required inspection. Deficiencies are corrected by the unit and rechecked by the inspection team. Once the inspection sequence is complete, the deploying unit arranges its vehicles, pallets, and equipment into load or chalk sequence. A final briefing is provided to deploying troops and the CRE reviews all manifests for accuracy.

4. Loading Ramp Area

The loading ramp area, including ready line area, is controlled by the CRG. At this point, control of units for movement purposes passes to AMC.

The chalk commander:

- Follows directions of load team chief
- Monitors and controls aircraft passengers; retains copy of the final manifests
- Provides assistance in loading and securing the load as required
- Ensures vehicle/equipment operators follow instructions

The load team:

- Receives loads at the ready line
- Loads and secures vehicles and equipment under supervision of the loadmaster
- Provides the loadmaster with manifests
- Informs CRE of load completion time

V. Arrival/Departure Airfield Control Group (A/DACG) Responsibilities

Ref: ATP 3-35 (w/Chg 2), Army Deployment and Redeployment (Mar '15), app. G.

Contingency Response Element (CRE)

The USAF normally exercises overall control of airlift and loading/unloading operations at the departure and arrival airfields. Airlift resources will at all times, remain under the operational control of the USAF. The USAF will establish a CRE at both departure and arrival airfields. The CRE provides the commander, command and control staff, and communications required to support AMC's worldwide refueling and airlift operations. The CRE may also include additional contingency support elements such as aircraft maintenance, aerial terminal, weather, intelligence, flight surgeon, etc. These contingency support elements are under the direct command of the CRE commander and are organizationally subordinate to the CRE.

The A/DACG will coordinate and control loading and offloading of units for deployment or redeployment.

Arrival/Departure Airfield Control Group (A/DACG)

The A/DACG is an ad hoc Army organization established to control and support arrival at the APOD and departure from the APOE. Elements of a movement control team and an inland cargo transfer company typically operate the A/DACG however the mission can be performed by almost any unit with the properly trained personnel and equipment. The organizational elements of an A/DACG, command, operations, joint inspection, and loading/unloading, remain constant but the size and capabilities are mission dependent. In CONUS the A/DACG will normally work for the DOL however OCONUS a sustainment brigade may be assigned the responsibility to oversee the establishment and operation of the A/DACG. As a minimum the following steps should take place:

- Exercise overall operational planning, supervision and management of the A/DACG activities within installation as pertains to safety, training and operations
- Develop a SOP outlining safety, training, and operations procedures
- Ensure all civilian personnel involved in A/DACG operations are sufficiently manned, trained and funded for mission accomplishment. Ensure A/DACG personnel receive annual training
- Ensure all A/DACG personnel with a need to drive on the flight line are trained and certified by the appropriate authority. This certification should be annotated on the individual's U.S. Government Motor Vehicle Operator's ID Card, OF 346.
- Establish and provide guidance on security requirements
- Brief all personnel engaged in A/DACG operations
- Establish communications
- Obtain parking and flow plan from the mobility force
- Coordinate MHE with the mobility force
- Ensure sufficient loading team personnel and vehicles are available to accomplish the mission
- Advise the deploying units of the airflow and expected arrival of aircraft
- Receive passenger and cargo manifests from the loadmaster
- Supervise off-loading the aircraft (normally accomplished by the aerial port squadron), including removal of shoring and dunnage
- Ensure communications between CRE and A/DACG and all functional areas of CRE
- Inform unit of any change in operations

III. Reception, Staging, Onward Mvmt, Integration

Ref: ATP 3-35 (w/Chg 2), Army Deployment and Redeployment (Mar '15), chap. 4.

The purpose of RSOI is to build the combat power necessary to support the CCDR's concept of operation. Force closure is that point which the combatant commander determines that an adequate combat-ready force is available. Force closure requires well-defined criteria by which unit commanders can measure their readiness. Assessment of combat power begins with established standards for readiness and is based on unit capability, rather than simple tallies of vehicles and weapon systems on hand. Readiness and reporting are inherently operational matters, normally handled through operations channels.

RSOI Segments

- I Reception
- II Staging
- III Onward Movement
- IV Integration

Ref: ATP 3-35, Army Deployment and Redeployment, chap. 4.

Reception is the unloading of personnel and equipment from strategic transport, marshaling them, transporting them to staging areas, and if required, providing life support services.

Staging is the assembling, holding, and organizing arriving of personnel, equipment, and basic loads into units; preparing the units for onward movement; and providing life support until the unit becomes self-sustaining.

Onward Movement is moving units from reception facilities and staging areas to TAAs or other theater destinations; placing arriving nonunit personnel to gaining commands; and providing sustainment to distribution sites.

Integration is the synchronized transfer of authority of units to a designated component or functional commander for employment in the theater of operations.

RSOI Infrastructure

RSOI operations are the responsibility of the GCC and his designated headquarters, generally a theater sustainment command (TSC). The TSC controls the physical facilities and collaborates with the advanced echelon of the arriving headquarters to establish the throughput rate it can handle. The deploying forces have a responsibility for their own security, organization, and movement through the RSOI process to the extent possible. The process is supported by outside entities such as host nation operators and contractors.. The complex RSOI system is usually composed of several elements, each contributing to the process—

- In-place command and control forces
- Advanced echelon of the deploying units
- Deploying forces
- Host nation and multinational support elements
- Contractor support
- Army prepositioned stocks

The RSOI infrastructure also includes some of the theater's distribution nodes. Nodes are a location in a distribution system where a movement requirement is originated, processed for onward movement, or terminated.

RSOI Execution

RSOI effectiveness is dependent upon proper TPFDD development. For example, the combatant commander places rapid port clearance capabilities early in the TPFDD and coordinates personnel and equipment flows on the TPFDD so they can be united without delay at ports or staging areas. Decisions on force mix and sequence are critical, because adjustments after deployments begin become difficult to implement. Moreover, changes cause ripple effects and may seriously disrupt the flow of forces into the JOA.

Communication is necessary at all levels, and across all modes and nodes. The communication system must link the combatant commander, the supporting combatant commands, the deploying units, the RSOI providers, and the tactical commanders who will integrate the deploying force into their structures. Mission, enemy, terrain and weather, troops and support available-time available and civilian considerations (METT-TC) may cause certain units to be in high demand or necessary for immediate employment. Critical resources like heavy equipment transporters, fuel support, or buses to move personnel may have to be diverted to rapidly move these units. Secure, assured, compatible, and reliable means of relaying deployment data is essential.

RSOI Coordination

The Army operates in diverse environments and conducts a variety of operations as part of joint, multinational, or interagency teams. This fact increases the difficulty of RSOI and reaffirms the need for established procedures, mutually understood relationships, and robust liaison. Army commanders need to understand how best to integrate their forces into the various organizations under which they will operate.

- **Joint**. Joint integration of planning and execution is key to successful RSOI.
- **Multinational**. Major differences in logistics doctrine, mobility, resources, interoperability, and language create problems in coordinating the use of highways, rail lines, seaports, and airfields, as well as providing support and services for multinational RSOI operations.
- **Host Nation Support.** Host nation support is civil and military assistance rendered by a nation to foreign forces within its territory during peacetime, crises or emergencies, or war. This assistance is normally outlined in host nation support agreements mutually concluded between nations.
- **Interagency**. In the course of joint and multinational operations, the Army operates alongside US and non-US government agencies, non-governmental agencies. To promote unity of effort and assess the impact of these agencies and organizations on the RSOI effort, the JFC can establish a civil-military operations center.
- **Liaison**. Liaison with forces of each Service, nation, and higher and adjacent headquarters is a prerequisite for effective RSOI and timely transfer of critical information.

Principles of RSOI

Ref: ATP 3-35 (w/Chg 2), Army Deployment and Redeployment (Mar '15), pp. 4-2 to 4-3.

1. Unity of Command

One commander should control and operate the RSOI process - adjusting resources based upon deployment flows, controlling movements in the area of operations, and providing life support to arriving personnel.

2. Synchronization

Synchronization occurs when the right units, equipment, supplies, and capabilities arrive in the correct order at the appropriate locations, and supporting activities are coordinated to operate with one another to ensure the tempo of deployment is uninterrupted.

3. Unit integrity

Moving unit personnel and equipment on the same strategic lift platform provides distinct advantages for units and the force closure process. It leverages the strength of the chain of command, simplifies force tracking, and increases training opportunities. While it is impossible to put an entire armored battalion on one airplane, the increased sealift of the LMSR allows movement of all the battalion equipment on a single ship. Maintaining unit integrity during strategic lift can simplify the RSOI challenge of incrementally building combat power.

4. Balance

Defining the size of the required support structure required is essential to effectiveness. The goal is to avoid burdening strategic lift, infrastructure, and the commander with more support than is necessary, yet deploy minimum assets necessary to optimize throughput of units and materiel. Supporting assets must be deployed in a properly timed sequence to leverage their capabilities and may be increased to reduce vulnerability of the overall force. Increasing the RSOI capability to clear backlogs in ports and staging areas can be a tool to reduce force vulnerability.

Optimization

The commander's planning and operational dilemma is balancing the need for early deployment of combat forces against the requirement to deploy tailored logistical units that maximize throughput of sustainable combat forces. To resolve this dilemma, the commander must have the ability to see, understand, and balance the flow. The combatant commander defines force requirements in terms of size, location, and time while the TPFDD defines the force flow needed to meet these requirements. Knowledge of the RSOI infrastructure present in the theater, coupled with assets arriving via the TPFDD, is critical to understanding the flow.

The relationship between throughput volume and RSOI infrastructure is important to commanders trying to optimize force closure capacity. Accelerating the arrival of combat forces in the TAA requires an increased deployment of RSOI forces. Deploying additional RSOI forces costs space on strategic lift and requires additional positions in the TPFDD. The combatant commander applies the necessary command and control to ensure unity of command, and establishes communications for a seamless flow of information to manage and influence the force buildup.

The Army operates in diverse environments and conducts a variety of operations as part of joint, multinational, or interagency teams. This fact increases the difficulty of RSOI and reaffirms the need for established procedures, mutually understood relationships, and robust liaison. Army commanders need to understand how best to integrate their forces into the various organizations under which they will operate. Joint integration of planning and execution is key to successful RSOI.

I. Reception

As the initial step in introducing combat power, reception can determine success or failure of the RSOI operation. Reception from strategic lift is implemented at or near designated air and seaports of debarkation, normally under control of the GCC. It must be thoroughly planned and carefully executed. While the reception plan for each theater may vary, reception capacity should at least equal planned strategic lift delivery capability.

For the initial period of deployment, the aerial port is the lifeline to the front-line. All that is not pre- positioned or available from the host nation comes through the aerial terminal. Then the first surge of sealift ships begins to arrive, dramatically increasing forces. Airlift remains a critical element regarding delivery of personnel, but most unit equipment to build the combat power arrives through seaports.

Synchronizing transportation reception activities are critical to facilitating throughput at the ports of debarkation. They include command, staff oversight, movement control, and port operations.

Port Operations

The RSOI command and control headquarters must control the deployment flow so that reception capabilities are not overwhelmed. APODs and SPODs should be considered integral parts of a single reception complex, unless the distance separating them precludes mutual support. Reception capacity depends on—

- Port and airfield infrastructure, condition, and characteristics
- Availability of host nation labor and port services
- Off-loading and holding space
- Weather
- Enemy situation

II. Staging

Staging is that part of the RSOI operation that reassembles and reunites unit personnel with their equipment and schedules unit movement to the tactical assembly area, secures or uploads unit basic loads, and provides life support to personnel. These activities occur at multiple sites in controlled areas called ISBs that are required because space limitations normally preclude reassembly of combat units at seaports of debarkation. In general, there will be at least one intermediate staging base (ISB) for each SPOD/APOD pairing.

Theater with Mulitple Ports and ISB

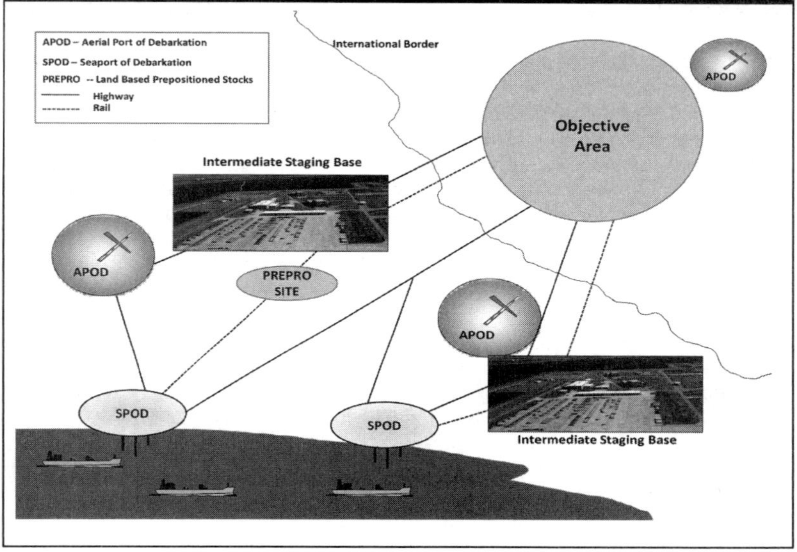

Ref: ATP 3-35, Army Deployment and Reployment, fig. 4-1, p. 4-8.

Intermediate Staging Base (ISB)

An ISB is a secure staging base established near but not in the area of operations. ISBs are temporary staging areas enroute to an operation and also may be used to sustain forces in the area of operations. ISB tasks and capabilities are contingent on the operational situation and are located where they can best support the force.

No two ISBs will be alike; some will be in operation for a few days while others will operate for an extended period. Although we refer to an ISB as if it were a single unit it is in reality a collection of brigade combat teams, sustainment, signal, military police, engineer, and support units brought together for a specific purpose. ISBs must deploy early to be prepared to receive deploying forces and to operate the nodes inherent in the theater distribution plan.

See p. 9-37 for discussion of ISB functions. Refer to DA Pamphlet 700-33 for guidance on establishing and operating an ISB.

Force Closure

In order to meet the force closure requirements, the time that units spend in the ISB must be minimized. Staging should not be a lengthy process, but inefficiencies can cause delays, for example, personnel arriving before their equipment, equipment arriving before its personnel, frustrated cargo, and gaps in matching troops with proper equipment. In fact, a battalion-sized unit should strive to spend no more than two days in the ISB.

ISBs should be located in areas convenient to both the SPOD and APOD, with good lines of communication back to ports of debarkation and forward to designated TAAs. In addition, the ISB should have sufficient space to accommodate the largest force scheduled to stage through it, together with facilities for vehicle marshalling, materiel handling, equipment maintenance and calibration, and possibly bore sighting and test firing of weapons. All of these are needed if the ISB is to fulfill its function of converting personnel and equipment into mission-ready combat units.

RSOI Port Selection

Ref: ATP 3-35 (w/Chg 2), Army Deployment and Redeployment (Mar '15), pp. 4-5 to 4-7.

Seaport and airfield throughput capacities significantly influence the speed, order, and, to a large extent, the types of units that can deploy through them. Consequently, before thought is given to actual deployment of forces, planners must evaluate available airfield and port facilities within the area of operations, as well as the transportation networks linking them with each other and to the interior. As was the case during Operation Desert Storm, it may be better to use a world class port hundreds of miles away from TAAs rather than conduct an in-stream discharge operation or use a smaller, degraded port facility with limited capacity and throughput. Diplomatic and military contacts should be made at the earliest possible opportunity with the host nation controlling key facilities and rights of way.

The combatant commander in conjunction with USTRANSCOM selects the PODs that will be used for deployment. METT-TC considerations and the theater transportation infrastructure will drive the sequence, type, size of forces, and materiel arriving at ports of debarkation. These decisions impact the speed of combat power buildup and continued development of the theater. Ports of debarkation may need improvement and repair to accommodate high throughput rates required for rapid force closure. Thus, the early entry of units such as cargo transfer companies, Army watercraft, causeway detachments, and engineer assets can be critical to off-loading materiel, clearing ports and consequently speeding deployment.

Aerial Port of Debarkation (APOD)

Deployment by air is often constrained by the capabilities of the arrival airport more often than a shortage of aircraft. Issues such as concurrent civilian use, competition for landing and takeoff slots, ramp space, number of aircraft on the ground at one time, and political restrictions limit its use to military aircraft. Consequently, maximum throughput at limited airports is paramount. The APOD is by its very nature a joint facility and will likely be a multinational facility. It is a POD for deploying forces, and a POE for forces moving to other theaters and noncombatant evacuation. The host nation may limit the APOD to coalition military use, or the military may be sharing the facility with commercial activities. Governmental, non-governmental, and private organizations will likely be competing for use of the APOD along with military forces.

The APOD serves as the primary port of entry for all deploying personnel, as well as for early entry forces normally airlifted into theater together with their equipment. USTRANSCOM through AMC is the DOD-designated single port manager (SPM) for all common user APODs worldwide. The SPM performs those functions necessary to support the strategic flow of forces and sustainment supplies through the APOD. The SPM is responsible for providing deployment status information to the supported CCDR and clearing the airfield in accordance with the CCDR's priorities. Responsibility for APOD functions is divided between the USAF and the Army, with the USAF responsible for airfield operations including air terminal control, loading, unloading, and servicing of aircraft. The Army is responsible for clearing personnel and cargo off the tarmac and for required logistics support for transiting units. USAF/Army interface occurs between the USAF contingency response group (CRG)/CRE and the Army arrival/departure airfield control groups (A/DACG) and MCTs.

Necessary communication, personnel, and cargo handling equipment must be in place to facilitate rapid movement out of the airport. Both the CRE and the A/DACG must be included in the lead elements of the deploying force. The CRE controls all activities at the off-load ramp area and supervises aircraft offloading. The A/DACG escorts loads and personnel to holding areas, thus clearing the airfield and ensures airfield operations and strategic airflow are not obstructed and limited due to the accumulation of cargo.

With responsibility divided between the USAF and the Army, and sometimes multinational forces, multiple chains of command exist within the aerial terminal, which may result in a variety of unforeseen challenges. Given this potential command relationship, potential for conflicting priorities necessitates careful planning and coordination during the reception process. For example, something straightforward as security responsibilities becomes complicated when there are two chains of commands at the same site. Special attention must be paid to ensure that airfield security (a USAF responsibility) and area security (an Army responsibility) are well coordinated among themselves as well as with multinational forces and the host nation.

Seaport of Debarkation (SPOD)

Activities at seaports are normally joint, multinational, and intermixed with commercial operations. Seaports can serve as ports of debarkation for arriving forces and simultaneously as ports of embarkation for forces deploying to other theaters of operations. USTRANSCOM through SDDC is the DOD-designated SPM for all common user ports worldwide. The SPM (normally a transportation terminal group) performs those functions necessary to support the strategic flow of the deploying forces' equipment and sustainment supplies through the SPOD. The SPM is responsible for providing strategic deployment status information to the GCC and to workload the SPOD port operator based on the GCC's priorities and guidance.

Theater planners must consider several factors when assessing the port's capacity to receive the planned strategic flow. Some of the factors include state of repair, commercial utilization, the port's capability, congestion, and throughput capacity. Throughput capability is based on the port's ability to receive, process, and clear personnel and equipment. The cargo reception function is based on the number and size of the berths, MHE and water depth; the cargo process function is based on staging area and the time it takes to marry units with their respective equipment; the cargo clearing function is based on truck and rail out loading facilities; gate capacity, and links to the theater transportation networks.

Seaport operations are similar to airport operations; once the vessels are off-loaded, unit equipment is moved to temporary holding areas within the port to be configured into convoys, rail loads, or watercraft loads. Unit equipment clearing the port moves to an intermediate staging base (ISB), an inland water terminal, or directly to the tactical assembly area.

The volume of cargo arriving in the theater in a small window of time can drive the need for multiple seaports to meet deployment timelines. The physical size of roll-on/roll-off ships and depth of water required to bring vessels of this class alongside a pier may also present a challenge. If world-class port facilities are available, off-loading can be rapidly accomplished. If facilities are less than world class or austere, then multiple ports and slower in-stream operations may be required.

The ability to receive forces in an operational area, despite degraded or austere ports is essential to the Army's force projection strategy. Army watercraft is the primary enabler in this process; they allow ships that cannot get to a pier to be off-loaded in-stream. Additional watercraft then moves the cargo to smaller coastal ports or directly over the shore. In-stream discharge operations are sensitive to weather and seas conditions and generally require a protected anchorage.

Refer to JP 4-01.6 for additional information on logistics-over-the-shore operations.

Other factors affecting selection of an ISB include geography and terrain and availability of organic and host nation assets. These factors, together with the size of the deploying force, may often necessitate multiple ISBs. The requirement for multiple staging bases is most evident in the urban sprawl of Europe and Korea, particularly around seaport facilities. In many cases, it is tremendously difficult to find even one square mile of open terrain much less the total space requirement for an ISB.

Under normal circumstances, troops deploy by air, while equipment deploys by sea and or rail. The speed differential between air and sea surface transportation is the fundamental cause of complexity and potential difficulties in the staging process. Troops and equipment must be sequenced in the TPFDD so that both arrive (nearly) simultaneously, expeditiously unite, and ready themselves for onward movement.

Troops arriving too early must be provided with meals and quarters while waiting for their equipment to arrive. The TSC or one of its sustainment brigades would be expected to accommodate these needs. Moreover, the mass of immobile, unprotected troops presents an inviting and vulnerable target. On the other hand, if equipment arrives much earlier than the troops, ports of debarkation can become congested, and space management becomes critical.

Early deployment of essential support units at the expense of combat units pays dividends later by speeding the flow of the entire force, enhancing the GCC's ability to build combat power and increase operational flexibility. Conversely, front loading the TPFDD with combat forces may jeopardize the GCC's ability to build up forces as rapidly as required and reduce his flexibility.

III. Onward Movement

Personnel and equipment reassembled as combat-ready units must be moved to the TAA based on the CCDR's priorities. Onward movement is a joint/multinational effort using capabilities and organizational structures of other Services, allies, host nation and other governmental entities. It is an iterative activity in which units advance from one LOC node to another. Onward movement occurs when units move from ports to theater staging bases or forward to the TAA. Three primary factors affecting onward movement are movement control, transportation infrastructure, and security.

Movement Control

Movement control is defined as planning, routing, scheduling, and control of forces and sustainment over lines of communication, while maintaining in-transit visibility and force tracking. This is not a passive activity. Successful movement control requires continual analysis of requirements, capabilities, shortfalls, alternatives, and enhancements. Bottlenecks within the theater must be identified and potential interruptions to the flow minimized. One of the biggest challenges of movement control is rapidly adjusting to changes in the operational environment and the commander's priorities. The challenge of a theater movements program is to merge the CCDR's concept of operations and priorities in a movement plan and execute them. This challenge can be met by employing an adequate number of movement control resources, appropriately enabled by communications, to anticipate and improvise. Efficient movement control enables the commander to redirect forces and rapidly overcome disruptions in the LOC.

See pp. 3-16 to 3-17 and p. 7-30 for further discussion. Refer to ATP 4-16, Movement Control, tor more information on movement control.

Transportation Infrastructure

The transportation infrastructure routes, control factors, host nation support, and specialized handling requirements must be coordinated to maximize speed of movement. Capabilities of the transportation network must be balanced against movement requirements, so that modes and routes are neither saturated nor under used.

Intermediate Staging Base (ISB) Functions

Ref: ATP 3-35 (FM 3-35), Army Deployment and Redeployment (Mar '15), pp. 4-8 to 4-9.

An ISB is a secure staging base established near but not in the area of operations. ISBs are temporary staging areas enroute to an operation and also may be used to sustain forces in the area of operations. Functions occurring at an ISB may vary widely depending on the specifics of the operation, but the following are some critical functions that should be considered for every ISB. *See related discussion on p. 1-29.*

Communications

Reliable, secure, and compatible communications are essential to operations in the theater staging base. The GCC must know when forces are combat-capable and prepared for onward movement to give him the capability to control and employ these forces at the decisive point and time.

Force tracking provides situational awareness of combat-ready units within the operational area. While in transit visibility begins at home station, the process of force tracking begins in the staging area, where equipment and personnel are reassembled into combat-ready units. Staging operations must have the communications, data processing equipment, and personnel assets to provide and manage force tracking data. Efficient movement control can provide force tracking information but it must be able to communicate directly with operational commanders.

ITV acts as a staging enabler by providing commanders with clear pictures of locations of units and materiel in RSOI and deployment. For the ISB commander, ITV provides an awareness of the scheduled arrival of personnel and equipment, so the resources required to support them, as well as time required to assemble the unit in a mission-ready configuration, are available.

Life Support

Regardless of time actually spent in the ISB, troops staging through it will require support, including housing, sustenance, sanitation, and health care. RSOI planners must ensure that the force provider units are sequenced early in the TPFDD to be in place and functioning by the time the first units arrive. Even if this requires displacement of some combat capability, it pays dividends later in the operation in the form of higher throughput, faster buildup of combat power, and earlier force closure. The Army Force Provider units, each designed to provide base camp support to 550 people, as well as USAF's Prime Beef and Prime Rib units, are viable options for providing support to transient forces.

Arming, Fueling, and Fixing

Equipment arriving at the ISB may require maintenance before it becomes combat ready. This includes calibration of equipment, bore sighting of weapons, replacement of parts damaged in transit, painting, and re-fueling. The ISB should provide adequate facilities to support these activities, including marshalling areas, maintenance shelters, fuel and ammunitions storage, a test-driving loop, and range areas.

Preparation of Units for Onward Movement

In addition to preparing equipment, units at the ISB undergo training and reorganization. Communications networks are established, vehicle loads are reconfigured and RFID tags are updated so that tracking systems allow senior commanders to monitor the buildup of combat power. Commanders must participate in planning the onward movement including route planning, unit tracking, and movement control.

Security

ISBs are high-value targets. Their destruction or damage results in serious delays in force closure and disruption of the GCC's concept of operations. Maintaining the flow of forces through the ISB can be the best means of reducing their vulnerability to attack.

The other services and allied forces will be competing for the same networks as the Army and congestion will result if proper coordination is not accomplished. Planners should anticipate simultaneous demands on limited infrastructure, difficulties with communications, and differences in transportation capabilities.

During onward movement, mode selection determines whether the commander of the unit in transit maintains control during the move or whether control of the move is exercised by elements of the TSC. Ideally, tracked vehicles should be moved by rail or heavy equipment transporters and wheeled vehicles should move in a convoy. Inland and coastal waterways should be used when available if they afford useful solutions.

Establishment of convoy support centers and trailer transfer points along main supply routes and other support centers at temporary airfields, rail sites and waterway drop off points, further aids onward movement. These allow units and line haul drivers to rest, eat, perform vehicle maintenance, and contact unit/movement control personnel to receive operational updates, revised priorities, and when necessary diversions.

Security

The onward movement phase can provide the enemy with numerous opportunities to inflict serious losses and delay the build-up of combat power by exploiting vulnerability of units in transit from the ISB to the TAA. Security consists of those actions taken by the unit to protect it against all acts designed to destroy or impair its effectiveness.

Enemy interdiction of onward movement presents special challenges that can be partially overcome by using alternative routing and mode substitution when feasible but all units must be prepared to defend themselves. Organized and tightly controlled convoys afford a higher degree of security. Moreover, hardened gun trucks may escort the convoys and additional armed personnel ride in the vehicles to immediately engage threats as required.

Security is an important component of warfighting. Security is the responsibility of the moving force itself and forces of the GCC. Care must be taken to avoid or neutralize explosive devices and attacks on the movement with direct or indirect fires.

IV. Integration

During integration, combat-ready units are transferred to the operational commander and merged into the tactical plan. The transfer may require interaction and familiarization among units and that arriving units meet certain standards before being completely integrated into the combat plan. Consequently, requirements for integration planning and coordination must occur early in the force projection process and modified according to METT-TC until force closure is achieved.

The time required for integration may vary, depending upon the size of the total force, contingency conditions, and amount of predeployment and ongoing planning and coordination. Rapid integration, however, is critical to the success of combat operations, and adequate planning and coordination can reduce integration time.

Thorough integration has to be completed before a unit is operational and can perform its mission. Integration is complete when the CCDR establishes positive command and control over the arriving unit, usually in the TAA, and the unit is capable of performing its assigned mission.

Control measures, such as liaison officers or movement control teams can reduce confusion between integrating units, RSOI forces, and receiving headquarters. These measures act as guardians of the commander's intent and focus effort on force integration. These measures should be established immediately as part of the planning process and be maintained throughout the RSOI process.

IV. Redeployment

Ref: ATP 3-35 (w/Chg 2), Army Deployment and Redeployment (Mar '15), chap. 5.

Redeployment involves the return of personnel, equipment, and materiel to home and/or demobilization stations and is considered as an operational movement critical in reestablishing force readiness. The GCC defines the conditions for redeployment. The same elements that operate and manage the theater distribution system during deployment and sustainment will usually perform support roles during redeployment. Redeployment planning is an integral part of employment planning and should be coordinated with mission termination or transition plans.

I. Redeployment Planning

Commanders plan for redeployment within the context of the overall situation in the theater. The phases of redeployment are redeployment planning, pre-redeployment activities, movement, and reception and integration at the home or demobilization station.

Redeployment planning by the theater Army, ASCC, or Army Forces Commander normally precedes the actual issuance of an order and tentatively outlines information about the support network, follow-on operations, security requirements, and movement limitations imposed by infrastructure and resources. Redeployment operations must be conducted at a pace that does not disrupt the ability of the GCC to execute continuing missions, including deployment of replacement forces.

There are several considerations that should be accounted for during redeployment planning. If the redeploying unit is being replaced by another unit, then time may be required to effect the change in responsibilities. The possible presence of multiple units in one space will have an impact on space, support, and services requirements.

The ASCC or Army Forces Commander is responsible for planning, organizing, and coordinating theater redeployments by providing guidance, facilities, training, and personnel to assist redeploying units.

The redeploying unit's plan is nested within the plans (IDSP or equivalent) of their higher headquarters. The redeployment plan conveys the commander's intent and includes responsibilities,priorities, and guidance for movement of forces, individuals, and materiel. Issues that must be addressed in the plan are——

- Scheduling of redeployment activities
- Personnel accountability
- Cleaning of equipment
- Transfer of equipment
- Ammunition turn-in
- APS procedures
- Security of the force
- Availability of theater transportation assets
- Availability of strategic lift

II. Preredeployment Activities

Ref: ATP 3-35 (FM 3-35), Army Deployment and Redeployment (w/Chg 2), pp. 5-2 to 5-4.

When a unit is identified for redeployment, the GCC issues a redeployment operations order releasing units from their missions and authorizing movement. Redeploying forces move to designated assembly areas. Redeployment operations at the assembly areas are under the control and supervision of the TSC commander and include actions necessary to prepare the unit for movement. In some cases, a fragmentary order to a deployment order may be used instead of a separate redeployment order.

Redeployment priorities are outlined in the OPLAN by the supported GCC. During redeployment preparation units must update movement data to reflect gains and losses to the OEL. The changes are normally attributed to combat losses, maintenance, or supply. Moreover, the redeploying force is directed to leave equipment and materiel behind for use by a follow on force or by host nation or other multinational forces. Subordinate organizations and component commands must verify unit movement data to the supported combatant commander. USTRANSCOM develops the redeployment strategic movement schedule after receiving the validated requirements from the supported combatant commander.

A movement order may be issued sequentially for each movement or may be contained in one movement order designating the timing and means of transport to the POE. The theater movement control element issues movement tables that give detailed movement instructions to redeploying units. The TSC usually manages the redeployment support that can be performed by a subordinate organization.

The unit begins the redeployment process by identifying requirements and determining current unit status. Other actions include—

- Submitting personnel and pay actions.
- Maintaining personnel accountability.
- Conducting medical screening.
- Performing equipment checks and services.
- Conducting equipment inventory.
- Refining the UDL.

The redeploying unit conducts the following pre-redeployment activities in the second phase of the redeployment process—

- Redeployment site surveys of tactical assembly areas, APOE, A/DACG, wash racks, and marshalling areas.
- Receive equipment disposition instructions (theater provided equipment and/or transfer to USAMC for induction into RESET).
- Meet with theater redeployment organization to finalize Redeployment, theater provided equipment list, destination DOD Activity Address Code, address and phone numbers in accordance with theater policy.
- Clear all Life Cycle Management Command, DOD Activity Address Codes which will no longer be used, and/or Property Book loans.
- Meet with theater redeployment organization to update UDL and enter JOPES level VI data in TC-AIMS in accordance with theater policy.
- Validate UMD accuracy prior to submission to FORSCOM or ASCC, as appropriate.
- Provide UDL and DD Form 1384 in accordance with theater policy.
- Plan and arrange unit movement from forward operating locations to Tactical Assembly Areas and onward movement to POEs.
- Arrange and coordinate for customs inspection of unit equipment prior to loading into boxes, crates, containers, sea vans, or similar receptacles.
- Burn RFID tags and print MSLs for equipment, vehicles and containers in accordance with The DTR and theater policy.

- Apply RFID tags and MSLs to vehicles and containers IAW DTR and theater policy.
- Determine requirements for blocking and bracing materials at POD rail movement.
- Determine Passenger Travel & Baggage Movement requirements to the APOE and submit movement requests.
- Determine Personnel Requirements for assisting at Port Support Activity POE and at A/DACG
- Document and mark all hazardous material.
- Obtain a separate ULN for sensitive unit equipment redeploying separate
- Inventory and account for contents packed in containers.

Movement

Redeployment planning results in a network of transit areas designed to efficiently move forces from their area of operations to their final destinations. Use of these areas may vary with the situation. The redeploying unit will need to establish a redeployment command post at the tactical assembly area to track the flow of personnel and equipment as they move from the tactical assembly area and onward to SPOE or APOE to include the management of departing personnel and the various teams needed to support operations at the SPOE and APOE. Other responsibilities of the redeploying unit are-

- Submittal of departure and arrival reports through the Chain of Command within one hour of wheels up or down for airlift and, as required, for surface modes.
- Track air missions that are allocated to ULNs.
- Meet Port Call windows.
- Maintain communications with the SPOE to ensure convoy and rail departure and arrival times are known.
- Provide liaison between the seaport and redeploying units.
- Provide support personnel to assist PSA if requested.
- Track redeploying unit equipment from the redeployment Port of Embarkation to Port of Debarkation, to installation, and onto home station.

The ASCC or Army Forces Commander is responsible for the following activities of subordinate and supported redeploying units during the movement phase-

- Supervise the execution of the redeployment plan.
- Operate theater redeployment transportation nodes
- Operate A/DACG and/or PSA.
- Operates theater staging and marshalling areas.
- Controls all activity in the staging area.
- Coordinate and supervise marshalling and out load of the unit.
- Support unit movements at railheads airfields, and seaports.
- Transport unit equipment, containers, and pallets to the staging area.
- Inspect unit vehicles, trailers containers, and pallets.
- Check redeploying equipment to ensure RF tags are properly affixed and operating
- Provides liaison between the airport and redeploying units.
- Ensure an electronic manifest list is built and the manifest is sent in accordance with DOD criteria to the IGC.
- Establishes and operates SRP and personnel holding area site.
- Inspect unit passenger and cargo manifests for accuracy.
- Maintain communications with the SPOE to ensure convoy and rail departure and arrival times are known.
- Provide liaison between the seaport and redeploying units.
- Ensure sealift equipment data is provided to IBS at SDDC.
- Supervise execution of unit rail load plans.

III. Movement to and Activities at the Point of Embarkation (POE)

Ref: ATP 3-35 (FM 3-35), Army Deployment and Redeployment (w/Chg 2), pp. 5-4 to 5-6.

Redeployment planning results in a network of transit areas designed to efficiently move forces from their area of operations to their final destinations. Use of these areas may vary with the situation.

Assembly Area

Units move to an assembly area to prepare for redeployment after being relieved from their operational mission. The assembly area should be away from the immediate employment area. Movement to, and within, the area is under control of a redeployment coordination cell. Units in the assembly area inventory, inspect and process equipment for turn-in or transfer; load containers; prepare documentation; conduct US Customs inspections; finalize unit movement data; and plan rail loads, bus movements, barge movements, and convoys for movement to a POE or APS turn-in site for movement to the port holding area based on movement instructions. Units update UDLs, generate documentation, RFID tags, and MSLs using TC-AIMS II. Equipment moving from the assembly area to the POE must have RFID tags and MSLs applied prior to loading.

Units wash major end items to satisfy US Department of Agriculture standards. Customs and agricultural inspection standards are based on the destination and types of equipment being redeployed. Units should make plans to perform the activities necessary to meet these standards. The time required to wash vehicles can be considerable and likely will be the overriding factor in redeployment scheduling. For example, a M1098 high mobility multi-purpose wheeled vehicle (HMMWV) can take approximately 12 hours to wash to meet the agricultural standards, and larger equipment can take a day or more to wash. Considerations in computing the estimated time to wash unit equipment should include the equipment density, estimated time for each piece of equipment, the number of wash points, and the staffing at each location. Once the equipment is cleared by customs inspectors it will be held in a secure sterile area until it is moved to the POE.

Customs Procedures

All DOD-sponsored cargo is inspected at the overseas point of origin by customs and border clearance agents. Military equipment is inspected at the time it is placed in boxes, crates, or containers for movement and secured until departure from the overseas area. Vehicles and similar items to be shipped are inspected and secured immediately prior to loading on the departing aircraft or vessel. After the inspection is completed a DD Form 2855 (U.S. Military Agriculture and Customs Preclearance Program) is prepared and securely affixed to the container or vehicle.

Inspectors normally check a minimum of 10 percent of all baggage 24 hours before the departure time. Once inspected, baggage is stored in a sterile area until transported and loaded at the APOE, approximately 4-6 hours prior to the scheduled departure. Soldiers process through customs with their carry-on bags and once cleared remain in the sterile area until they board the aircraft.

Detailed requirements of the military customs preclearance program can be found in Part V of the Defense Travel Regulation.

Activities at the SPOE

Units normally move to the SPOE staging area from assembly areas. Some SPOEs may not have total use of the port area. Port managers and operators must closely coordinate their activities with host nation authorities as well as joint and multinational elements. Joint-use facilities and limited real estate availability may require port authorities and redeploying forces to modify processes to accommodate port capabilities.

SDDC, as the single port manager, directs water terminal operations to include oversight of contracts, cargo documentation, security operations, and the overall flow of information. SDDC is responsible for providing strategic redeployment information to the GCC and to workload the port operator based on the GCC's priorities and intent.

See pp. 9-24 to 9-25 for discussion of activities associated with moving through SPOEs.

Activities at the APOE

The agencies and processes involved in moving Army units through an APOE during a deployment are similar to those at an APOE during redeployment. Customs and agricultural inspections are based on U.S. standards. Special emphasis should be placed on ensuring that baggage, passengers, and cargo be clean and free of dirt and organic matter, which may be difficult under field conditions.

See pp. 9-26 to 9-27 for discussion of activities associated with moving through APOEs.

POE Staging Area

Intratheater transportation assets may move units directly to a POE staging area or to an intermediate staging area. These movements are largely determined by the distance to be traveled, the size of the redeploying force, and theater capabilities. Units that were issued APS equipment usually turn it in at a separate location prior to moving to the POEs. Procedures for return of APS to storage locations are established during redeployment planning.

Refer to FM 3-35.1 for additional information regarding APS.

SPOE staging operations prevent congestion within the terminal area and provide space for segregating vehicles for vessel loading. This is the final en route location for preparation of unit equipment for strategic movement prior to the equipment entering the port holding area. The redeployment coordination cell monitors the flow of vehicles and equipment into the port and notifies the theater movement element when there is a backlog. The TSC establishes and operates the SPOE staging area and assists with opening the staging area at the SPOE.

Movements into the POE staging area must be carefully managed to preclude congestion and to avoid exceeding the capacity of the facility. Early planning in the assembly area ensures that units arrive at the POE on time and fill scheduled modes of transportation. Instructions directing movement to the port will come in the form of a call forward message from SDDC and is based on the availability of space in the port and the TPFDD timelines.

The theater human resources manager, in coordination with the TSC is responsible for personnel accountability at the theater processing centers. The unit remains responsible for conducting strength accounting through manpower and personnel channels. The human resources element at the processing centers verify unit manifests, coordinate manifest changes with the USAF, and transmit final flight manifests to the appropriate commands, human resources agencies, and destination installation commanders.

Movement to POD

The combination of strategic airlift and sealift provides the capability to redeploy forces, albeit in different timeframes and along separate routes. Personnel are transported by strategic airlift to the destination APOD and then moved by bus to the destination installation. Vehicles, unit equipment, and containers are moved by strategic sealift to the designated SPOD, unloaded, and transported by convoy, commercial truck, or rail to the destination installation.

It is extremely important for the redeploying unit, assisted by their home station ITO, to maintain visibility of their vehicles and unit equipment. A small investment in maintaining visibility throughout the redeployment pipeline can be rewarded by having vehicles and equipment delivered to the right place at the right time.

IV. Reception and Integration at Home or Demobilization Station

Ref: ATP 3-35 (FM 3-35), Army Deployment and Redeployment (w/Chg 2), pp. 5-6 to 5-7.

The destination for active component units is normally their home station whereas reserve component units return through a demobilization station. Typically the demobilization station is the same installation that served as the unit's mobilization station.

The supporting installation is responsible to assist returning forces until they reach their destination. The installation coordinates the support for the arrival ports and airfields and establishes en route sites as required by the redeployment plan. Once the unit vehicles and equipment arrive at the SPOD, the destination installation has the primary role of coordinating with SDDC for onward movement. The unit is responsible to provide load/unload teams and drivers at the POD and railhead. The supporting installation has the following responsibilities at the POD—

- Stage equipment for movement to the final destination
- Coordinate for customs clearance inspections
- Complete equipment inspections and process movement documentation
- Notify owning unit of equipment arrival.
- Operate the railhead and line-haul reception site to receive and download redeploying unit equipment.
- Sign for arriving equipment from commercial carriers.
- Operate the installation staging location for all downloaded equipment.
- Confirm cargo arrival to the installation back to SDDC for closure reporting.

Units returning by air to an APOD are off-loaded under the control of the CRE and moved to the holding area where they are released to the A/DACG. The unit remains in the holding area briefly to ensure that they have accounted for their personnel and equipment and then moves to the marshalling area where they are loaded on appropriate transportation for movement to home station, demobilization site, or other destination. The installation will—

- Maintain a central control and inspection point
- Provide a security area for sensitive items
- Coordinate life support facilities

The unit will—

- Ensure that all aircraft pallets and nets are returned to the CRE or A/DACG.
- Perform required maintenance checks and refuel equipment.

In most instances vehicles and unit equipment are transported to their destination by commercial transportation, contracted by SDDC in coordination with the destination ITO. The ITO receives the movement documents for all equipment flowing through their areas of responsibility. They receive the commercially delivered assets, process all paperwork, and release the equipment to the unit.

Closure Reporting

Redeploying units must also be tracked back to their home stations and reported to the original force provider and Headquarters, Department of the Army (HQDA). This facilitates reintegration into the Army force generation process. Once redeployment begins, force tracking is conducted until the force has completed movement through the redeployment pipeline and has emerged intact at the destination. Actions taken to track and then report the closure of units back to their home stations is a command responsibility.

(SMFLS5) Index